Intelligent System and Computing

Edited by Yang (Cindy) Yi

Published in London, United Kingdom

IntechOpen

Supporting open minds since 2005

Intelligent System and Computing
http://dx.doi.org/10.5772/intechopen.84928
Edited by Yang (Cindy) Yi

Contributors
Emmanouil Psarakis, Nikos Aspragathos, Panagiotis Koutmos, Vasileios Moulianitis, Georgios Tsiourlis, Georgios Thanellas, Eleftherios Dogkas, Pavlos Konstantinidis, Nefeli Lamprinou, Georgios Paterakis, Evangelos Sartinas, Konstantinos Souflas, Nikitas Xanthopoulos, Panteleimon Xofis, Eduardo Avendaño Fernández, Giovanni Serafino, Antonella Bogoni, Ana María Cardenas Soto, Neil Guerrero Gonzalez, Paolo Ghelfi, Kaung Htet Myint, Na-Young Ahn, Himanshukumar Patel, Vipul Shah, Wenwen Tu, Shiya Liu, Yang Yi, Gowrishankar Subrahmanyam, Ishwari G, Fadi Alnaimat, Mohammed Ziauddin, Bobby Mathew, Jayabrata Goswami, Anuva Ganguly, Anirudhha Ghosal, J.P Banerjee, Carlo Ferri, Prasanna Lakshmi M, Pushparaj Shetty D, Orken Mamyrbayev, Nurzhamal Oshanova, Nurbapa Mekebayev, Mussa Turdalyuly, Kian Hamedani, Zhou Zhou, Lingjia Liu, Kangjun Bai, Sandesh Shrestha, Sandrine Prom Tep, Perrine Ruer, Alexandra Nemery

First published in London, United Kingdom, 2020 by IntechOpen
IntechOpen is the global imprint of INTECHOPEN LIMITED, registered in England and Wales, registration number: 11086078, 7th floor, 10 Lower Thames Street, London, EC3R 6AF, United Kingdom
Printed in Croatia

British Library Cataloguing-in-Publication Data
A catalogue record for this book is available from the British Library

Additional hard and PDF copies can be obtained from orders@intechopen.com

Intelligent System and Computing
Edited by Yang (Cindy) Yi
p. cm.
Print ISBN 978-1-83880-522-7
Online ISBN 978-1-83880-523-4
eBook (PDF) ISBN 978-1-78985-722-1

For EU product safety concerns:
IN TECH d.o.o., Prolaz Marije Krucifikse Kozulić 3, 51000 Rijeka, Croatia, info@intechopen.com or visit our website at intechopen.com.

Meet the editor

Yang (Cindy) Yi is an associate professor in the Bradley Department of Electrical and Computer Engineering (ECE) and an associate director of the Multifunctional Integrated Circuits & Systems Center (MICS) at Virginia Tech (VT), USA. She obtained a MS and BS in Electrical Engineering from Shanghai Jiao Tong University, and a PhD in Electrical and Computer Engineering from Texas A&M University, USA. Dr. Yi has more than 100 publications in international journals and conference proceedings. Eight of her papers have been selected as Best Paper Award and Best Paper Award Finalist at leading conferences. She has received numerous awards and fellowships, including the VT College of Engineering Faculty Fellow and ICTAS Junior Faculty Award (JFA) in 2019; NSF CAREER Award in 2018; United States Air Force (USAF) Summer Faculty Fellowship; Miller Professional Development Award for Distinguished Research in 2016; National Science Foundation (NSF) EPSCoR First Award; KU Miller Scholar Award; KU New Faculty General Research Award in 2015; Air Force Research Lab (AFRL) Visiting Faculty Research Fellowship; University Academic Program Faculty Scholar; and University Outstanding Service Award 2014.

Contents

Preface XI

Section 1
Machine Learning and Deep Learning 1

Chapter 1 3
Machine Learning Techniques to Mitigate Nonlinear Phase Noise in Moderate Baud Rate Optical Communication Systems
by Eduardo Avendaño Fernández, Ana María Cárdenas Soto, Neil Guerrero Gonzalez, Giovanni Serafino, Paolo Ghelfi and Antonella Bogoni

Chapter 2 31
Deep Learning Based Prediction of Transfer Probability of Shared Bikes Data
by Wenwen Tu

Chapter 3 49
Voice Identification Using Classification Algorithms
by Orken Mamyrbayev, Nurbapa Mekebayev, Mussa Turdalyuly, Nurzhamal Oshanova, Tolga Ihsan Medeni and Aigerim Yessentay

Chapter 4 63
Parallel Processing for Range Assignment Problem in Wireless Sensor Networks
by M. Prasanna Lakshmi and D. Pushparaj Shetty

Section 2
Intelligent Systems 77

Chapter 5 79
Vehicle Secrecy Parameters for V2V Communications
by Na-Young Ahn and Dong Hoon Lee

Chapter 6 95
The Novel Applications of Deep Reservoir Computing in Cyber-Security and Wireless Communication
by Kian Hamedani, Zhou Zhou, Kangjun Bai and Lingjia Liu

Chapter 7 113
SMS Security on Android Using RC4 Algorithm
by Kaung Htet Myint

Chapter 8 125
From Pillars to AI Technology-Based Forest Fire Protection Systems
by Nikos Aspragathos, Eleftherios Dogkas, Pavlos Konstantinidis, Panagiotis Koutmos, Nefeli Lamprinou, Vassilis C. Moulianitis, Georgios Paterakis, Emmanouil Z. Psarakis, Evangelos Sartinas, Konstantinos Souflas, Georgios Thanellas, Georgios Tsiourlis, Nikitas Xanthopoulos and Panteleimon Xofis

Chapter 9 147
Vehicle Tracking Using Video Surveillance
by Sandesh Shrestha

Chapter 10 167
Intelligent Routing Mechanisms in IoT
by Gowrishankar Subhrahmanyam and Ishwari Ginimav

Section 3
Emerging Devices for Intelligent Systems 181

Chapter 11 183
Graphene Nanowire Based TFETs
by Jayabrata Goswami, Anuva Ganguly, Anirudhha Ghosal and Jyoti Prakash Banerjee

Chapter 12 191
Wax Deposition in Crude Oil Transport Lines and Wax Estimation Methods
by Fadi Alnaimat, Mohammed Ziauddin and Bobby Mathew

Section 4
Computational Intelligence Methods 209

Chapter 13 211
Comparative Study of Interval Type-2 and Type-1 Fuzzy Genetic and Flower Pollination Algorithms in Optimization of Fuzzy Fractional Order $PI^{\lambda}D^{\mu}$ Controllers
by Himanshukumar R. Patel and Vipul A. Shah

Chapter 14 235
Uncertainty in Measurement
by Carlo Ferri

Chapter 15 253
Quantized Neural Networks and Neuromorphic Computing for Embedded Systems
by Shiya Liu and Yang Yi

Chapter 16 273
Persuasion in Mobile Financial Service: A Case Study with a Bank Savings Mobile Application
by Prom Tep Sandrine, Ruer Perrine and Nemery Alexandra

Preface

The book "Intelligent Systems and Computing" will potentially revolutionize a variety of industries, including automotive, communication, aerospace, semiconductor, health care, economics, and manufacturing. These systems cover a wide range of technologies, including machine learning and deep learning, neural networks, fuzzy logic systems, pattern recognition, social intelligence, the Internet of Things, and many others.

This book provides the state of the art in theory, mathematical models, algorithms, design methods, and applications of intelligent systems and computing, along with a vision for future research. It covers various disciplines including computer and information science, electrical and computer engineering, natural sciences, economics, neuroscience, and so on.

I would like to thank the contributing authors for participating in this project. Without their support, this book would not have been possible. I would also like to thank Ms. Dolores Kuzelj and her team at IntechOpen for their persistent efforts in moving the project forward.

Yang (Cindy) Yi
Bradley Department of Electrical and Computer Engineering,
College of Engineering,
Virginia Tech,
Virginia, United States of America

Section 1

Machine Learning and Deep Learning

Chapter 1

Machine Learning Techniques to Mitigate Nonlinear Phase Noise in Moderate Baud Rate Optical Communication Systems

Eduardo Avendaño Fernández, Ana María Cárdenas Soto, Neil Guerrero Gonzalez, Giovanni Serafino, Paolo Ghelfi and Antonella Bogoni

Abstract

Nonlinear phase noise (NLPN) is the most common impairment that degrades the performance of radio-over-fiber networks. The effect of NLPN in the constellation diagram consists of a shape distortion of symbols that increases the symbol error rate due to symbol overlapping when using a conventional demodulation grid. Symbol shape characterization was obtained experimentally at a moderate baud rate (250 MBd) for constellations impaired by phase noise due to a mismatch between the optical carrier and the transmitted radio frequency signal. Machine learning algorithms have become a powerful tool to perform monitoring and to identify and mitigate distortions introduced in both the electrical and optical domains. Clustering-based demodulation assisted with Voronoi contours enables the definition of non-Gaussian boundaries to provide flexible demodulation of 16-QAM and 4+12 PSK modulation formats. Phase-offset and in-phase and quadrature imbalance may be detected on the received constellation and compensated by applying thresholding boundaries obtained from impairment characterization through statistical analysis. Experimental results show increased tolerance to the optical signal-to-noise ratio (OSNR) obtained from clustering methods based on k-means and fuzzy c-means Gustafson-Kessel algorithms. Improvements of 3.2 dB for 16-QAM, and 1.4 dB for 4+12 PSK in the OSNR scale as a function of the bit error rate are obtained without requiring additional compensation algorithms.

Keywords: nonlinear phase noise, clustering, Voronoi, decision boundary

1. Introduction

Novel transceiver architectures for the 5th Generation (5G) communication networks front- and backhaul must make use of the digital signal processing (DSP)-assisted transmission capabilities to optimize the trade-off among signal processing complexity, spectral efficiency and reach in the short access radio networks [1]. One key element in hybrid optical-wireless next-generation technologies is the capability to identifying the noise and distortions, their impact on the signal, and how the modulation format is adapted to that channel quality conditions [2]. For this reason, an important disadvantage of analog Radio-over-Fiber (A-RoF) systems using high-order modulation formats is its inherent propensity to transmission impairments risen from the electrical and optical domain. The immediate effect on the signal is a limitation in the dynamic range (DR) [3], in terms of reach as a function of the bit rate. In the radiofrequency (RF) communications context, oscillators are key devices of transmission chain whose main purpose is to deliver stable reference signals to perform frequency translation from baseband to pass-band and to provide in-phase (I)/quadrature (Q) components synchronization. As inherent in electronic devices, practical RF oscillators exhibit nonideal behavior and characteristics leading to random phase noise (PN) on the received signal. The spectral regrowth at the output of a nonlinear amplifier and IQ imbalance, combined with the effect of PN, are the major RF impairments that degrade the performance of communication systems in the electrical domain [4]. PN arises mainly by inherent passive and active components, which introduce thermal noise inside the circuitry of oscillators; therefore, due to transistor transient state's, aspects, as a rule of thumb, the higher the frequency of an oscillator, the worst is its performance in terms of PN. For coherent systems, the impact of PN over data symbols may be observed in the constellation diagram, where the main effect is a random rotation of symbols, causing their distribution over rings with elongated shapes. Because of this shape distortion, some immediate effects are (i) increased detection of erroneous symbols, (ii) spectral broadening, which can lead to interusers interference, (iii) in-band signal injection of out-of-band emission, and (iv) loss of orthogonality among multi-carrier waveforms. These effects over signal imply tackling synchronization issues leveraging on dedicated blocks of the DSP chain at the receiver [5]. The main task of the synchronization stage is to match the frequency and phase of transmitter and receiver oscillators, in both the electrical and optical domains. This process includes time recovery and tracking of the local oscillator (LO) frequency mismatch, to compensate for impairments caused by PN.

In this scenario, a critical block in the DSP chain of optical communication systems is the carrier phase recovery (CPR) block, which is required before the symbol decision takes place. It has the role of compensating the PN, directly related to the frequency noise of the lasers employed as optical carriers in the system. The noise of lasers consist of short-term random fluctuations of the optical phase and other different output parameters. Besides, the baud rate, the transmission length, and, in general, the DR of optical communication systems are limited by noise issues. The key parameter related to the noise in lasers is the linewidth, defined for a monochromatic source as the width of its optical spectrum, i.e., the power spectral density (PSD) of the emitted electrical field, oscillating at optical frequencies, as a function of frequency, wavenumber, or wavelength. An ideal electronic or optical oscillator would be, strictly speaking, monochromatic, exhibiting a zero linewidth. In other words, all the energy it emits is completely concentrated on its nominal frequency. Real oscillators are not exactly monochromatic and have a finite, non-zero linewidth. Besides, before the first laser was experimentally demonstrated, Schawlow and Townes showed that a laser linewidth could not have values lower than a fundamental quantum limit [6]. Nowadays, the most widely employed lasers in optical communications are semiconductor lasers; Henry demonstrated that this kind of lasers exhibits a linewidth enhancement factor [7], concerning to the Schawlow-Townes quantum limit. The line broadening is caused by the coupling of phase and amplitude noise, due to the variations of carrier density (i.e., of the refractive index) in the laser cavity. Further analysis of these phenomena is out of the scope of this chapter, and the interested reader is referred to [6] and [7]. There exists a strong relationship between the laser linewidth (LW) and the temporal coherence, characterized by the coherence time or coherence length, i.e., the measure of temporal coherence expressed as the time over which the field correlation decays. The linewidth can be used for estimating the coherence time as a function of the optical spectral shape, and the bandwidth can be approximated just by the inverse of the coherence time for typical spectral shapes, such as the Lorentzian or the Gaussian shape. In this chapter, the proposed methods mitigate the PN introduced by the mismatch between the optical carrier and the transmitted RF signals that are not correlated within the coherence time [8] of the local oscillator for the specific baud rate (250 MBd) used in the experimental setup.

Due to electrical-to-optical conversion, optical fiber propagation, and optical-to-electrical conversion, the signal is distorted and impaired due to the impact of system devices in the conversion stages, combined with distortive effects in the optical fiber segment, such as chromatic dispersion (CD), the Kerr effect, and, to a lesser extent, polarization mode dispersion (PMD). The experimental setup was conceived to realize a Radio-over-Fiber (RoF) system with the particular characteristics explained above, including severe impairments in both the optical and the electrical domains. A typical issue that must be compensated for is the frequency mismatch between the transmitter frequency carrier and the LO used for down-conversion at the receiver, either direct or coherent detection. This mismatch is due to PN and produces fluctuations in the constellation point locations concerning to their ideal positions, causing the outer symbols to shift away from their natural position, leading to a broadening of constellation data symbol. In [9], the received constellation exhibits data symbol (or clusters) with quasi-elliptical (elongated) shape in the case of modulation formats with 16 data symbol points. Some effects were observed on cluster symbols like the imbalance gain, direct current (DC) offsets, time skews, and PN, creating an irregular distribution in data symbols, resulting in overlaps among adjacent neighbor clusters when the conventional rectangular grid is used as demodulation rule. This same distortion appeared in our experimental scenario, corresponding to 16-Quadrature-Amplitude Modulation (QAM) and 4 + 12 Phase Shift Keying (PSK) constellations. To overcome PN impairments, well-known post-equalization techniques have been proposed, such as (i) detection methods based on the symbol error probability (SEP) in which minimum Euclidean distance (ED) decisions are used to define decision regions (for example, the rectangular grid for 16-QAM); (ii) demodulation methods based on maximum likelihood criterion applied for decision-making rules under conditions of uncertainty [10]; (iii) machine learning techniques applied to clusters, to classify and demodulate data symbol points, and mainly to improve the error system tolerance against nonlinear effects [11, 12].

On the other hand, a novel method for clustering was applied using Voronoi diagrams [13], inspiring the proposal of generic methods that can be extended to coherent receivers, avoiding the carrier frequency offset (CFO) and carrier phase offset (CPO) equalization algorithms at DSP stage [14]. Besides, in optical systems using coherent receivers, it is observed that the received constellation, even after applying DSP algorithms, remains influenced by enhanced noise originated from the PN added by the system components, and frequency noise from the LO [15]. In [16], a guideline for a joint design using forward error correction (FEC) and coded modulation (CM) techniques show the trade-off among electronic computational complexity, optical transparent reach, and hardware constraints. CM schemes with signal shaping and rate adaptation capabilities are combined to adapt the optical data transmission architecture under different signal qualities.

2. Nonlinear phase noise (NLPN)

As demonstrated by Gordon and Mollenhauer [17], the linear phase noise associated with ASE noise from inline amplifiers interacts with the transmitted signals via the Kerr effect, generating a strong distortion over symbols, referred to as NLPN. In an optical transmission system, amplification is an essential operation to compensate for signal propagation losses. Commonly, erbium-doped fiber amplifiers (EDFAs) are employed as optical boosters in the transmitters, as well as in-line optical amplifiers and pre-amplifiers in the receivers. Their benefit in preserving the signal level among the receiver sensitivity comes with the undesired amplified spontaneous emission (ASE) noise. The main effect of NLPN consists of creating a nonlinear distortion over the shape of external symbols, which degrades the symbol error performance. To compensate for this effect, the demodulation process requires non-Gaussian decision boundaries that, in practice are not simple to realize.

In the considered scenario for the validation of the proposed method, NLPN was observed in both the 16-QAM and 4 + 12 PSK constellation diagrams. In our case, the NLPN effects are very strong, due to the low adopted baud rate, which translates to a long symbol duration, that must be compared to the coherence time t_c between the local oscillator and the incoming signal from the transmitter (ideally, a baud rate of 250 MBd needs a t_c around 4 ns). Usually, this effect on symbols is due to higher linewidths of the laser [18], but, in our case, the linewidth used was 1 kHz. In **Table 1**, a state-of-the-art synthesis and review of different methods to compensate impairments using the conventional equalization techniques through DSP techniques and the design and optimization of constellations tolerant to noises and distortions is introduced. This table presents columns with a brief explanation of the proposed method showing the metrics used to evaluate the performance, the modulation scheme, and the authors and year of publication.

For multilevel advanced modulation formats, especially for m-ary phase-shift keying (m-PSK) and m-ary quadrature amplitude modulation (m-QAM), the symbols in the constellation diagram suffer from linear amplitude noise and PN. In an ideal oscillator, the phase variance over a given time interval remains small, and the output signal is nearly periodic. However, as already mentioned, in practical oscillators, the phase random variation or phase jitter is a random variable associated with the PN random process, being the instantaneous deviation of phase from the ideal value caused by PN [28]. Either in homodyne or heterodyne detection, the beating of the received signal with a LO with the same frequency of the optical carrier enables the detection and demodulation process; however, since the phase is not stable in time, frequency will not be the same between LO and carrier, causing a lack of synchronization that translates into the degradation of the received signal. The constellation symbol will not be in its ideal place (for example, in QAM the ideal location of a symbol point i.e., $1 + i$), and it will be rotated by an angle given by the phase difference between the phasor of the transmitted signal (i.e., symbol) and the phasor of the LO. If this phase difference is large, the symbols fall outside the right decision region, tracing an elongated shape in the constellation, and the symbol error performance is degraded. In long-haul scenarios, the ASE noise from inline high-gain amplifiers becomes the main source of the noise [29]. The NLPN can be compensated electronically by subtracting from the received phase a correction that is proportional to the received intensity. The optimal scaling factor is approximately equal to half of the ratio of the mean NLPN and the mean received intensity [30]. Thus, the phase jitter, i.e. the standard deviation of residual phase noise is halved, and the transmission distance is doubled. In [31], the minimization of the total noise variance is performed by using a power profile along with the link that is linear with distance.

Method	Description	Mod. format	Authors and year
Digital signal processing and equalization algorithms	Digital backpropagation (DBP) with coherent detection are proposed to mitigate dispersion effects and nonlinear impairments through noniterative asymmetric split-step Fourier method (SSFM). An optimal decision map and a heuristic for the step size and sampling rate requirements for RZ-QPSK were obtained by numerical simulation.	RZ-QPSK	Ezra and Kahn [19]
	PN tolerance of circular multilevel quadrature amplitude modulation (C-mQAM) constellations employing different carrier phase recovery (CPR) algorithms is studied and evaluated using a differential decoding and bit-mapping scheme. Carrier phase recovery achieves similar performance than the blind phase search (BPS) algorithm at lower computational complexity level, obtaining 3.8e-3 and 1e-2 BER levels for C-16QAM and C-64QAM modulation formats, respectively.	16-QAM 64-QAM	Rodríguez Navarro et al. [20]
	Laser PN and NLPN limit and degrade performance at high launch power level, the phase-modulated coherent optical communication systems. Authors propose the use of a Kalman filter to estimate, track, and mitigate both impairments in decision-directed mode filtering.	QPSK	Jain and Krishnamurthy [21]
	Two-stage extended Kalman filtering (EKF) technique for the joint tracking of frequency offset (FO), laser PN, fiber nonlinearity, and amplitude noise is numerically evaluated for polarization multiplexed 16-QAM at 28 GBd. Results show tolerance over SSMF in a span length of 3000 km, obtaining a BER of 2.4e-2.	16-QAM	Pakala and Schmauss [22]
	An effective *k*-nearest neighbor (*kNN*) detector to further improve the performance of the optical phase conjugate (OPC) system by mitigating non-Gaussian impairments caused by NLPN is numerically evaluated.	QPSK, 16-QAM	Jiang et al. [23]
Constellation design and optimization	The authors propose an optimal signal constellation design (OSCD) algorithm suitable for the phase noise channels using the cumulative log-likelihood (LLR) function as the optimization criterion.	DPSK	Liu et al. [24]
	Joint coding and modulation format optimization using bit-interleaved coded modulation (BICM) mutual information shows that circular constellation 8-QAM is suitable for high-coding rates while rectangular 8-QAM provides the best performance for low-coding rates, and both schemes outperform start 8-QAM.	8-QAM Circular and star 8-QAM	Ríos Müller et al. [25]
	Prototype nonconventional constellations generated using digital pre-distortion and demodulated at the receiver using clustering were validated experimentally. ONSR improvements of 1.2 dB for nonconventional	16-QAM, 4 + 12 PSK	Fernández et al. [26]

Method	Description	Mod. format	Authors and year
	ring 6-2-8 constellation and up to 2.9 dB using the same constellation demodulated with fuzzy c-means Gustafson-Kessel was obtained comparing to 16-QAM.		
	A new geometric shaping method is proposed, leveraging autoencoder unsupervised machine learning (ML) to optimize the constellation design. The learned constellation mitigates nonlinear effects with gains up to 0.13 bit/4D when trained with a simplified fiber channel model.	64-QAM 256-QAM	Jones et al. [27]

Table 1.
Nonlinear fiber impairment mitigation techniques.

3. Machine learning (ML) techniques

The main challenge of ML algorithms consists of mitigating nonlinearities, where several performance parameters and error estimation techniques have been applied to linear and nonlinear impairments (chromatic dispersion, polarization mode dispersion, Kerr effect, etc.) that distort the shape of symbol points in the constellation diagram. Under adverse conditions during propagation, complex highly unpredictable analytical models should be better implemented with flexible techniques that take advantage of feature extraction and exploit the knowledge of historical data to create direct input–output relations between monitored parameters and desired outputs. The immediate effect of these ML methods is to decrease the error rate by enhancing the distance transmission or increasing the optical signal to noise ratio (OSNR) tolerance against these linear and nonlinear effects. An additional key issue is that ML in optical communication systems can work where is not possible to obtain a closed-form expression that model the optical channel [32]. Parameter estimation or system monitoring must be performed before signal demodulation, but is not a trivial task because of linear and nonlinear transmission impairments, especially when the signal crosses through different channels like the hybrid optical-wireless considered here (the Radio-over-Fiber system) has not a defined mathematical model including channel condition variations and transient impairment contributions during operation. Therefore, it is appropriate to use versatile methods that learn from data to predict i.e., the conditional probability or maximum likelihood function or to use the classical compensation method where to have a constrained channel model is possible to apply training sequences to estimate and equalize the channel response and improve the system performance. In this sense, to apply different machine learning approaches is attractive to characterize, to train and learn a mapping function between the signal features in the optical-wireless scenario, to and define non-Gaussian boundaries in constellation diagram before signal demodulation.

Machine learning has been successfully applied to a wide variety of problems in the context of telecommunication networks, mainly in wireless networks, ranging from opportunistic spectrum access, to channel estimation and signal detection in orthogonal-frequency division multiplexing (OFDM) to multiple-input-multiple-output (MIMO) communications. Many techniques within the previous categories can be addressed at the physical layer of optical transmission for optical performance monitoring (OPM) [33–40], modulation format recognition (MFR) [41], or nonlinearity mitigation. Ideally, OPM should be implemented just with a single photodiode and machine learning algorithms that can learn the mapping between the detected signal and optical fiber channel parameters and finally predict optical fiber channel parameters from energy constellation diagrams or power eye diagrams. Moreover, machine learning algorithms enable the MFR [39] of an incoming signal to perform accurate constellation demodulation, also, to apply signal processing and detection at the receiver.

In **Table 2**, novel architectures and techniques under the artificial intelligence umbrella are reviewed, mainly from perspective of optical communication systems to mitigate nonlinear effects and perform signal demodulation. Machine learning techniques based on clustering or support vector machine (SVM) may assist the

Method	Description	Mod. format	Authors year
Clustering	Burst optical signal detection based on a modified *k*-means clustering algorithm is used to establish a threshold, and the effects of the unbalanced ratio of bits zero and one to improve detection performance in terms of SNR without preamble field for amplitude recovery. In addition, a data-aided feedforward symbol-timing recovery method based on a polynomial interpolation and maximum-likelihood estimation theory are developed.	N.A.	Zhao et al. [43]
	Experimental demonstration of successful coherent detection and digital demodulation for 3 × WDM of 2.5 Gbit/s QPSK/8-PSK phase-modulated RoF optical link after transmission over a 78.8 and 40 km of field-deployed fiber, respectively. The iterative clustering approach based on the *k*-means algorithm was applied to perform phase recovery on QPSK signal successfully.	QPSK, 8-PSK	González et al. [37, 44]
	A novel phase offset compensation by dual-stage ring fragmentation and clustering for 16-QAM and 4 + 12 APSK constellation is demonstrated experimentally in RoF scenario. OSNR improvements up to 3.3 and 1.4 dB, for 16-QAM and 4 + 12 PSK, respectively, were obtained.	16-QAM, 4 + 12 PSK	Fernández et al. [9]
	Unsupervised *k*-means and Gaussian mixture model (GNN) algorithms are proposed in different sections of coherent DSP to perform pre-decisions and enhance the performance of multi-modulus (MMA) algorithm and carrier phase recovery (CPE). Experiments demonstrate more than 3 and 1 dB improvements in OSNR sensitivities for dual-polarization 64-QAM at BER thresholds of 4.5e-3 and 1.6e-2, respectively.	64-QAM	Xu et al. [45]
	A *k*-means clustering algorithm is proposed and experimentally demonstrated in visible light communication (VLC) systems to mitigate nonlinear distortion effects. Improvements up to 5.5 dB in Q-factor are obtained for 400 Mb/s Nyquist 4-PAM signal at BER threshold of 2.4e-2.	4-PAM	Ma et al. [46]
Support vector machine (SVM)	A SVM-based detector is proposed for coherent optical fiber amplitude phase shift keying (APSK) constellation to mitigate NLPN. Comparing with 16-QAM, the SVM detector increases the nonlinear system tolerance by 4.88 dB at a BER of 1e-3.	16-APSK	Han et al. [11]

Method	Description	Mod. format	Authors year
	A machine learning-based classifier, namely SVM, is introduced to create the nonlinear decision boundary in M-ary PSK-based coherent optical system to mitigate NLPN. The maximum transmission distance and launch power dynamic range (LPRD) tolerance are improved by 480 km and 3.3 dBm for 8-PSK.	8-PSK	Wang et al. [35]
	A proposal of a bit-based SVM as a non-parameter nonlinear mitigation approach in the millimeter-wave Radio-over-Fiber (mm-RoF) system for different modulation formats is demonstrated. Experimental results outperform the *k*-means algorithm and show improvements of 1.2 dB for 16-QAM, 1.3 dB for 64-QAM, 1.8 dB for 16-APSK, and 1.3 dB for 32-APSK at BER of 1e-3 with the SVM detector, respectively.	QPSK, 16-QAM	Cui et al. [47]
Convolutional neural network (CNN)	An intelligent constellation diagram analyzer is proposed to implement both modulation format recognition (MFR) and OSNR estimation by using a CNN-based deep learning technique. The experimental results showed that the OSNR estimation errors for all the signals were less than 0.7 dB and the accuracy of MFR was 100%, proving the feasibility of the proposed scheme.	m-PSK m-QAM	Wang et al. [48]
	Maximization of capacity over deployed links require operation regime estimation based on precise understanding of transmission conditions through linear and nonlinear SNR from the received signal. The extraction of NLPN and second-order statistical moments by a neural network is trained to estimate SNR from extensive realistic fiber transmissions. Measured performance of 0.04 and 0.2 dB of standard error for the linear and nonlinear SNRs, respectively, are obtained.	16-QAM WDM	Vaquero Caballero [49]
Supervised machine learning	A machine learning detector based on *kNN* algorithm overcomes non-Gaussian symmetric noise coming from laser PN and NLPN in the zero-dispersion link and dispersion-managed link (DML). The method increases the linewidth tolerance by 180 kHz, and the nonlinear tolerance for dispersion neglected, managed, and unmanaged links by 1.7, 1 and 1.4 dBm, respectively.	16-QAM	Wang et al. [12]
	A nondata (ND)-aided *k*-nearest neighbor (*kNN*) equalization technique can provide efficient nonlinear compensation at a low computational cost and zero data redundancy in coherent optical communication systems. 0.5 dB of BER improvement in 800 km standard single-mode Fiber (SSMF) using 16-QAM and 2 dB of BER in 80 km of SSMF for 64-QAM.	16-QAM 64-QAM	Zhang et al. [50]

Table 2.
Machine learning approaches used in optical fiber communications.

definition of nonlinear boundaries for demodulation, avoiding additional stages for phase estimation and correction, extraction of features from constellations of real transmission scenarios, demonstrating the potential of artificial intelligence algorithms in next-generation optical-wireless technologies. In [42], a description of the mathematical background of machine learning techniques from a signal processing perspective applied to nonlinear transmission systems, OPM, and cross-layer network optimizations for software-defined networks (SDN) is reviewed as an alternative, unique, and powerful set of tools to conventional compensation techniques in the wireless-optical fiber context. In this chapter, clustering techniques like *k*-means and fuzzy c-means are used to classify data symbols of the classical 16-QAM constellation and 4 + 12 PSK modulation formats. Here, we report on an experimental RoF testbed to demonstrate the impact of techniques based on statistical analysis and advanced clustering methods for the mitigation of NPLN-induced impairments. The definition of such RoF testbed includes a clustering-based detector that allows identifying the distortive effects over symbols in the constellation diagram of both channels, the electrical and the optical fiber channel that, together, introduce particular distortions including NLPN. The processing that applies Voronoi contours before signal demodulation, defines non-Gaussian boundaries and increases noise tolerance improving the system error performance.

4. Oscillator phase noise model

The general model of a noisy signal in the time domain is represented by

$$v(t) = [V_c + \varepsilon(t)] \cos [2\pi f t_0 + \phi(t)] \quad (1)$$

where $\varepsilon(t)$ is the amplitude modulation (AM) noise, and $\phi(t)$ is the phase modulation (PM) noise, also known as phase jitter. The normalized carrier power spectrum without noise is given by

$$P_c(f_0) = V_c^2/2 \quad [W] \quad (2)$$

The power spectrum of Eq. (1) is shown in **Figure 1** with and without noise. The PN causes the instantaneous frequency to "jitter" around the carrier frequency, f_c. This leads to the noise sidebands on either side of f_c as shown in the picture. An important relation between the coherence time and the PN must be considered here. The coherence time is the time over which a propagating wave may be considered correlated (coherent). In simple words, it is the interval within which the wave phase is, on average, predictable. Therefore, the coherence time can be expressed as

$$t_c = 1/\Delta v \quad (3)$$

where Δv is the spectral width (Hz) of the source or, in other words, the range within which the frequency "jitters."

The PN is a random process that is a function of time, but, commonly, its monolateral power spectrum density is considered and represented versus the offset frequency from the carrier, as depicted as an example in **Figure 2**. The phase jitter $\Delta\varphi$ can be calculated from the power of the PN random process $S_\phi(f)$ by the following formula:

$$\Delta\varphi = \sqrt{\int_{f_l}^{f_u} S_\phi(f) df} \quad (4)$$

usually expressed in radians. The lower and upper integral bound f_l and f_u can be arbitrarily set depending on the way the phase jitter is calculated, as it will be explained in the following. It is worth noticing that the value of the integral corresponds to the PN process variance, and besides, the process to be non-zero only in the range f_l to f_u.

Another important random variable for the analysis of the stability of an oscillator is the time jitter $\Delta\tau$, which can be calculated from the phase jitter

$$\Delta\tau = \frac{\Delta\varphi}{2\pi f_c} = \frac{1}{2\pi f_c}\sqrt{\int_{f_l}^{f_u} S_\phi(f)df} \tag{5}$$

where f_c is the carrier frequency. The timing jitter, usually expressed in seconds (or fractions of seconds), represents the mean variation in time of the wave period, which can change since the wave frequency fluctuates.

For this reason, it can also be expressed in parts per million (ppm) of the period. The timing jitter is divided by the carrier frequency. The integral lower bound of Eq. (4) is f_l, i.e., the closest offset frequency from considered to calculate the jitter. Therefore, the inverse of the lower bound frequency $1/f_l$ can be considered as a good approximation, the coherence time t_c, because it represents the longest time over which that curve of PN ensures that particular value of jitter. To illustrate the relation between the coherence time t_c and the PN, consider **Figure 2**. As an example, for a baud rate of 250 MBd (equivalent to 1 Gb/s in 16-QAM), the needed coherence time corresponds is $t_c \approx 4$ ns, while for 10 GBd it is $t_c \approx 100$ ps because it is the duration of one symbol. In other words, the required signal stability should be kept for at least one symbol duration.

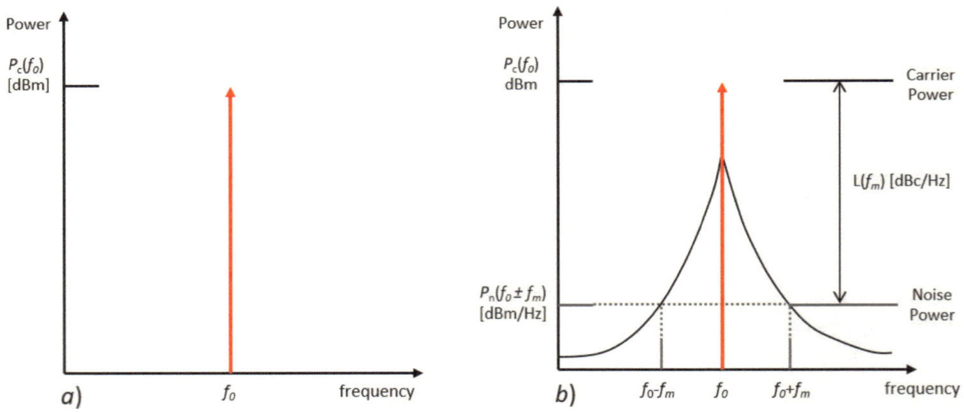

Figure 1.
Power spectrum density for (a) an ideal noise-free oscillator at frequency f_o and (b) a real oscillator with noise bands and non-zero linewidth.

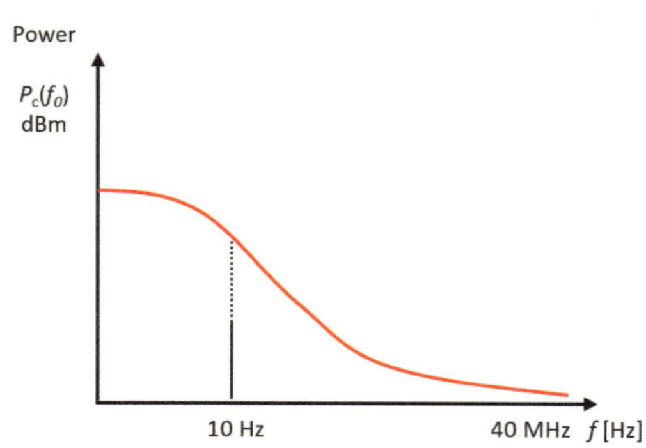

Figure 2.
Example of a phase noise spectrum profile.

PN manifests as the jitter on a signal zero crossings. The variance of such (random) zero-crossing jitter is related to the variance of the phase noise itself [51]. One way to deal with the PN effect on the BER consists in measuring the parameter $f_{\text{sym}}/f_{\text{c}}$. For large values of this ratio, the noise effect on BER becomes smaller as $f_{\text{sym}}/f_{\text{c}}$ increases [52]. For that range, one symbol time is short compared with the phase fluctuation period caused by the PN. Hence, the phase fluctuation in one symbol time delayed signal is virtually the same as the fluctuation in the signal before the onset of the delay. Conversely, in small $f_{\text{sym}}/f_{\text{c}}$ range, the PN effect on BER becomes smaller as $f_{\text{sym}}/f_{\text{c}}$ decreases. This is because the PN integration value over the receiver noise-equivalent bandwidth becomes smaller due to the filter at the receiver.

As already mentioned, a way to introduce PN in a system to measure its effect is by using larger laser linewidth on the light source, despite this is not this case because the experimental setup that will be reported here used a laser LW of 1 kHz. Therefore, from the laser's LW point of view, the non-spectral purity (non-monochromatic) of the lasers employed for signal modulation/demodulation in coherent optical systems generally translates into PN impairment in the detected signal. The normalized output of a transmitting laser can be modeled as

$$x_{laser}(t) = E_0 e^{j(w_{tx}(t)+\phi_{PN})} \tag{6}$$

w_{tx} and ϕ_{PN} are the central emitting frequency and PN of the laser. The PN can be modeled as a Wiener process

$$\phi_{PN}(k) = \phi_{PN}(k-1) + \Delta\phi_{PN}(k) \tag{7}$$

With $\Delta\phi_{PN}$ being an independent and identically distributed (*i.i.d.*) random variable with a mean $\mu = 0$ and variance $\sigma^2_{\Delta\phi_{PN}} = 2\pi(\Delta f \cdot Ts) \cdot \Delta v$ is the laser LW, and T_s is the symbol period. When transmission uses higher m-ary modulation formats, the effect of the laser LW must be as low as possible to avoid PN and other impairments arise from combined effects introduced from devices in the system.

Clock signals are required in almost every integrated circuit and some desired characteristics related to their quality must be considered. Digital signal processing is used to perform conversions between analog and digital signals and to process and transmit data at higher data rates and also higher resolutions. The clock signal quality can be described by timing jitter or PN measurements. Therefore, the typical clock PN measurement explores the power density spectrum (PSD) of a clock signal through the above defined time (or period) jitter as the often used jitter measurement parameter.

Through the Fourier series expansion, it can be shown that a square-wave clock signal has the same jitter behavior as its base harmonic sinusoid signal. For this case, a clock sinusoid signal with PN [51] can be written by

$$s_{clk}(t) = A\sin\left(2\pi f_c t + \phi(t)\right) = A\sin\left(2\pi f_c\left\{t + \frac{\phi(t)}{2\pi f_c}\right\}\right) \tag{8}$$

And the period jitter being given by

$$J_{PER}(t) = \frac{\phi(t)}{2\pi f_c} \tag{9}$$

The signal in Eq. (9) is a phase-modulated sinewave by the PN, $\phi(t)$. Since the PN is usually much smaller than $\pi/2$, Eq. (8) can be approximated as

$$s_{clk}(t) = A\sin(w_c t) + A(\phi)\cos(w_c t) \tag{10}$$

From an operational point of view, therefore, it is possible to set to set a method measuring the PN process observing that the signal $s_{clk}(t)$ mixed with a $\cos(2\pi f_c t)$ and filtered by a low-pass filter in the frequency domain allows to obtain the PN spectrum $S_\phi(f)$ and its logarithmic version $L(f)$, related by the following formula:

$$S_\phi(f) = 10^{\frac{L(f)}{10}} \tag{11}$$

The PN spectrum $L(f)$ defined as the attenuation in dBc/Hz from the peak value $S_{clk}(f)$ at the clock frequency, f_o, to a value of $S_{clk}(f)$ at f_m, as shown in **Figure 1b**. The mathematical expression that represents the PN spectrum is given by

$$L(f_m - f_o) = 10\log\{S_{clk}(f_m)/S_{clk}(f_o)\} \quad [dBc] \tag{12}$$

Phase noise testing characterizes the spectral purity of a LO by calculating the ratio of the desired energy being delivered by the oscillator at the specified output frequency to the amount of undesired energy being delivered at adjacent frequencies. It is a measure of noise power per unit bandwidth appearing at a given offset from the carrier frequency [53]. From Eq. (12), the mean square (MS) of $\phi(t)$ is equivalent to the PN variance, since it is a zero-mean process, and it can be calculated by

$$\langle\phi^2(t)\rangle = \Delta\varphi^2 = 2\int_0^\infty S_\phi(f)df = 2\int_0^\infty 10^{\frac{L(f)}{10}}df \tag{13}$$

that, finally, shows the relation between the timing jitter $\Delta\tau$ and the PN spectrum, $L(f)$, as

$$\Delta\tau = \frac{\Delta\varphi}{2\pi f_o} = \frac{1}{2\pi f_o}\sqrt{2\int_0^\infty 10^{\frac{L(f)}{10}}df} \tag{14}$$

The real integral bounds are set to be f_l (that is not 0, but with an offset from the carrier) and f_u (usually corresponding to the frequency where the PN spectrum enters the noise floor of measurement instrument). An approximated linear piecewise function using a log scale for the $L(f)$ spectrum can be used to estimate the $\Delta\tau$ as presented in [51].

5. Clustering algorithms

According to state of the art, clustering algorithms are presented in this section to understand how we apply the algorithms to identify the change of shape and distortions and to classify each data symbol affected by PN from statistical analysis. Also, like each symbol is surrounded by a decision region, we use the Voronoi diagram to estimate the contour of each symbol, and hence, the last section describes how to apply Voronoi algorithm to obtain flexible thresholding decision boundaries.

5.1 *k*-means clustering algorithm

k-means is a well-known clustering algorithm applied to perform phase recovery and demodulation processes [54] in a phase-modulated RoF system. It is one unsupervised learning algorithm, whose start point is a predefined and fixed grid on IQ plane. For the modulation formats 16-QAM and 4 + 12 PSK to be evaluated, the partition corresponds to 16 clusters, where each observation is assigned to one of the k clusters defined by centroids according to the nearest mean. The algorithm aims to minimize an objective function known as squared error J and works as follows: for a cluster centroid $v_i = \{v_1, \ldots, v_c\}$ of the data-set $\{x_1, \ldots, x_N\}$, consisting of N observations, the algorithm [55] first groups randomly selected centroids, which are used as the beginning points for every cluster, and then performs iterative calculations to optimize the positions of the centroids, according to

$$J = \sum_{i=1}^{c} \sum_{k=1}^{N} \|x_k - v_i\|^2 \quad (15)$$

A data-set of N observations assigned according to this algorithm (7), create different clusters with irregular shapes. **Figure 3** shows the general way in which the k-means algorithm works when no vector of centroids is predefined. The first step (**Figure 3a**) is to place centroids v at random locations or referenced locations, i.e., the ideal location of symbols according to the m-ary level of the modulation format used. Then, in the second step (also as in **Figure 3a**), each point from x_k is assigned to the nearest centroid of the cluster i, by applying some distance metric, e.g., Euclidean distance (ED) between sample x_k and cluster centroid v_i. After this, the third step (**Figure 3b**) consists of updating each cluster by averaging all the points assigned to cluster i in the previous step. Finally, at the fourth step (also in **Figure 3b**), all the points are assigned to the closest cluster, and the centroids are updated until convergence.

5.2 Fuzzy c-means (FCM)

The FCM is an algorithm for clustering that allows to observations belonging to one, two, or more clusters with some membership degree [55]. FCM is based on the minimization of the following objective function

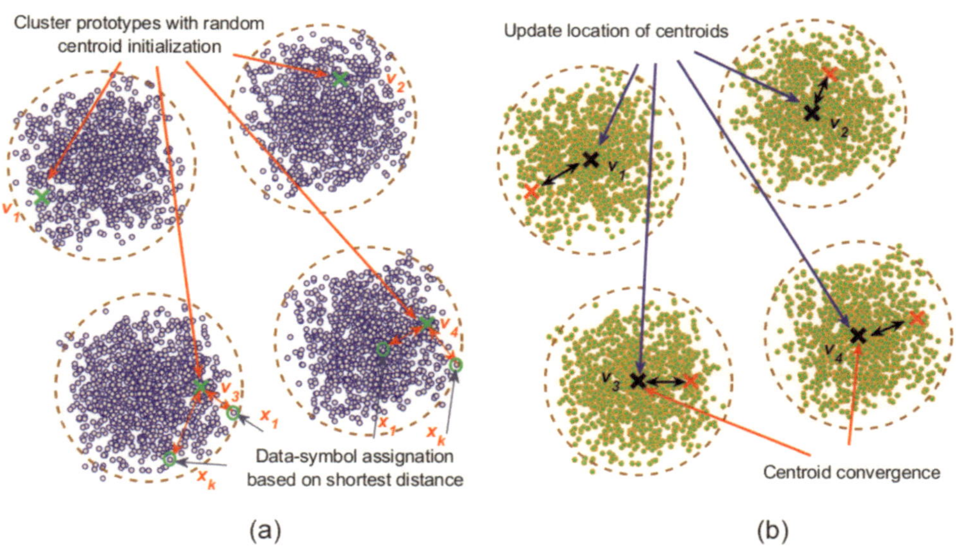

Figure 3.
k-means algorithm principle operation. (a) Cluster prototype and data symbol assignation. (b) Centroid location updating until convergence.

$$J_m = \sum_{i=1}^{c}\sum_{k=1}^{N} \{\mu_{ik}\}^m \|x_k - v_i\|^2 \tag{16}$$

where k is the number of clusters, N is the total number of observations. The squared norm represents the sum of squares of the distances of each data point to its assigned centroid vector c_k, for a set of binary indicator variables $\mu_{ik} \in \{0,1\}$, known as the degree of membership of the data points x_n associated to the cluster centroids c_k. The centroid of the cluster prototype is calculated as

$$v_i = \frac{\sum_{k=1}^{N} (\mu_{ik})^m x_k}{\sum_{k=1}^{N} (\mu_{ik})^m} \tag{17}$$

The goal is to find an assignment value of membership for each data point to the cluster where belongs (it means, the belonging degree per symbol for each cluster in the constellation diagram), as well as, the weighted sum of distances of each data point to its closest vector $\|c_k - v_i\|^2$ that minimizes J. In (8) and (9), $m > 1$ is a real number that governs the influence of individual fuzzy sets in fuzzy partition, and besides, when m goes toward infinity, all clusters tend to converge to the centroids of the data-set X.

5.3 Fuzzy c-means Gustafson-Kessel (FCM-GK)

The Gustafson-Kessel algorithm [55, 56] associates each sample point with its centroid and its covariance. The main feature of this clustering algorithm is the local adaptation of the distance matrix to identify clusters of different geometrical shapes in one data-set. Each cluster has its own norm-inducing matrix A_i, which yields the following inner product norm:

$$D_{ikA}^2 = (x_k - v_i)^T A_i (x_k - v_i), \quad 1 \le i \le c, \quad 1 \le k \le N \tag{18}$$

where v_i is the mean for that points over cluster i (v_i is referred as the cluster prototype), and the matrices A_i are used as optimization variables in the c-means functional, besides, allowing each cluster to adapt the distance norm to the local topological structure of the data. The objective functional J_m of GK algorithm is defined as

$$J_m = \sum_{i=1}^{c}\sum_{k=1}^{N} \mu_{ik}^m D_{ikA_i}^2 \tag{19}$$

However, the objective function cannot be directly minimized concerning A_i, because it is linear in A_i, and then to obtain a feasible solution, it is necessary to constrain the determinant of A_i. This procedure allows the matrix A_i to vary with its determinant fixed corresponding to the optimization of the cluster's shape while its volume remains constant:

$$\|A_i\| = \rho_i, \quad \rho > 0 \tag{20}$$

where ρ_i is fixed for each cluster. Using the Lagrange multiplier method, the following expression for A_i is obtained:

$$A_i = [\rho_i \ det(F_i)]^{1/n} F_i^{-1} \tag{21}$$

Being F_i the fuzzy covariance matrix of the *ith* cluster defined by:

$$F_i = \frac{\sum_{k=1}^{N} (\mu_{ik})^m (x_k - v_i)(x_k - v_i)^T}{\sum_{k=1}^{N} (\mu_{ik})^m} \tag{22}$$

The substitution of Eqs. (13) and (14) in Eq. (10) gives a generalized squared Mahalanobis distance norm between x_k and the cluster mean v_i, where the covariance is weighted by the membership degrees in the partition matrix U = $[\mu_{ik}]$. Besides, whereas the original FCM algorithm makes the implicit assumption that the clusters are spherical, the Gustafson-Kessel algorithm is not subject to this constraint and can identify ellipsoidal clusters, due to a size restriction on the covariance matrix whose determinant must be the unit.

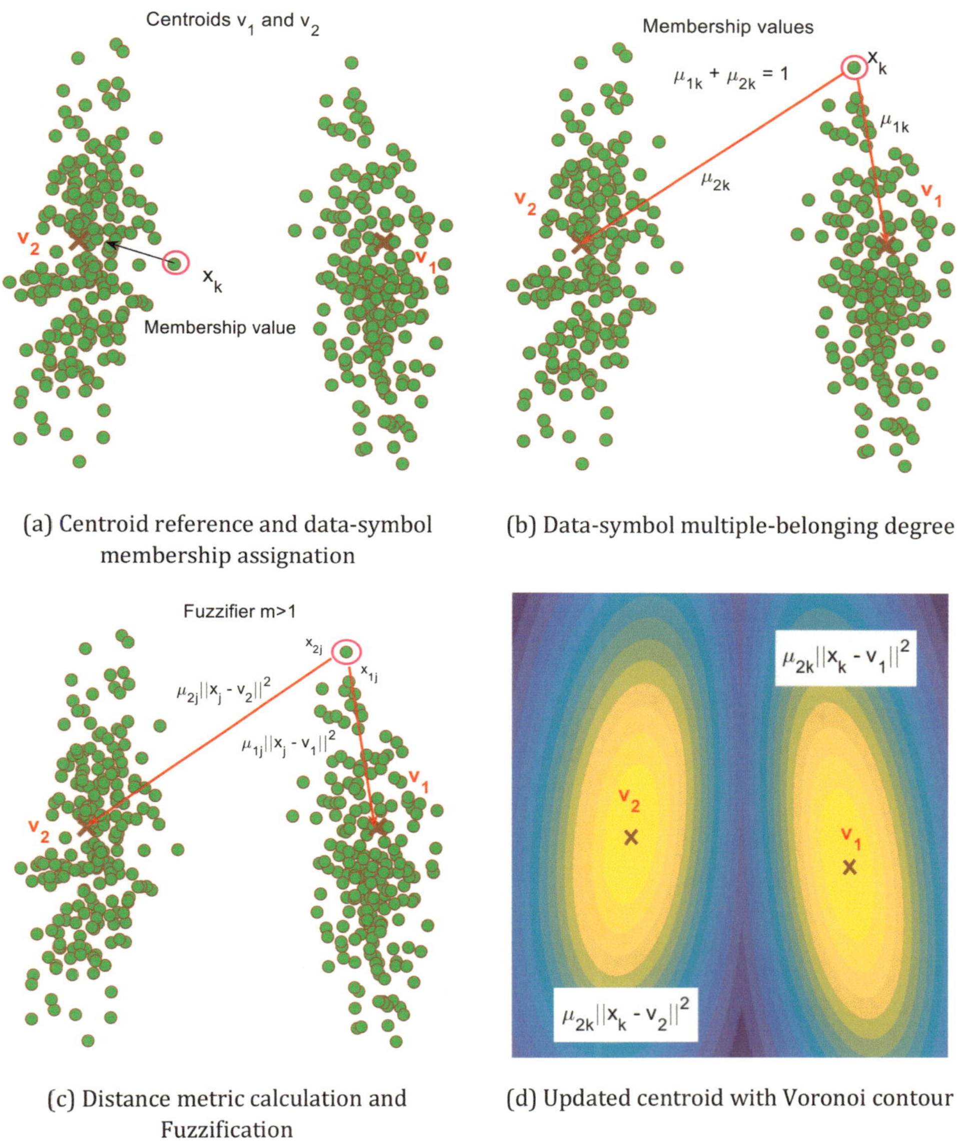

(a) Centroid reference and data-symbol membership assignation

(b) Data-symbol multiple-belonging degree

(c) Distance metric calculation and Fuzzification

(d) Updated centroid with Voronoi contour

Figure 4.
Fuzzy c-means GK step-by-step algorithm. (a) Centroid of cluster initialization at predefined locations and membership value estimation, (b) the association of each data point to the clusters, (c) distance norm computation and fuzzification stage, and as a final step, (d) Voronoi contour to define the non-Gaussian boundaries for each cluster distorted by PN.

In **Figure 4**, a graphical description of the FCM-GK algorithm is shown. The general setting for our clustering-based demodulation (for simplicity, we only use two clusters) is applied as follows: (i) the FCM-GK (in general, also k-means and FCM procedures are similar) algorithm requires as inputs, the data, and, the number of clusters v_i (predefined value for 16-QAM modulation format). The data symbols correspond to bidimensional (2-D) vectors $x(k) = [x_i(k)\ x_q(k)]$, where x_i and x_q are the in-phase and quadrature (IQ) component samples projected on complex plane. From this data-set, the weighted exponent m, the tolerance criteria $\varepsilon \leq 0.001$, and the partition matrix U (randomly) are all initialized; (ii) then centroid prototypes are estimated as indicated in (8) and, as shown in **Figure 4a**; (iii) after that, membership values are obtained for each data x_k with respect to the closest centroid (see **Figure 4b**), calculating the distance norm (k-means and FCM uses Euclidean norm, and, FCM-GK uses Mahalanobis distance norm, but constrained to fixed determinant of A equal to 1); (iv) the matrix fuzzy partition U is updated (see **Figure 4c**); and, finally, (v) the criterion for termination is calculated as $||U(l) - U(l-1)|| < \varepsilon$, if convergence is achieved, the algorithm stops, if not, returns to step *iii*. Besides, we obtain Voronoi contours [57] to define the nonlinear decision boundaries without considering the shape or orientation of the

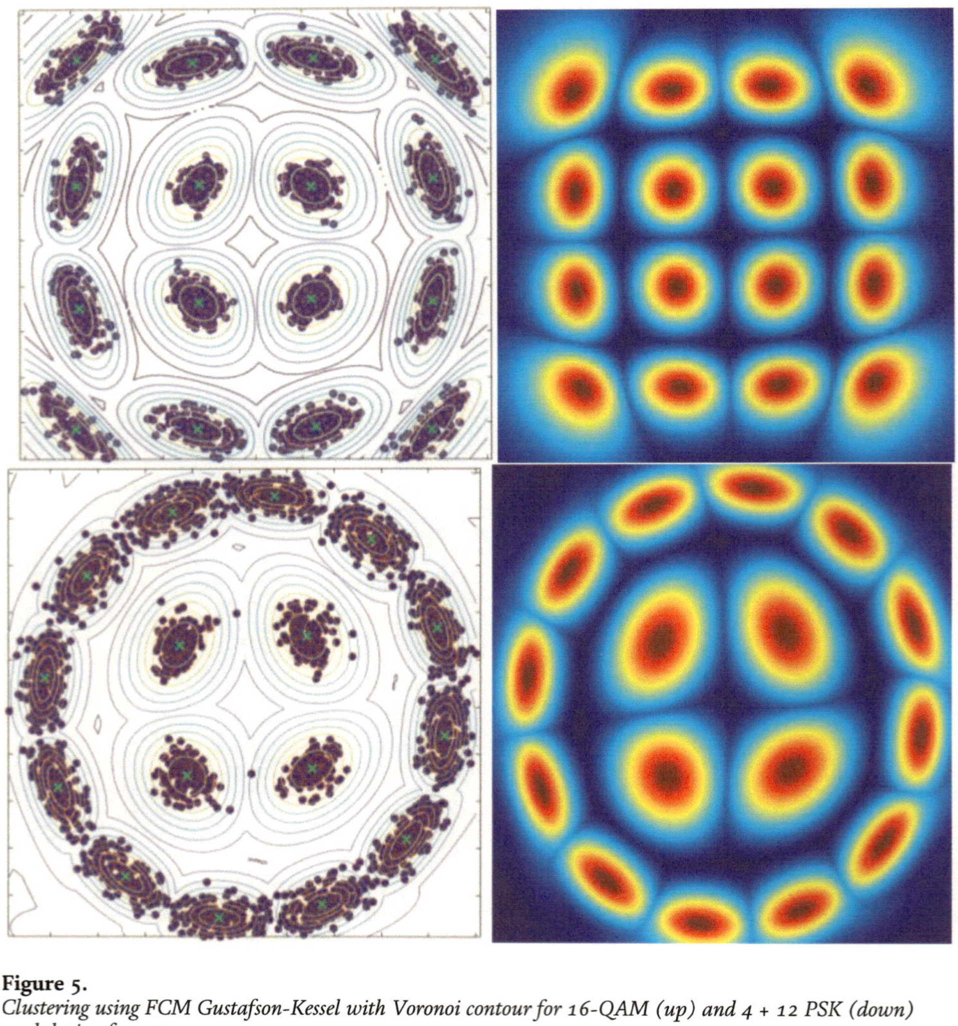

Figure 5.
Clustering using FCM Gustafson-Kessel with Voronoi contour for 16-QAM (up) and 4 + 12 PSK (down) modulation formats.

symbol within the constellation diagram. This effect is relevant especially for the symbols in the corners, which are the most severely impacted by PN, as it is shown in **Figure 4d**, and explained as follow.

5.4 Voronoi contours for flexible thresholding

The Voronoi diagram by contours has been used to assist the clustering-based demodulation stage when symbol distortion is severe [11]. An approximation to its definition may be introduced as the proximity regions to a set of objects in a bidimensional plane. As an example, **Figure 5** shows a Voronoi diagram for a 16-QAM and 4 + 12 PSK constellation defining the boundaries of the data symbols. This technique adds flexibility to the demodulation process because non-Gaussian boundaries caused by PN are not trivial to estimate and require nonlinear equalizers highly time-consuming and computationally expensive. In the 4 + 12 PSK modulation scheme, due to the lower ED between symbols in the external ring, the overlapping on the symbol clusters is evident, and as a direct consequence, the symbol error rate is increased at the demodulation stage. Then, an arbitrary symbol point is assigned to the site (single point—centroid—of a region inside a Voronoi partition that meets certain properties [58]) with the minimal Euclidean distance. The set of all symbol points with the same assigned site builds the Voronoi region (VR). In our case, for both modulation schemes, we have 16 sites leading to more Voronoi edges (VEs), these VEs are defined by the points which hold the same ED for two sites. But, in general, there are more than two sites requiring at least one Voronoi vertices (VV), which represents a symbol point with the same minimum Euclidean distance toward the 16 sites. In this way, the Voronoi diagram resulting as can be seen in **Figure 5** on the right, defining the contours properly for all the sites corresponding to the 16 symbol points of the constellation.

6. Experimental setup

Figure 6 shows the schematic diagram for the RoF system setup as single channel and single polarization. A pseudorandom binary sequence (PRBS) with length 2^{18} at a baud rate of 250 MBd was generated to deliver 1 Gb/s in a single channel and single polarization case. The pulse shaping was implemented using a finite impulse response (FIR) root-raised-cosine (RRC) filter with eight taps and a commonly used roll-off factor α = 0.2. After that, the modulation process was performed encoding 4 bits per symbol for both 16-QAM and 4 + 12 PSK. This signal was digitized through a Fujitsu LEIA board with a sampling frequency of 64 GSa/s with an embedded up-conversion stage that translated the baseband signal to a RF

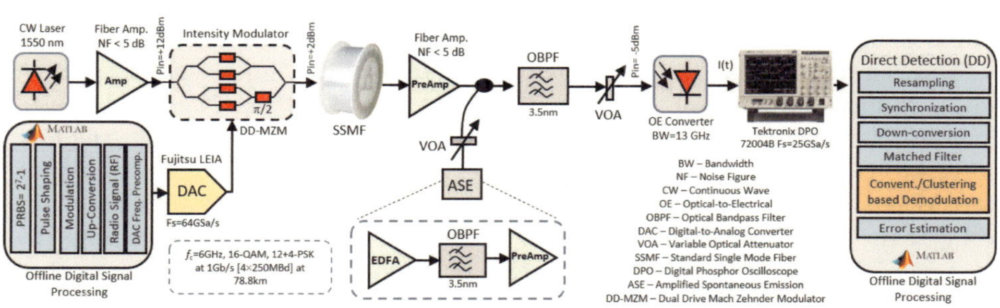

Figure 6.
RoF experimental setup.

carrier at 6 GHz. The RF signal was fed into the Mach-Zehnder modulator (MZM) RF inputs to modulate a continuous wave (CW) laser. The obtained optical signal was propagated over a spool of fiber with length 78.8 km. The employed optical source was a distributed feedback (DFB) laser (LW of 100 kHz) emitting at 1550 nm, which was amplified by an EDFA before the electro-optical MZM. To test the channel performance against noise, ASE noise was generated over a bandwidth of 3.5 nm centered around the CW wavelength and injected in the channel before the receiver. The received signal was pre-amplified and filtered by an optical bandpass filter (OBPF) with a bandwidth of 3.5 nm and photodetected by a Nortel PP-10G photoreceiver with an electrical bandwidth of 13 GHz. A variable optical attenuator (VOA) was adjusted to set the power at the input of the photodetector at −5 dBm. Then, the electrical signal was entered into a digital oscilloscope with a sampling rate of 25 GSa/s, and the captured data were stored for offline processing. Homodyne detection was performed at the receiver, and the frequency difference at the transmitter between the LO and the optical carrier was not compensated. This mismatch introduces PN distortion on data symbols, mainly on outer symbols like the one shown in **Figure 5**. After the data acquisition, the down-conversion stage translated the RF signal to baseband; and then, the matched filtering with the same prototype characteristics of the transmitter side was applied. Besides, in the demodulation stage, the conventional rectangular grid demodulations, i.e., *k*-means, FCM, and FCM-GK clustering-based methods are evaluated varying the OSNR from 16 to 36 dB, as a function of bit error rate (BER). The FCM-GK demodulation process is aided with the Voronoi diagrams to estimate the non-Gaussian boundary per symbol using different step-size per contour line.

7. Analysis of the results

Observing the 16-QAM constellation of **Figure 5**, the shape of symbols mainly in the closest to corners is quasi-ellipsoidal: this non-Gaussian shape is due to the hybrid channel nonlinearities jointly combined with NLPN, conversely, the inner clusters of symbols follows a circular behavior. It is important to highlight that the main distortions introduced by the PN are proportional to the noise variance (power) of each symbol, being this effect consistent with the higher amplitude of the external symbols as shown in **Figure 7**. In this figure, three rings can be clearly identified. Here, a ring is defined as the circumference along which symbols are distributed, and it is associated with an energy level on a circumference where symbols are distributed. Considering this non-Gaussian shape, and taking advantage of the clustering technique, it was possible to perform an individual symbol characterization through statistical analysis. In **Figure 8**, we show a 16-QAM constellation, where fragmentation per ring and symbol has been applied. We perform an individual analysis of each cluster by estimating the mean, standard deviation, variance, histogram, and contour. For example, for symbol "1001" (Gray coding), we obtain a mean of −0.9530 and a variance of 0.1986; this inner symbol grows with a circular shape like observed in an additive white Gaussian Noise (AWGN) channel. But for the symbol at the corner "0000," the shape is ellipsoidal, and the normal distribution has a higher variance, i.e. dispersion, and variance of the symbols. With these features, we can choose the appropriate clustering algorithm to mitigate the PN impact on specific symbols, i.e., by applying FCM-GK for detecting clusters with ellipsoidal shape, and *k*-means or conventional demodulation method based on a rectangular fixed grid in inner symbols, where the distortion is minor. As observed in the histograms of **Figure 8**, the symbol toward close to the center of the constellation diagram (i.e., the symbol 1111) has a Gaussian shape with lower

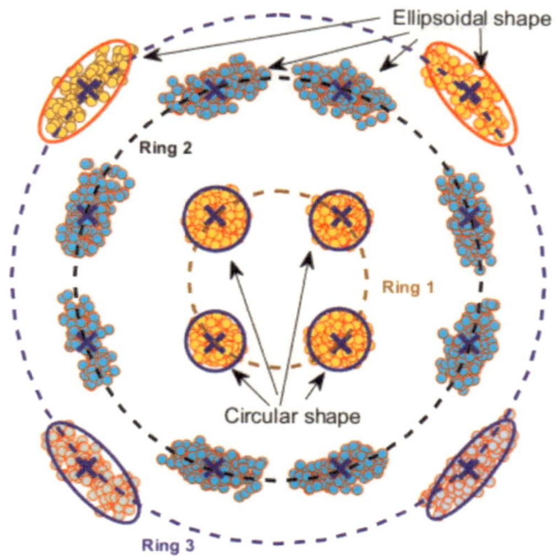

Figure 7.
PN impact over symbols distributed over rings for a 16-QAM constellation.

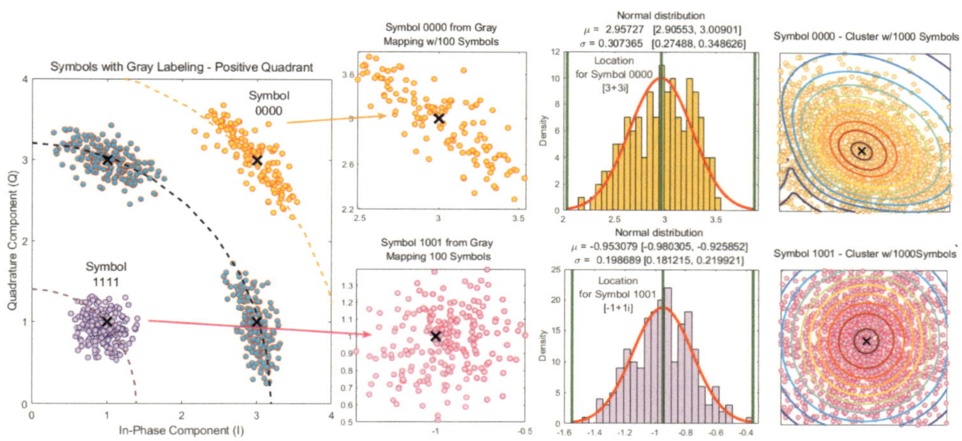

Figure 8.
Independent statistical analysis on each symbol of constellation diagram.

standard deviation (σ = 0.1986) than the symbol on the corner. In the same way, the symbol 0000 spreads the Gaussian bell on the top and has a higher standard deviation value of 0.3073. This analysis allows to conclude that lower variance (or, equivalently, lower standard deviation) values correspond to more compact clusters that have circular shape, comparable with the cluster that would be obtained when only AWGN is present.

The Voronoi contour shows a more defined Gaussian shape for the inner symbol and an elongated non-Gaussian shape for the external symbol. This statistical analysis is performed for the different OSNR levels measured in the RoF experimental setup, and the combined effect of data symbol clustering with Voronoi contour enhances the demodulation stage, either using a direct detection [26] or coherent receiver. Besides, by observing the constellation characterization, a residual phase offset was detected on external symbols of constellations, then using the clustering fragmentation a the dual-stage phase offset compensation (DS-PS-RF) proposed method in [9], an additional improvement of BER can be obtained when conventional demodulation is applied with rectangular decision boundaries.

In **Figure 9**, we plot the BER performance vs. OSNR for 16-QAM at the error-free condition after the forward error correction (FEC) encoding that sets the error-free threshold at 2×10^{-2}. The line with blue color and circle marker corresponds to the conventional demodulation method; the error performance for FCM is the curve in orange color with a diamond marker. The other curve, with a square red marker, corresponds to the k-means algorithm, and, finally, with the asterisk green marker, the FCM-GK method is plotted. k-means and FCM-GK perform a similar gain of 2.1 dB in the OSNR scale comparing with conventional demodulation of 16-QAM; however, beyond 26 dB, the gain margin for FCM-GK over k-means holds around 1 dB for the rest of the curve up to 36 dB. Besides, for FCM-GK assisted with Voronoi contour, the OSNR achieves a gain of 3.2 dB.

The constellation inset shows the data symbol distribution for FCM-GK obtaining an error vector magnitude (EVM) of 20.1% for 16-QAM. On the other hand, in **Figure 10**, the BER performance results are presented for 4 + 12 PSK modulation format. An improvement in the OSNR scale of 1.4 dB for FCM-GK compared with the conventional demodulation method is observed in **Figure 10**, besides in the inset, an EVM of 20.9% for the 4 + 12 PSK is obtained at 22 dB of OSNR. The curve for the FCM algorithm shows the worst performance compared with the other methods. These effects are possibly due to distortions on the shape of the clusters, which are ellipsoidal and have higher overlaps due to lower Euclidean distance among clusters. Similar performance curves are shown for k-means and conventional demodulation algorithms.

The decision boundaries obtained from Voronoi contours for each cluster define non-Gaussian regions; then, when the inner symbols grow circularly (following a Gaussian shape), the conventional fixed grid or *k*-means demodulation techniques must be used but, in contrast, when outer symbols take ellipsoidal shape, the FCM-GK provides higher tolerance to distortions and improves the BER and OSNR. Also, it is evident that to optimize the geometric shaping is a non-trivial task because the Euclidean distance must be maximized by performing a nonconventional

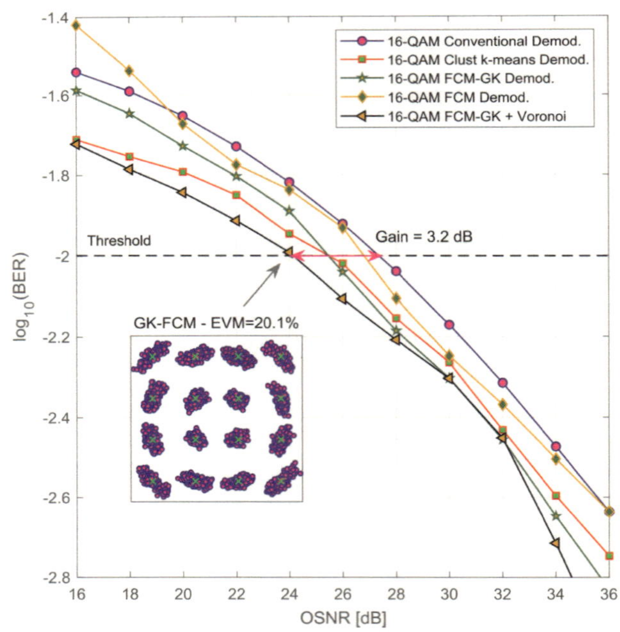

Figure 9.
BER performance for 16-QAM at 1 Gb/s after the 78.8 km span.

distribution of symbols, besides limiting the symbol energy level to avoid power penalties concerning conventional demodulation formats. However, to have an idea of the non-Gaussian behavior due to NLPN, **Figure 11a** and **b** and represent the 3D behavior of the received data symbols showing how the distortions impact and change the shape of the symbols. Besides, a residual phase offset of centroids mainly over external clusters shows that ideal centroids are shifted to newer locations over the plane as demonstrated experimentally in [34]. Additionally, it is possible to improve error performance using a second stage to estimate amplitude changes and phase offset to compensate these residual shifts observed in the constellation. The a priori initialization of centroids with a fixed number of clusters avoids introducing a

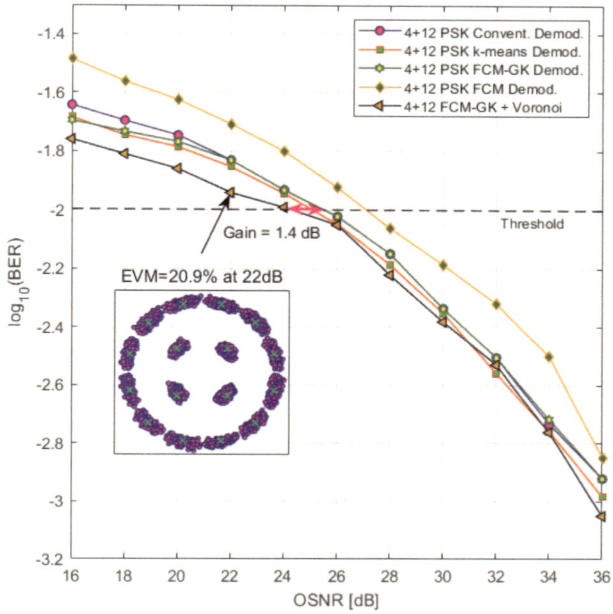

Figure 10.
BER performance for 4 + 12 PSK at 1 Gb/s after the 78.8 km span.

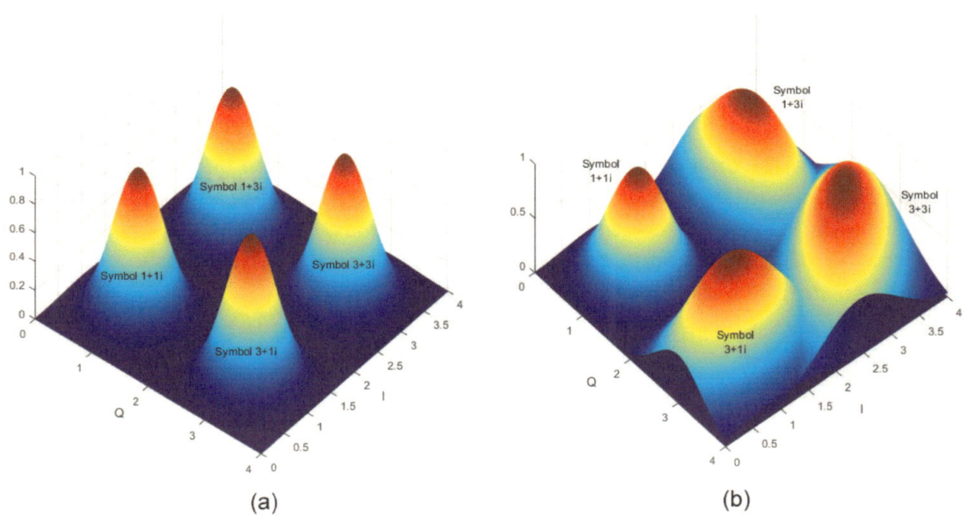

Figure 11.
(a) Data symbols under AWGN, and (b) data symbols under non-Gaussian distortion.

higher computational time of the evaluated algorithms. Besides, computational complexity in comparison with linear and nonlinear equalizers is an issue to be further explored in a future work. Additional specific experimental studies should be conducted to validate with more data the EVM/SNR estimation, and to define the impact of key aspects such as the shape of the clusters, or the performance for different distance norms, and the evaluation of optimized algorithms derived from conventional k-means and FCM.

Finally, the lower time-computing because of centroid initialization compensates the time spent in the calculation of inverse matrices required during the execution of Gustafson-Kessel algorithm. However, further studies on different optimization techniques must be conducted, e.g. for smaller OSNR values and using nonconventional modulation formats, implementing them in real-time scenarios, for a deeper analysis of the processing time issue.

8. Conclusions

We proposed and experimentally evaluated a constellation-based tool for demodulation, aided by Voronoi thresholding, clustering fragmentation, and signal demodulation, validated with 16-QAM and 4 + 12 PSK constellations, distorted especially with nonlinear phase noise. Fuzzy c-means Gustafson-Kessel was evaluated as an alternative to performing the demodulation-based clustering, taking advantage of the improved identification of clustering with non-Gaussian shapes enabled by the FCM-GK. The non-Gaussian boundaries for symbol demodulation were estimated using contour Voronoi diagrams, obtaining improvements of 3.2 dB for 16-QAM and 1.4 dB for 4 + 12 PSK using the fuzzy c-means Gustafson-Kessel algorithm. We compared the error performance for different OSNRs as a function of the BER at a threshold level of 2×10^{-2}. Our fragmented constellation in combination with Voronoi thresholding method mitigates distortive effects over data symbols, performing localized transmission impairments characterization by applying the individual analysis of each cluster of symbols or subconstellations (quadrant in the constellation diagram), without requiring frequency and phase carrier recovery algorithms. Besides, this method can be extended to the coherent receiver avoiding the time-consuming post-compensation algorithms to equalize the received signals. However, if during the characterization stage the distortions of the received constellation have Gaussian behavior, employing clustering-based demodulation using k-means or the conventional demodulation algorithm with fixed grid provides similar error performance as demonstrated, but the trade-off is focused on the computational effort. But if non-Gaussian shapes are identified over the constellation, then a better approach to improve the system performance is to use the FCM-GK method aided with a Voronoi diagram. Finally, a new proposal for estimating the EVM reduces the number of symbols required to calculate the error metric; this reduces the computational time because only external symbols are required to approach the result.

Acknowledgements

This work has been supported by the Grant 617 of Colciencias and the Universidad de Antioquia. A special thanks to the Universidad Pedagógica y Tecnológica de Colombia (UPTC) for their financial support to encourage my doctoral studies by agreement 041. The reported activity has resulted from experimental setups conducted at the Photonics and Microwave Laboratory of Scuola

Sant'Anna at Pisa, Italy, and are supported by the EU ITN project FiWin5G No. 642355. Project PREVENTION with the contribution of the Italian Ministry of Foreign Affairs, General Directorate for the Country Promotion.

Author details

Eduardo Avendaño Fernández[1*], Ana María Cárdenas Soto[2], Neil Guerrero Gonzalez[3], Giovanni Serafino[4], Paolo Ghelfi[5] and Antonella Bogoni[6]

1 Universidad Pedagógica y Tecnológica de Colombia, Sogamoso, Colombia

2 Universidad de Antioquia, Medellín, Colombia

3 Universidad Nacional de Colombia, Manizalez, Colombia

4 Scuola Superiore Sant'Anna and National Inter-University Consortium for Telecommunications (CNIT), Pisa, Italy

5 National Inter-University Consortium for Telecommunications (CNIT), Pisa, Italy

6 Scuola Superiore Sant'Anna and National Inter-University Consortium for Telecommunications (CNIT), Pisa, Italy

*Address all correspondence to: eduardo.avendano@uptc.edu.co

References

[1] Beygi L, Agrell E, Kahn JM, Karlsson M. Coded modulation for fiber-optic networks: Toward better tradeoff between signal processing complexity and optical transparent reach. IEEE Signal Processing Magazine. 2014;**31**(2):93-103

[2] Chagnon M. Optical communications for short reach. Journal of Lightwave Technology. 2019;**37**(8):1779-1797

[3] Karim A, Devenport J. High dynamic range microwave photonic links for RF signal transport and RF-IF conversion. Journal of Lightwave Technology. 2008; **26**(15):2718-2724

[4] Valkama M, Springer A, Hueber G. Digital signal processing for reducing the effects of RF imperfections in radio devices — An overview. In: Proceedings of 2010 IEEE International Symposium on Circuits and Systems. 2010. pp. 813-816

[5] Agrell E et al. Roadmap of optical communications. Journal of Optics. 2016;**18**(6):63002

[6] Schawlow CH, Townes AL. Infrared and optical maser. Physical Review. 1949;**112**:1940-1949

[7] Henry C. Theory of the linewidth of semiconductor lasers. IEEE Journal of Quantum Electronics. 1982;**18**(2): 259-264

[8] Li L, Zhang G, Zheng X, Li S, Zhang H, Zhou B. Phase noise suppression for single-sideband modulation radio-over-fiber systems adopting optical spectrum processing. IEEE Photonics Technology Letters. 2013;**25**(11):1024-1026

[9] Fernández EA, Serafino G, Bogoni A, Cárdenas Soto AM, Fresi F, González NG. Adaptive phase-offset compensation on fragmented constellation diagrams in radio-over-fiber systems. In: Latin America Optics and Photonics Conference. 2018. p. Tu5E.4

[10] Alvarado A, Agrell E, Brännström F. Asymptotic comparison of ML and MAP detectors for multidimensional constellations. IEEE Transactions on Information Theory. 2018;**64**(2): 1231-1240

[11] Han Y, Yu S, Li M, Yang J, Gu W. An SVM-based detection for coherent optical APSK systems with nonlinear phase noise. IEEE Photonics Journal. 2014;**6**(5):1-10

[12] Wang D et al. Nonlinearity mitigation using a machine learning detector based on *k*-nearest neighbors. IEEE Photonics Technology Letters. 2016;**28**(19):2102-2105

[13] Reddy D, Jana PK. Initialization for K-means clustering using Voronoi diagram. Procedia Technology. 2012;**4**: 395-400

[14] Savory SJ. Digital coherent optical receivers: Algorithms and subsystems. IEEE Journal of Selected Topics in Quantum Electronics. 2010;**16**(5): 1164-1179

[15] Savory SJ. Digital filters for coherent optical receivers. Optics Express. 2008; **16**(2):804-817

[16] Kakkar A et al. Impact of local oscillator frequency noise on coherent optical systems with electronic dispersion compensation. Optics Express. 2015;**23**(9):11221-11226

[17] Minzioni P, Pusino V, Cristiani I, Marazzi L, Martinelli M, Degiorgio V. Study of the Gordon–Mollenauer effect and of the optical-phase-conjugation compensation method in phase-modulated optical communication

systems. IEEE Photonics Journal. 2010; **2**(3):284-291

[18] Xu D et al. Laser phase and frequency noise measurement by Michelson interferometer composed of a 3 x 3 optical fiber coupler. Optics Express. 2015;**23**(17):22386-22393

[19] Ip E, Kahn JM. Compensation of dispersion and nonlinear impairments using digital backpropagation. Journal of Lightwave Technology. 2008;**26**(20): 3416-3425

[20] Navarro JR et al. Carrier phase recovery algorithms for coherent optical circular mQAM systems. Journal of Lightwave Technology. 2016;**34**(11): 2717-2723

[21] Jain A, Krishnamurthy PK. Phase noise tracking and compensation in coherent optical systems using Kalman filter. IEEE Communications Letters. 2016;**20**(6):1072-1075

[22] Pakala L, Schmauss B. Two stage extended Kalman filtering for joint compensation of frequency offset, linear and nonlinear phase noise and amplitude noise in coherent QAM systems. In: 2017 19th International Conference on Transparent Optical Networks (ICTON). 2017. pp. 1-4

[23] Jiang L, Yan L, Yi A, Pan Y, Pan W, Luo B. K-nearest neighbor detector for enhancing performance of optical phase conjugation system in the presence of nonlinear phase noise. IEEE Photonics Journal. 2018;**10**(2):1-8

[24] Liu T, Djordjevic IB, Wang T. Optimal signal constellation design for ultra-high-speed optical transport in the presence of phase noise. In: 2014 Conference on Lasers and Electro-Optics (CLEO) - Laser Science to Photonic Applications. 2014. pp. 1-2

[25] Rios-Müller R, Renaudier J, Schmalen L, Charlet G. Joint coding rate and modulation format optimization for 8QAM constellations using BICM mutual information. In: 2015 Optical Fiber Communications Conference and Exhibition (OFC). 2015. pp. 1-3

[26] Fernández EA, GranadaTorres JJ, Cárdenas Soto AM, González NG. Geometric constellation shaping with demodulation based-on clustering to mitigate phase-noise in radio-over-fiber systems. In: Latin America Optics and Photonics Conference. 2018. p. Tu5E.3

[27] Jones RT, Eriksson TA, Yankov MP, and Zibar D. Deep Learning of Geometric Constellation Shaping Including Fiber Nonlinearities. In: Proceedings of the 44rd European Conference and Exhibition on Optical Communications (ECOC 2018). IEEE; 2018. DOI: 10.1109/ECOC.2018.8535453

[28] Hajimiri A, Lee TH. A general theory of phase noise in electrical oscillators. IEEE Journal of Solid-State Circuits. Feb 1998;**33**(2):179-194. DOI: 10.1109/4.658619

[29] Demir A. Nonlinear phase noise in optical-fiber-communication systems. Journal of Lightwave Technology. 2007; **25**(8):2002-2032

[30] Ho K-P, Kahn JM. Electronic compensation technique to mitigate nonlinear phase noise. Journal of Lightwave Technology. 2004;**22**(3): 779-783

[31] Lau APT, Kahn JM. Power profile optimization in phase-modulated Systems in Presence of nonlinear phase noise. IEEE Photonics Technology Letters. 2006;**18**(23):2514-2516

[32] Hueda MR, Crivelli DE, Carrer HS, Agazzi OE. Parametric estimation of IM/DD optical channels using new closed-form approximations of the signal PDF. Journal of Lightwave Technology. 2007;**25**(3):957-975

[33] Khan FN, Lu C, Lau APT. Optical performance monitoring in fiber-optic networks enabled by machine learning techniques. In: 2018 Optical Fiber Communications Conference and Exposition (OFC). 2018. pp. 1-3

[34] Skoog RA et al. Automatic identification of impairments using support vector machine pattern classification on eye diagrams. IEEE Photonics Technology Letters. 2006; **18**(22):2398-2400

[35] Wang D et al. Nonlinear decision boundary created by a machine learning-based classifier to mitigate nonlinear phase noise. In: 2015 European Conference on Optical Communication (ECOC). 2015. pp. 1-3

[36] Wang D et al. KNN-based detector for coherent optical systems in presence of nonlinear phase noise. In: 2016 21st OptoElectronics and Communications Conference (OECC) Held Jointly with 2016 International Conference on Photonics in Switching (PS). 2016. pp. 1-3

[37] Guerrero Gonzalez N, Caballero Jambrina A, Amaya Fernández FO, Zibar D, Tafur Monroy I. Experimental 2.5 Gbit/s QPSK WDM coherent phase modulated radio-over-fibre link with digital demodulation by a K-means algorithm. In: IEEE, pp. 1-2 BT-35th European Conference on Optical Comm. 2009

[38] Zibar D, Piels M, Jones R, Schäeffer CG. Machine learning techniques in optical communication. Journal Lightwave Technology. 2016;**34**: 1442-1452

[39] Zhou T, Liu J, Guo C, Wu X, Zhong K, Yu S. Optical modulation format identification based on Gaussian mixture model of signal amplitude distribution. In: 2016 Asia Communications and Photonics Conference (ACP). 2016. pp. 1-3

[40] Wang Z et al. Failure prediction using machine learning and time series in optical network. Optics Express. 2017;**25**(16):18553-18565

[41] Zhang W et al. Identifying modulation formats through 2D stokes planes with deep neural networks. Optics Express. 2018;**26**(18): 23507-23517

[42] Khan FN, Fan Q, Lu C, Lau APT. An optical communication's perspective on machine learning and its applications. Journal of Lightwave Technology. 2019; **37**(2):493-516

[43] Zhao T, Nehorai A, Porat B. K-means clustering-based data detection and symbol-timing recovery for burst-mode optical receiver. IEEE Transactions on Communications. 2006; **54**(8):1492-1501

[44] Gonzalez NG, Zibar D, Yu X, Monroy IT. Optical phase-modulated radio-over-fiber links with k-means algorithm for digital demodulation of 8PSK subcarrier multiplexed signals. In: 2010 Conference on Optical Fiber Communication (OFC/NFOEC), Collocated National Fiber Optic Engineers Conference. 2010. pp. 1-3

[45] Xu M et al. Multi-stage machine learning enhanced DSP for DP-64QAM coherent optical transmission systems. In: 2019 Optical Fiber Communications Conference and Exhibition (OFC). 2019. pp. 1-3

[46] Ma J, He J, Shi J, He J, Zhou Z, Deng R. Nonlinear compensation based on K-means clustering algorithm for Nyquist PAM-4 VLC system. IEEE Photonics Technology Letters. 2019; **31**(12):935-938

[47] Cui Y, Zhang M, Wang D, Liu S, Li Z, Chang G-K. Bit-based support vector machine nonlinear detector for millimeter-wave radio-over-fiber

mobile fronthaul systems. Optics Express. 2017;**25**(21):26186-26197

[48] Wang D et al. Intelligent constellation diagram analyzer using convolutional neural network-based deep learning. Optics Express. 2017; **25**(15):17150-17166

[49] Caballero FJV et al. Machine learning based linear and nonlinear noise estimation. Journal of Optical Communications and Networking. 2018;**10**(10):D42-D51

[50] Zhang J, Gao M, Chen W, Shen G. Non-data-Aidedk-nearest neighbors technique for optical fiber nonlinearity mitigation. Journal of Lightwave Technology. 2018;**36**(17):3564-3572

[51] Pulikkoonattu R. Oscillator Phase Noise and Sampling Clock Jitter (PDF), Tech Note, Bangalore, India: ST Microelectronics, retrieved Oct. 10, 2019. 12 Jun 2007. Available from: http://documents.epfl.ch/users/p/pu/pulikkoo/private/report_pn_jitter_oscillator_ratna.pdf

[52] Nakagawa T, Araki K. Effect of phase noise on RF communication signals. In: Vehicular Technology Conference Fall 2000. IEEE VTS Fall VTC2000. 52nd Vehicular Technology Conference (Cat. No.00CH37152). Vol. 2. 2000. pp. 588-591

[53] Kester W. Converting Oscillator Phase Noise to Time Jitter (PDF), Tech Note, Analog Devices, retrieved Oct. 10, 2019. Available from: https://www.analog.com/media/en/training-seminars/tutorials/MT-008.pdf

[54] Gonzalez NG, Zibar D, Caballero A, Monroy IT. Experimental 2.5-Gb/s QPSK WDM phase-modulated radio-over-fiber link with digital demodulation by a *k*-means algorithm. IEEE Photonics Technology Letters. 2010;**22**(5):335-337

[55] Bezdek JC. Pattern Recognition with Fuzzy Objective Function Algorithms. 1st ed. New York: Springer Verlag; 1981

[56] Gustafson DE, Kessel WC. Fuzzy clustering with a fuzzy covariance matrix. In: 1978 IEEE Conference on Decision and Control Including the 17th Symposium on Adaptive Processes. 1978. pp. 761-766

[57] Atsuyuki Okabe SNC, Boots B, Sugihara K. Spatial Tessellations: Concepts and Applications of Voronoi Diagrams. 2nd ed. New York: John Wiley & Sons; 2000

[58] Okabe DG, Boots A, Sugihara B, Chiu K, Kendall SN. Poisson Voronoi Diagrams, in Spatial Tessellations: Concepts and Applications of Voronoi Diagrams. 1st ed. New Jersey: John Wiley & Sons, Ltd; 2000

Chapter 2

Deep Learning Based Prediction of Transfer Probability of Shared Bikes Data

Wenwen Tu

Abstract

In the pile-free bicycle sharing scheme, the parking place and time of the bicycle are arbitrary. The distribution of the pile does not constrain the origin and destination of the journey. The travel demand of the user can be derived from the use of the shared bicycle. The goal of this article is to predict the probability of transition for a shared bicycle user destination based on a deep learning algorithm and a large amount of trajectory data. This study combines eXtreme Gradient Boosting (XGBoost) algorithm, stacked Restricted Boltzmann Machines (RBM), support vector regression (SVR), Differential Evolution (DE) algorithm, and Gray Wolf Optimization (GWO) algorithm. In an experimental case, the destinations of the cycling trips and the probability of traffic flow transfer for shared bikes between traffic zones were predicted by computing 2.46 million trajectory points recorded by shared bikes in Beijing. The hybrid algorithm can improve the accuracy of prediction, analyze the importance of various factors in the prediction of transfer probability, and explain the travel preferences of users in the pile free bicycle-sharing scheme.

Keywords: deep learning, restricted Boltzmann machines, support vector regression, eXtreme gradient boosting, shared bikes data

1. Introduction

Bicycle sharing is a new type of transportation with low energy consumption and emissions. It serves short-distance travel and helps solve the "last mile" problem [1]. With the rapid development of the mobile Internet, the pileless bicycle began to replace the pile station bicycle [2]. In the pile-free bicycle sharing scheme, the parking place and time of the bicycle are arbitrary. The distribution of the pile does not constrain the origin and destination of the journey. The travel demand of the user can be derived from the use of the shared bicycle. The distribution of destinations for shared bike users is a valuable study. However, the large number of shared bicycle tracks requires a lot of computation time. This paper sets up different traffic areas and studies the law of shared bicycle flow transfer between the traffic areas. On this basis, we predict the ratio of the traffic flow of shared bicycles between traffic areas. It can be considered as the probability that the shared bicycle user selects the traffic area A as the origin and the traffic area B as the destination.

The pile-free shared bicycle system puts a large number of bicycles in the city. The amount is much higher than the number of traditional piled public bicycles. Therefore, when dealing with massive volume of trajectory data volume as a data set, classical statistical methods and traditional neural network algorithms would have limited processing capabilities.

As a newly developed travel method, the algorithms for the destination prediction of trips based on shared bikes need to be researched in depth [3–5]. In Deep neural networks (DNN), the model with multi-hidden layers can be developed based on the artificial neural network. The hidden layers of DNN convert the input data into a more abstract compound representation [6–10].

The Restricted Boltzmann Machine (RBM) is an algorithm that can be used for dimensionality reduction, classification, regression, and feature learning problems. RBM reconstructs data in an unsupervised algorithm and adjusts the weight through the process of reverse transfer and forward transfer. The RBM gradually approaches the original input and learns the probability distribution on the input set [11–15].

In this paper, a stacked RBM-SVR algorithm is constructed by combining support vector regression (SVR) [16] and stacking RMB algorithm. RBM-SVR is used to predict continuous output values. The error penalty factor c_s and kernel function parameter γ_s are the basic parameters of the radial basis function of the SVR model. The value of c_s and γ_s will directly affect the fit and generalization ability of the SVR [17–19]. In order to improve the accuracy of prediction, this paper needs to introduce intelligent algorithms to optimize the selection of parameter values.

In machine learning algorithms, Mirjalili et al. [20] proposed Gray Wolf Optimizer (GWO) as a meta-heuristic algorithm for solving many multi-modal functions in 2014. In addition, Storn and Price [21] proposed a differential evolution (DE) algorithm. The DE algorithm is an optimization algorithm based on modern intelligent theory. DE intelligently optimizes the direction of search through groups generated by cooperation and competition among individuals. Based on the above theory, an algorithm called the differential evolution Gray Wolf Optimizer algorithm (DEGWO) is used to optimize c_s and γ_s. DEGWO generates initial populations, subpopulations, and variant populations for each iteration, and then uses the GWO's capabilities for global searching to optimize the c_s and γ_s parameters.

After processing by RBM algorithm, the input data are transformed into a sparse matrix containing a high amount of information. It can reduce the computation time of SVR prediction, but it may also increase the number of outliers in the SVR algorithm, increase the complexity of the SVR model and reduce the stability of the fitting process. To solve this problem, this paper proposes a hybrid algorithm that combines the eXtreme Gradient Boosting (XGBoost) algorithm with the stacked RBM-SVR algorithm. XGBoost is an optimized distributed gradient boosting library designed to be highly efficient, flexible, and portable [22]. XGBoost uses the Exclusive Feature Bundling method to transform several sparse features into a dense matrix [23]. In the process, the approximate greedy algorithm is used to find the best combination of merged features when the number of bundles is the smallest. The XGB algorithm realizes the partition of nodes by the second order gradient and optimizes the greedy algorithm of splitting nodes by the local approximation algorithm [24].

The principal purpose of this paper is to build a hybrid model that combines the XGBoost model, the stacked RBM-SVR network, and the DEGWO optimization algorithm. This paper analyzes the trajectory data of shared bicycles, extracts the cell information, and predicts the probability of user destination selection in the traffic area, that is, predicts the transfer probability of shared bikes.

2. Background

Artificial intelligence (AI) is a domain of computer science that studies how to apply computings to simulate the fundamental theories, methods, and techniques of human knowledge. As the mainstream algorithm of artificial intelligence, deep learning is considered capable of solving many challenges in the field of computer vision, prediction, and optimization. It realizes the automatic positioning of targets and automated learning of target features, which improves the speed and accuracy of target detection. Artificial intelligence is mainly used to share bicycles in the following aspects.

First, the user's travel behavior and the law of spatial movement can be obtained through machine learning algorithms and statistical theory analysis. The user's travel preferences can be quantitatively analyzed. Researchers can discuss the impact of various influencing factors on shared bicycle usage, such as the mix of land use, the degree of convergence with public transport facilities, the sharing of bicycle infrastructure, rainfall and high temperatures [25, 26].

Second, through the deep learning algorithm of AI technology, the dynamic demand and parking demand of the shared bicycle users can be predicted. The focus of this paper is on this issue. This paper uses a deep learning algorithm to predict the probability of user destination selection. Xu et al. [27] prove that the long short-term memory neural networks (LSTM NNs) in deep learning algorithms are superior to traditional statistical metrology algorithms and advanced machine learning algorithms. LSTM NNs can better predict the riding demand dynamically. Besides, based on the distribution of road networks and travel needs, researchers can predict parking demand and develop better layout strategies for electronic fences [28].

Finally, according to the deep reinforcement learning algorithm in AI technology, a shared bicycle scheduling model can be constructed. Deep reinforcement learning combines the perception of deep learning with the decision-making ability of reinforcement learning. It can be directly controlled according to the original input data. It is an artificial intelligence method that is closer to human thinking. Based on this algorithm, the dynamic scheduling model can efficiently optimize goals such as improving user satisfaction and reducing system cost [29–31].

3. A stacked RBM_SVR deep learning algorithm

RBM_SVR is a deep learning model that connects three stacked RBM models and one SVR model. First, in RBM_SVR, the bottommost RBM is trained with the original input data, and the top RBM takes the feature extracted by the bottom RBM as input and continues training. RBM_SVR repeats this process until the topmost RBM model is trained. Secondly, RBM_SVR fine-tunes the network through the traditional global learning algorithm (BP algorithm), so that the model can converge to the local best. Finally, RBM_SVR can efficiently train a deep network and output the predicted probability value according to the SVR model.

Each RBM model has a visible layer v and a hidden layer h. The neurons inside the RBM layer are unconnected, but the neurons between the layers are fully connected. The value of the RBM node variable is 0 or 1. The number of layers of visible layers and hidden layers of the RBM_SVR model are n and m, respectively. The energy equation of RBM_SVR is given by Eq. (1)

$$E(v,h|\theta) = -\sum_{i=1}^{n} a_i v_i - \sum_{j=1}^{m} b_j h_j - \sum_{i=1}^{n}\sum_{j=1}^{m} v_i w_{ij} h_j \tag{1}$$

where $\forall i, j \in \{0,1\}$, $\theta = \{w_{ij}, a_i, b_j\}$ involves the parameters of RBM; w_{ij} is a connection weight between the visible layer i and the hidden layer j; a_i represents the bias of the visible layer, and b_j denotes the bias of the hidden layer. $P(v,h|\theta) = \frac{e^{-E(v,h|\theta)}}{\sum_{v,h} e^{-E(v,h|\theta)}}$ is the joint probability distribution of this set of states (v,h) [32]. $P(v|\theta) = \frac{\sum_h e^{-E(v,h|\theta)}}{\sum_{v,h} e^{-E(v,h|\theta)}}$ is the marginal distribution of $P(v,h|\theta)$. Since each visible layer and each hidden layer are independent, the activation probability of the hidden layer j and the ith visible layer are shown in Eqs. (2) and (3), respectively [33].

$$P(v_i = 1|h,\theta) = \sigma\left(a_i + \sum_j w_{ij} h_j\right) \tag{2}$$

$$P(h_j = 1|v,\theta) = \sigma\left(b_j + \sum_i v_i w_{ij}\right) \tag{3}$$

where $\sigma(x) = \frac{1}{1+\exp(-x)}$. In RBM_SVR, the number of neurons per layer of RBM is 300. Based on the abstracted vector output from the stacked RBM model, the SVR model predicts the probability of the traffic transfer among traffic zones y_d, as shown in Eq. (4).

$$y_d = RBM.SVR(x) \tag{4}$$

where x is the input dataset, $RBM.SVR$ represents the RBM_SVR model.

4. An improved RBM-SVR algorithm

4.1 Principles of GWO algorithm

Assume that in a D-dimensional search space, the population size $X = [X_1, X_2, \ldots, X_{\tilde{N}}]$ is composed of N individuals. $X_{i_h} = \left(X_{i_h}^1, X_{i_h}^2, \ldots, X_{i_h}^D\right)$ is the location of the gray wolf i_h-the solution to the optimization problem. The top three wolves of the optimal solution of the objective function are wolf α, wolf β, and wolf δ, respectively. They are also the main wolves that guide the rest of the wolves to explore the optimal solution. The rest of the solution corresponds to the wolf as wolf ω. The parameters and explanations of the GWO algorithm are shown in **Table 1**.

The update process of $X(t)$ is given by Eq. (5). The first three obtained optimal values are saved to enforce other searching individuals (including ω) to constantly update their positions according to the position of the optimal value, and the calculation method is expressed as Eqs. (6)–(7).

$$X(t+1) = X_p(t) - A\left|CX_p(t) - X(t)\right| \tag{5}$$

$$X_\mu(t+1) = X_\pi(t) - A_\mu D_\pi \tag{6}$$

$$X_p(t+1) = \frac{1}{3}\sum_{\mu=1}^{3} X_\mu \tag{7}$$

Parameter	Interpretation
t	The number of current iterations
C	The swing factor $C = 2r_1$
$X_p(t)$	The position of the prey after the tth iteration
$X(t)$	The position of the gray wolf during the tth iteration
r_1	A random number within $[0, 1]$
A	The convergence factor, $A = 2a \cdot r_2 - a$
r_2	A random number uniformly distributed within [0,1]
a	a linearly decreases from 2 to 0 with the increase of the number of iterations
D_π	The distances between the individual gray wolves, $D_\pi = \lvert C_\mu X_\pi(t) - X(t) \rvert$

Table 1.
Parameters and explanations of the GWO algorithm.

where $\pi = \alpha, \beta, \delta$; $\mu = 1, 2, 3$. The distances between the other individual gray wolves and α, β, and δ, as well as the distances $D_\pi = \lvert C_\mu X_\pi(t) - X(t) \rvert$ between them and the updated position of the gray wolf are be determined by and (6). Then, the position of the prey can be determined by Eq. (7).

4.2 Principles of the DE algorithm

Assume that in the D-dimensional search space, in the population size NP, $Z(g)$ is the gth generation of the population, $Z(g) = \{Z_1(g), Z_2(g), \dots, Z_{np}(g)\}$. $Z_k(g)$ is the kth individual in the gth generation of the population, $Z_k(g) = [Z_{k,1}(g), Z_{k,2}(g), \dots, Z_{k,D}(g)]$, $k = 1, 2, \dots, NP$, $g = 1, 2, \dots, g_{\max}$, and $g_{\max}$ is the number of the last iteration.

4.2.1 Initialization of the population

Initially, the algorithm randomly generates the 0th generation of the population over the entire search space, and the value of the individual $z_{k,q}(0)$ in each dimension q is generated according to Eq. (8).

$$z_{k,q}(0) = z_{k,q}^L + rand(0,1)\left(z_{k,q}^U - z_{k,q}^L\right) \tag{8}$$

where $q = 1, 2, \dots, D$, $rand(0,1)$ is a random number, which is uniformly distributed within $[0, 1]$, $z_{k,q}^L$ is the lower threshold of the individual population, $z_{k,q}^U$ is the upper threshold of the individual population.

4.2.2 Mutation

Mutant individual is generated via Eq. (9).

$$\tau_{k,q}(g) = z_{p_1} + F(z_{p2} - z_{p3}) \tag{9}$$

where z_{p_1}, z_{p2}, z_{p3} are three different parameter vectors randomly selected from the current population, and $z_{p_1} \neq z_{p2} \neq z_{p3} \neq i$; F is an amplifying factor within [0,1].

4.2.3 Crossover

The crossover process in the DE algorithm is expressed as Eq. (10).

$$\mu_k(g+1) = \begin{cases} \tau_{k,q}(g), & rand(0,1) \leq CR \ \ or \ \ q = rand(0,1) \\ \tau_{k,q}(g), & rand(0,1) \geq CR \ \ or \ \ q \neq rand(0,1) \end{cases} \tag{10}$$

where CR is the crossover probability within $[0, 1]$, and $rand(0, 1)$ is a random number, which is uniformly distributed within $[0, 1]$ and used to guarantee that at least one-dimensional component comes from the target vector Z_k.

4.2.4 Selection

Selection operation compares the vector $\mu_k(g+1)$ and the vector $z_k(g)$ by an evaluation function, which is given by Eq. (11).

$$z_{k,q}(g+1) = \begin{cases} \mu_k(g+1), & f[\mu_k(g+1)] < f[z_k(g)] \\ z_k(g), & f[\mu_k(g+1)] \geq f[z_k(g)] \end{cases} \tag{11}$$

Therefore, this mechanism allows the populations of the offspring to evolve based on the current population. This optimization mechanism can improve the average optimization ability of the population and converge the optimal solution.

Algorithm 1. RBM_SVR _DEGWO

Input: $D_{train} = \{(\mathbf{x}_1, y_1), (\mathbf{x}_2, y_2), \ldots, (\mathbf{x}_{m'}, y_{m'})\}$, $D_{test} = \{(\mathbf{x}_1, y_1), (\mathbf{x}_2, y_2), \ldots, (\mathbf{x}_n, y_{n'})\}$, $\mathbf{S}_{degwo.dbn}$;
Output: $[\mathbf{r}_{test}, \mathbf{Z}_{parent.\alpha}]$

```
Initialize a, A, C, Z_parent and objective function V_parent
for each individual wolf k do
    V_parent ← RBM_SVR (D_train, D_test, Z_parent)
end for
sort V_parent
compute top three gray wolf individuals {X_α, X_β, X_δ}
for each generation g do
    update a ← 2 − g · (2/g_max)
    for each individual wolf k do
            Z_parent ← X_p
            V_parent ← RBM_SVR (D_train, D_test, Z_parent)
            compute mutant individuals τ_k,q ← z_p1 + F · (z_p2 − z_p3)
            compute children population Z_child ← μ_k,q
            V_child ← RBM_SVR (D_train, D_test, Z_child)
    end for
    for each individual wolf k do
            update Z_parent and V_parent
    end for
end for
update the parameters in DBN S ← Z_parent.α
RBM_SVR (D_train, D_test, S)
return [r_test, Z_parent.α]
```

Table 2.
The procedure of RBM_SVR _DEGWO algorithm.

4.3 DEGWO algorithm

In the DEGWO algorithm, $\mathbf{S}_{degwo.dbn} = (NP, g_{\max}, CR, D, ub, lb, F)$ where NP denotes population size, $g_{\max}$ denotes the maximum number of iterations, ub and lb are the search range. $\mathbf{r}_{test}$ and $\mathbf{r}_{train}$ denote the error in test and learning procedure respectively. **Table 2** is the specific procedure employing the DE and the GWO algorithms to optimize parameters c_s and γ_s in the RBM-SVR deep learning model.

5. A hybrid model based on RBM SVR DEGWO and XGBoost

5.1 XGBoost principle

5.1.1 Model function form

This paper assumes that the data set as the input sample is $D = \{(X_i, z_i)\}$ and the XGBoost model is a linear model (logistic regression, linear regression). The linear model is an additive model. The number of learning trees is n [34].The XGBoost model uses a pattern of linear superposition to calculate the predicted value, as shown in Eq. (12).

$$\hat{z}_i = \sum_{n}^{N} h_n(x_j) \tag{12}$$

Here, x_i is a feature vector and i is the number of data points. $h_n(x_i)$ is the regression tree function. $h_n \in H$, H is the set space of the regression trees.

$$H = \{h_n(z) = \alpha_{f(x)}\} \tag{13}$$

In Eq. (13), $f : R^m \to T$, $f(X)$ indicates that sample X is classified on a leaf node. T represents the number of leaf nodes of the tree. α is the score of the leaf node. $\alpha_{f(x)}$ represents the predicted value of the regression tree for the sample.

5.1.2 XGBoost learning objective function

The objective function based on the parameter space is shown in the following Eq. (14).

$$\kappa(\phi) = L(\phi) + \Omega(\phi) = \sum_{i=1}^{I} l(z_j, \hat{z}_i) + \sum_{n}^{N} \Omega(h_n) \tag{14}$$

where $\Omega(\phi)$ is a regularization term, indicating a penalty value for the complexity of the model. The regular term $\Omega(\phi)$ in the linear model includes: the regular term L_1, $\Omega(\alpha) = \lambda\|\alpha\|_1$, and the regular L_2, $\Omega(\alpha) = \lambda\|\alpha\|^2$. $L(\phi)$ is an error function that measures the fitting accuracy of the model. A can reduce model bias, such as square loss, exponential loss. Compared to GBDT, XGBoost adds a regular term to the objective function. XGBoost punishes the complexity of each regression tree and avoids overfitting during learning. XGBoost measures the complexity of the tree such as the number of internal nodes, the depth of the tree, the number of leaf nodes T, the leaf node score α, etc. XGBoost uses the regular term as shown in Eq. (15).

$$\Omega(h) = \gamma T + \frac{1}{2}\lambda\|\alpha\|^2 \tag{15}$$

5.1.3 Model optimization

In the model parameter optimization process, each iteration model is always added a new function on the optimal model obtained from the previous training. After the kth iteration, the prediction of the model is equal to the prediction function of the first $k-1$th model prediction function combined with the kth tree, as shown by Eq. (16).

$$\hat{z}_i^{(k)} = \sum_{i=1}^{I} h_n(x_j) = \hat{z}_i^{(k-1)} + h_i(x_j) \tag{16}$$

The objective function can be rewritten to Eq. (17).

$$\kappa(k) = \hat{z}_i^{(k)} = \sum_{i=1}^{I} l\left(z_i, \hat{z}_i^{(k-1)} + h_i(X_i)\right) + \Omega(h_i) \tag{17}$$

In formula (17), the model's goal is to learn the function of the kth tree. When the error function is replaced by a second-order Taylor expansion, the objective function can be rewritten as Eq. (18).

$$L^{(k)} = \sum_{i=1}^{I}\left[l\left(z_i, \hat{z}_i^{(k-1)} + v_i h_i(X_i) + \frac{1}{2} g_i h_k^2(X_i)\right)\right] + \Omega(h_k) \tag{18}$$

When $v_i = \partial_{\hat{z}_i^{(k-1)}} l\left(z_i, \hat{z}_i^{(k-1)}\right)$ a = 1 and $g_i = \partial^2_{\hat{z}_i^{(k-1)}} l\left(z_i, \hat{z}_i^{(k-1)}\right)$, the objective function is Eq. (19).

$$\tilde{L}^{(k)} = \sum_{i=1}^{I}\left[v_i h_i(X_i) + \frac{1}{2} g_i h_k^2(X_i)\right] + \Omega(h_k) \tag{19}$$

This objective function solves regression, classification, and sorting problems. Eqs. (20) and (21) are in the form of a tree structure of the regression tree function and the regular term. The objective function can be updated to Eq. (22).

$$h(X) = \alpha_{f(x)} \tag{20}$$

$$\Omega(h) = \gamma T + \frac{1}{2}\lambda\|\alpha\|^2 \tag{21}$$

$$\tilde{L}^{(k)} = \sum_{i=1}^{I}\left[v_i \alpha_f(x_i) + \frac{1}{2} g_i h_k^2 \alpha_{f(x)}^2\right] + \gamma T + \lambda \frac{1}{2}\sum_{j=1}^{T} \alpha_j^2 \tag{22}$$

This article defines the sample set on each leaf node as $J_j = \{i|f(x_i) = j\}$. The objective function based on the form of leaf node accumulation is Eq. (23).

$$\begin{aligned}\tilde{L}^{(k)} &= \sum_{j=1}^{T}\left[\left(\sum_{i\in G_j} f_i\right)\alpha_j + \frac{1}{2}\left(\sum_{i\in G_j} g_i + \lambda\right)\lambda\alpha_j^2\right] + \gamma T \\ &= \sum_{J=1}^{T}\left[\delta_j \alpha_j + \frac{1}{2}\left(\eta_j + \lambda\right)\alpha_j^2\right] + \gamma T\end{aligned} \tag{23}$$

This paper assumes that the structure of the tree is a certain value (i.e., $f(x_i)$ is determined). To solve the problem of minimizing the objective function, we can make the derivative of the objective function zero. The optimal predicted score for each leaf node is Eq. (24). The formula for the minimum loss function is Eq. (25), which can be thought of as a function that scores the tree structure. The tree structure is gradually optimized as the score is reduced.

$$A_j^* = -\frac{\delta_j}{\eta_j + \lambda} \tag{24}$$

$$\tilde{L}^{(*)} = -\sum_{j=1}^{T} \frac{\delta_j^2}{\eta_j + \lambda} + \gamma T = -\frac{1}{2}\sum_{j=1}^{T}\left(\left(\sum_{i \in G_j} f_i\right)^2 / \sum_{i \in G_j} g_i + \lambda\right) + \gamma T \tag{25}$$

5.1.4 Structure score

Eq. (25) is a function for scoring a tree structure, called structure score. The smaller the score, the better the tree structure is. The algorithm searches for the optimal tree structure by using Eq. (25). $\tilde{L}^*$ represents the contribution of the leaf node to the overall loss. The goal of the algorithm is to minimize the loss, so the larger part of $\frac{\delta_j^2}{\eta_j + \lambda}$ could be as good as possible. This article expands a leaf node and defines the gain as shown in Eq. (26).

$$gain = \frac{1}{2}\left[\frac{\delta_j^2}{\eta_L + \lambda} + \frac{\delta_R^2}{\eta_R + \lambda} - \frac{(\delta_L + \delta_R)_j^2}{\eta_L + \eta_R + \lambda}\right] - \gamma \tag{26}$$

In Eq. (26), $\frac{\delta_j^2}{\eta_L + \lambda}$ is the score of the left subtree, $\frac{\delta_R^2}{\eta_R + \lambda}$ is the score of the right subtree, $\frac{(\delta_L + \delta_R)_j^2}{\eta_L + \eta_R + \lambda}$ is the score without division, and γ is the cost of the complexity after introducing the new leaf node. The larger the value of *gain*, the more loss after splitting is reduced. Therefore, when segmenting a leaf node, we calculate the *gain* corresponding to all candidate features and select the segment with the largest *gain*.

5.1.5 Best branch

The core part of the XGBoost algorithm is to obtain the optimal node based on the maximum gain obtained. XGBoost looks for the best branch using a greedy algorithm. The greedy algorithm traverses all possible segmentation points of all features, calculating the Gain value and selecting the maximum value to complete the segmentation. The greedy algorithm is an algorithm that controls the local optimum to achieve global optimization. The decision tree algorithm can also be considered as a method of greedy algorithm. XGBoost is an integrated model of the tree. If each leaf is optimal, the overall generated tree structure is optimal. This avoids enumerating all possible tree structures. XGBoost uses the objective function to measure the structure of the tree, and then let the tree grow from depth 0. Each time a branch calculation is implemented, XGBoost calculates the reduction in the objective function. When the reduction is below a certain value, the tree will stop growing.

5.2 Hybrid model based on RBM_SVR_DEGWO and XGBoost

After the boosting tree is created, the XGBoost algorithm extracts the importance score for each attribute. The XGBoost importance score measures the value of

Algorithm 2. Hybrid Algorithm based on RBM_SVR_DEGWO and XGBoost

Input: $D = \{(X_i, z_i)\}, J_j = \{i|f(x_i) = j\}$
data sets normalization
Initialize *gain*
Initialize δ_L, η_L
for each k **do**
    $\delta_L \leftarrow 0, \eta_L \leftarrow 0$
    for j in sorted (J, by X_{jk}) **do**
        $\eta_L \leftarrow \eta_L + g_j, \delta_L \leftarrow \delta_L + f_j$
        $\eta_R \leftarrow \eta - \eta_L, \delta_R \leftarrow \delta - \delta_L$
        $score \leftarrow \max(score)$
    end for
end for
Split with max score
$y_d = RBM.SVR.DEGWO(X_i)$
$y_e = EXGBOOST\ (X_i)$
$y_{hybrid} = E[y_d, y_e]$
Output: y_{hybrid}

Table 3.
Hybrid algorithm based on RBM_SVR_DEGWO and XGBoost.

features in improving decision tree construction. The more an attribute is used to build a decision tree, the more important it is [35]. In order to further improve the accuracy of prediction and analyze the importance of feature quantity, this paper uses XGBoost to extract the feature quantity importance score. By combining the proposed RBM_SVR _DEGWO model prediction value, this paper proposes a hybrid prediction model, as shown in **Table 3**.

6. Experimental description and result analysis

This paper analyzes 2,468,059 trajectory data from Mobike's shared bikes. The data covers more than 300,000 users and 400,000 shared bikes. The data of each rental trip includes the start time, the end time, the Geohash code of the starting position, the Geohash code of the ending position, the bicycle ID and the user ID.

GeoHash is an algorithm for spatial indexing. In the GeoHash theory, the Earth is considered to be a two-dimensional plane that can be divided into multiple sub-regions. The latitude and longitude inside the sub-area will correspond to the same code. GeoHash-based spatial indexing can improve the efficiency of spatial data for latitude and longitude retrieval. In this paper, GeoHash encodes a square plane separated by a square of latitude and longitude of 0.001373. To improve the prediction accuracy, this paper combines nine adjacent areas into a square area with a length of 411.9873 meters. This paper divides Beijing into 10×10 traffic zones and numbers them from 1 to 100. Various indicators of the traffic area will be used as input data for the prediction model, as shown in **Table 4**.

The output of the model is the daily transfer probability of traffic flow among the traffic zones $p^t_{I,J}$, which is given by Eq. (27). In the cities of N interconnected traffic areas, $p^t_{I,J}$ indicates the transfer probability of the traffic flow with the original point I and the destination J in day t.

$$p^t_{I,J} = \frac{d^t_{I,J}}{\sum_{J=1}^{N} d^t_{I,J}} \tag{27}$$

Number	Variable name
1	Zone number of origin traffic zone
2	Zone number of destination traffic zone
3	Longitude of the origin point
4	Latitude of the origin point
5	Longitude of the destination point
6	Latitude of the destination point
7	Distance between the center points of the traffic area
8	Absolute value of the difference in the numbers of traffic zone
9	Number of the day

Table 4.
Input variables and interpretation.

	The first day of training data		The 2nd day of training data		Prediction data	
Test group	Date	Data amount (trajectories)	Date	Data amount (trajectories)	date	Data amount (trajectories)
1	2017/5/10	262,569	2017/5/11	272,210	2017/5/12	265,173
2	2017/5/13	225,281	2017/5/14	236,594	2017/5/15	279,554
3	2017/5/14	236,594	2017/5/15	279,554	2017/5/16	288,719
4	2017/5/15	279,554	2017/5/16	288,719	2017/5/17	322,201
5	2017/5/16	288,719	2017/5/17	322,201	2017/5/18	315,758
6	2017/5/10	262,569	2017/5/15	279,554	2017/5/16	288,719

Table 5.
Experimental group training data and test data.

where $I = 1, 2, 3, \ldots, N; J = 1, 2, 3, \ldots, N; d^t_{I,J}$ refers to the traffic flow with the original point I and the destination J in day d. $p^t_{I,J}$ represents the origin–destination (OD) probability distribution and reflects the distribution of demand in the city.

This paper builds a set of destinations that may correspond to the origin traffic zone of the test day. The calculated destination candidates can be used to predict the probability of the traffic flow among the traffic zones. In the experiment, we selected data of different adjacent days as 6 test groups (**Table 5**).

Based on data for the past 2 days as the training data, this paper predicts the subsequent third day of the transfer probabilities of bike-sharing traffic flow. **Figure 1** is the root mean square errors of a prediction result of transfer probabilities of bike-sharing traffic flow in Beijing based on the RBM_SVR_DEGWO algorithm.

Compared to the surrounding area, the central area of the city has higher shared bicycle usage and more bicycle trajectory data. Therefore, the Root Mean Square Error of the central region is smaller.

To illustrate the performance of the RBM_SVR_DEGWO algorithm, we calculated the predicted values of the SVR algorithm, the RBM_SVR algorithm, and the RBM_SVR_DEGWO algorithm based on the data from the experimental groups in **Table 5**. To ensure the fairness of the results, the data, network structure and parameter settings consistent. **Figure 2** shows the mean-square error bars of the predicted transfer probabilities of SVR, RBM_SVR, and RBM_SVR_DEGWO.

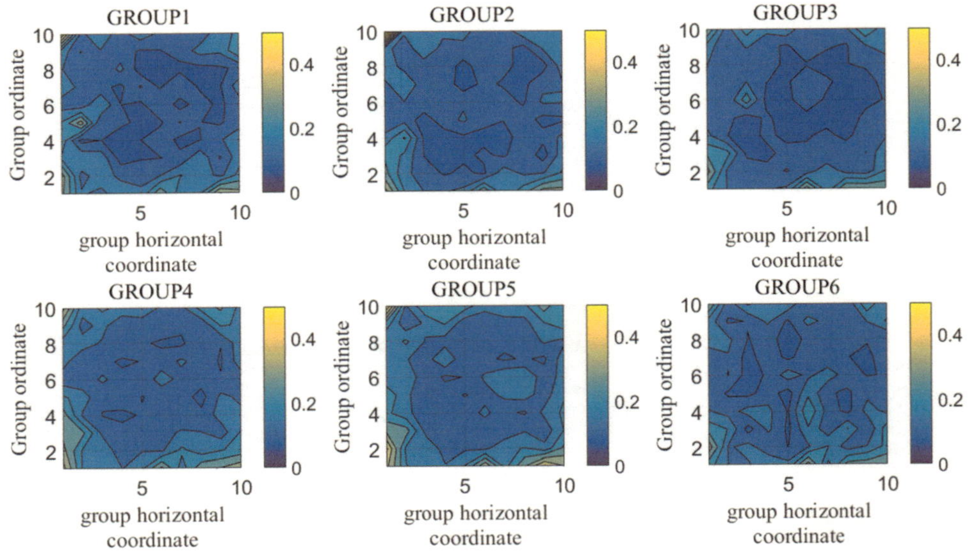

Figure 1.
The root-mean-square errors of the predicted transfer probabilities of bike sharing traffic flow.

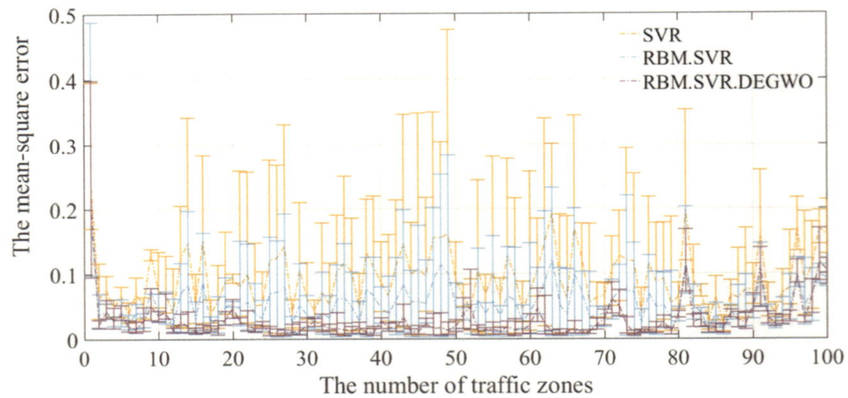

Figure 2.
The mean-square error bars of the predicted transfer probabilities of bike sharing traffic flow of six test groups for each type of comparison method.

The average values of the mean squared errors of the predicted values of the transfer probabilities of the algorithms SVR, RBM_SVR, and RBM_SVR_DEGWO are gradually reduced. The average mean square error of the SVR is 0.0916, the RBM_SVR is 0.0542, and the RBM_SVR_DEGWO is 0.0283. RBM improves the prediction accuracy of the model through the deep network structure. The DEGWO algorithm stabilizes the prediction value error to a lower value by optimizing the parameters of the RBM-SVR. Compared with SVR and RBM_SVR, RBM_SVR_DEGWO algorithm has better robustness.

According to the proposed hybrid algorithm of RBM_SVR_DEGWO and XGBoost, the value of transfer probabilities of bike-sharing traffic flow can be predicted. The data set for this experiment is from the grouped data of **Table 5**. The training data set, test data set, and feature variables are the same as those used in the previous experiments. **Table 6** is the parameters and explanation of the XGBoost model.

The root mean square error of the predicted values of the RBM_SVR_DEGWO algorithm, the XGBoost algorithm, and the hybrid algorithm is shown in **Figure 3**. In the six experimental groups, the mean, variance, kurtosis, maximum, minimum, and range of the predicted root mean square error of the RBM_SVR_DEGWO algorithm, the XGBoost algorithm, and the hybrid algorithm are shown in **Figure 3**.

The statistical characteristics of the proposed root mean square error of the algorithms are shown in **Figure 4**. The root-mean-square error of the predicted value of the mixed algorithm has a high kurtosis value. It indicates that the variance increases of root mean square error is caused by the extreme difference of low frequency greater than or less than the mean value. The plots of the minimum and variance indicate that RBM_SVR_DEGWO can achieve higher prediction accuracy than XGBoost. XGBoost is more stable than RBM_SVR_DEGWO in the prediction process. In the six experimental groups, compared with the RBM_SVR_DEGWO algorithm and the XGBoost algorithm, the mean, variance, maximum, minimum, and range of the root mean square error of the predicted value of the hybrid

Parameter	Value	Interpretation
early_stopping_rounds	200	If the loss function does not decrease after the model has been added to n trees continuously, the model will stop adding trees.
eval_metric	linear	Evaluation index
eta	0.3	The shrinking step size used in the update process. Eta is used to prevent overfitting.
min_child_weight	1	Min_child_weight refers to the sum of the weights of the smallest leaf nodes. If the sample weight of a leaf node is less than min_child_weight then the splitting process ends.
max_depth	6	Maximum depth of the tree
lambda	1	Penalty factor for L2 regular terms
alpha	0	Penalty factor for L1 regular terms
objective	linear	Loss function

Table 6.
Parameters and explanations of the XGBoost model.

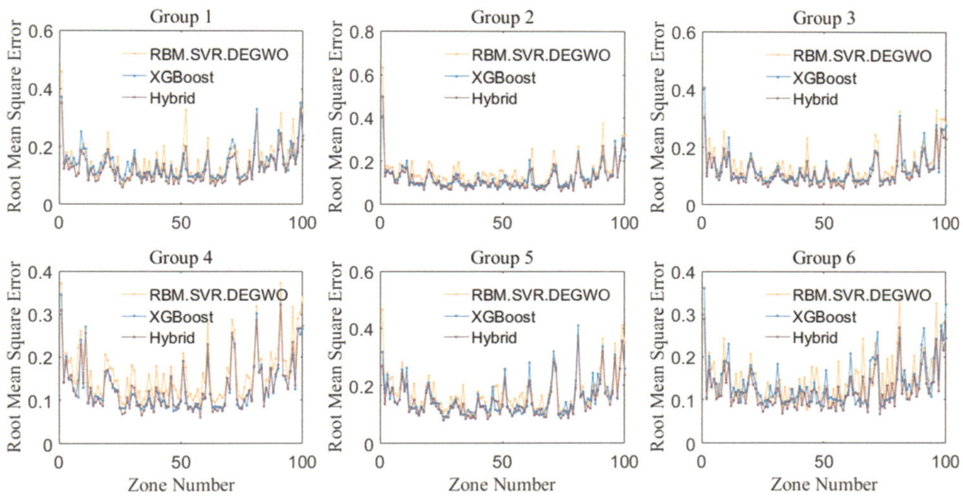

Figure 3.
The root mean square error of the predicted values of the RBM_SVR_DEGWO algorithm, the XGBoost algorithm, and the hybrid algorithm.

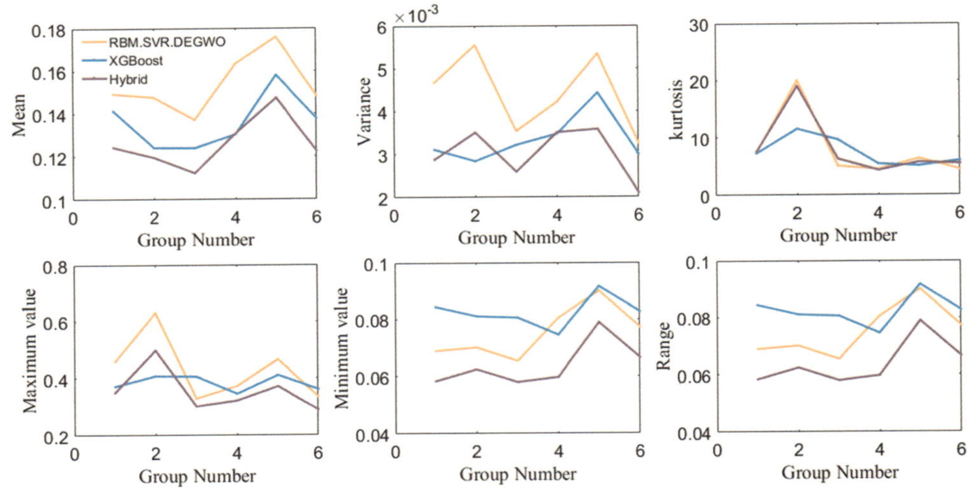

Figure 4.
Statistical characteristics of the root mean square error.

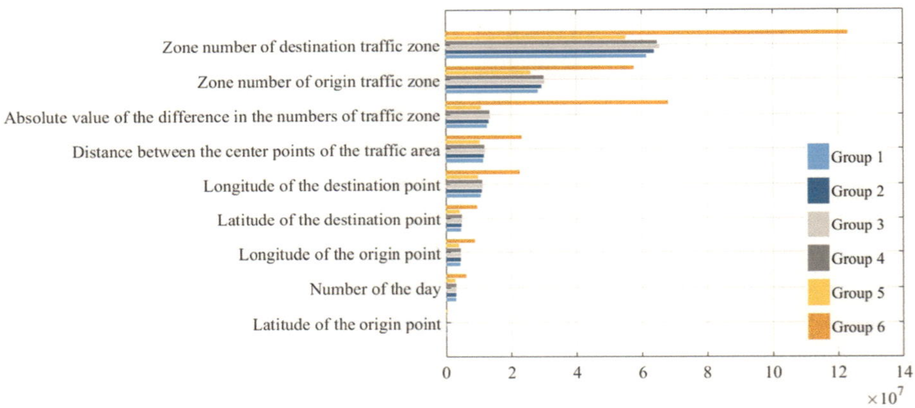

Figure 5.
Importance analysis of characteristic variables.

algorithm are lower. Therefore, by combining the prediction results of the RBM_SVR_DEGWO algorithm and the XGBoost algorithm, the hybrid algorithm improves the prediction accuracy and obtains a lower root mean square error of the predicted value. XGBoost scores the importance of each feature based on the number of times the feature is used to segment the sample in all trees and the average gain in all trees. In the six experimental groups, the ranking of each input feature variable is as shown in **Figure 5**.

The main factors affecting the transfer probabilities of bike-sharing traffic flow are the destination traffic zone number, the origin traffic zone number, and the absolute value of the difference between the numbers of traffic zone. It shows that the shared bike rider's choice of destination is usually affected by the starting point, the end position and the distance of the journey. The shared bicycle service is suitable for short trips. Travel destinations for shared bike riders are usually nearby business and lifestyle centers, and bus stops. The dates of the data for the six groups of experiments are within weekdays. On a normal weekday, for the riders in the same community, the main travel destinations are somewhat similar and fixed. Therefore, information such as the cell number of the origin and destination becomes a key factor for predicting the probability of travel destination.

7. Conclusions

The principal objective of this study is to predict the traffic flow transfer probability of shared bicycle by proposing a hybrid deep learning algorithm and accurately reflect the transfer probability of the user's OD demand. First, this paper constructs a deep-structured RBM model and connects it to the SVR model for predicting continuous probability values. Furthermore, we utilize the DEGWO optimization algorithm, named, to optimize the parameters c_s and γ_s in the stacked RBM-SVR algorithm. XGBoost improves the prediction accuracy and analyzes the importance of the feature variables in the input data.

Based on the comparison results, it demonstrates that the proposed hybrid algorithm outperformed the XGBoost model and RBM_SVR_DEGWO model. The XGBoost algorithm improves the stability of the prediction process and reduces the error of the RBM_SVR_DEGWO algorithm at extreme points. The deep-structured RBM algorithm simulates the probability distribution that best produces the training samples. In the case of massive training data, RBM improves the efficiency of algorithm calculation utilizing Gibbs sampling of small-batch data. In the DEGWO algorithm, the GWO algorithm guarantees the global search capability, and the DE algorithm avoids the fall into a local optimal through the mutant individual, crossover, and selection operations.

Author details

Wenwen Tu
Southwest Jiaotong University, Chengdu, China

*Address all correspondence to: tuvivimic@gmail.com

References

[1] Vogel P, Greiser T, Mattfeld DC. understanding bike-sharing systems using data mining: Exploring activity patterns. Procedia-Social and Behavioral Sciences. 2011;**20**:514-523

[2] Wu Y, Zhu D. Bicycle sharing based on PSS-EPR coupling model: Exemplified by bicycle sharing in China. Procedia CIRP. 2017;**64**:423-428

[3] Chemla D, Meunier F, Calvo RW. Bike sharing systems: Solving the static rebalancing problem. Discrete Optimization. 2013;**10**(2):120-146

[4] Contardo C, Morency C, Rousseau L-M. Balancing a Dynamic Public Bike-Sharing System. Montreal: Cirrelt; 2012

[5] Schuijbroek J, Hampshire RC, Van Hoeve W-J. Inventory rebalancing and vehicle routing in bike sharing systems. European Journal of Operational Research. 2017;**257**(3):992-1004

[6] Bengio Y, Lamblin P, Popovici D, Larochelle H. Greedy Layer-Wise Training of Deep Networks. Advances in Neural Information Processing Systems. Cambridge, MA, USA: MIT Press; 2007. pp. 153-160

[7] Hinton GE, Osindero S, Teh Y-W. A fast learning algorithm for deep belief nets. Neural Computation. 2006;**18**(7): 1527-1554. DOI: 10.1162/neco.2006.18.7.1527

[8] Hinton GE, Sejnowski TJ. Learning and Relearning in Boltzmann Machines, Parallel Distributed Processing: Explorations in the Microstructure of Cognition. Cambridge: MIT Press; 1986. pp. 282-317

[9] LeCun Y, Bengio Y, Hinton G. Deep learning. Nature. 2015;**521**(7553):436. DOI: 10.1038/nature14539

[10] LeCun Y, Boser B, Denker JS, Henderson D, Howard RE, Hubbard W, et al. Backpropagation applied to handwritten zip code recognition. Neural Computation. 1989;**1**(4):541-551. DOI: 10.1162/neco.1989.1.4.541

[11] Hinton GE. A Practical Guide to Training Restricted Boltzmann Machines, Neural Networks: Tricks of the Trade. NY, USA: Springer; 2012. pp. 599-619

[12] Larochelle H, Bengio Y. Classification using discriminative restricted Boltzmann machines. In: Proceedings of the 25th International Conference on Machine learning. NY, USA: ACM; 2008. pp. 536-543

[13] Le Roux N, Bengio Y. Representational power of restricted Boltzmann machines and deep belief networks. Neural Computation. 2008; **20**(6):1631-1649

[14] Nair V, Hinton GE. Rectified linear units improve restricted boltzmann machines. In: Proceedings of the 27th International Conference on Machine Learning. ICML-10. 2010. pp. 807-814

[15] Salakhutdinov R, Mnih A, Hinton G. Restricted Boltzmann machines for collaborative filtering. In: Proceedings of the 24th International Conference on Machine Learning. NY, USA: ACM; 2007. pp. 791-798

[16] Awad M, Khanna R. Support Vector Regression. Berkeley, CA: Apress; 2015

[17] Drucker H, Burges CJC, Kaufman L, Smola AJ, Vapnik V. Support vector regression machines. Advances in Neural Information Processing Systems. 1997;**28**(7):779-784

[18] Li L, Duan Y. A GA-based feature selection and parameters optimization for support vector regression. In: International Conference on Natural Computation. 2011. pp. 335-339

[19] Wu CH, Wei CC, Su DC, Chang MH. Travel time prediction with support vector regression. In: Proceedings of the 2003 IEEE International Conference on Intelligent Transportation Systems. Vol. 1432. 2004. pp. 1438-1442

[20] Mirjalili S, Mirjalili SM, Lewis A. Grey wolf optimizer. Advances in Engineering Software. 2014;**69**:46-61. DOI: https://doi.org/10.1016/j.advengsoft.2013.12.007

[21] Storn R, Price K. Differential evolution–A simple and efficient heuristic for global optimization over continuous spaces. Journal of Global Optimization. 1997;**11**(4):341-359

[22] Chen T, Guestrin C. Xgboost: A scalable tree boosting system. In: Proceedings of the 22nd ACM Sigkdd International Conference on Knowledge Discovery and Data Mining. NY, USA: ACM; 2016. pp. 785-794

[23] Chen T, He T, Benesty M, Khotilovich V, Tang Y. Xgboost: extreme gradient boosting. R package version 0.4–2. 2015. pp. 1-4

[24] Tu W, Liu H. Transfer probability prediction for traffic flow with bike sharing data: A deep learning approach. In: Science and Information Conference. NY, USA: Springer; 2019. pp. 71-85

[25] Li X, Zhang Y, Sun L, Liu Q. Free-floating bike sharing in Jiangsu: Users' behaviors and influencing factors. Energies. 2018;**11**(7):1664

[26] Shen Y, Zhang X, Zhao J. Understanding the usage of dockless bike sharing in Singapore. International Journal of Sustainable Transportation. 2018;**12**(9):686-700

[27] Xu C, Ji J, Liu P. The station-free sharing bike demand forecasting with a deep learning approach and large-scale datasets. Transportation Research Part C: Emerging Technologies. 2018;**95**:47-60

[28] Zhang Y, Lin D, Mi Z. Electric fence planning for dockless bike-sharing services. Journal of Cleaner Production. 2019;**206**:383-393

[29] Caggiani L, Camporeale R, Ottomanelli M, Szeto WY. A modeling framework for the dynamic management of free-floating bike-sharing systems. Transportation Research Part C: Emerging Technologies. 2018;**87**:159-182

[30] Li M, Wang X, Zhang X, Yun L, Yuan Y. A multiperiodic optimization formulation for the operation planning of free-floating shared bike in China. Mathematical Problems in Engineering. 2018;**2018**:1-11

[31] Liu Y, Szeto W, Ho SC. A static free-floating bike repositioning problem with multiple heterogeneous vehicles, multiple depots, and multiple visits. Transportation Research Part C: Emerging Technologies. 2018;**92**:208-242

[32] Shim VA, Tan KC, Cheong CY, Chia JY. Enhancing the scalability of multi-objective optimization via restricted Boltzmann machine-based estimation of distribution algorithm. Information Sciences. 2013;**248**:191-213

[33] Wen-juan X, Jian-feng L. Application of vision sensing technology in urban intelligent traffic control system. In: 2018 4th International Conference on Computer and Technology Applications (ICCTA). NY, USA: IEEE; 2018. pp. 74-77

[34] Zheng H, Yuan J, Chen L. Short-term load forecasting using EMD-LSTM neural networks with a Xgboost algorithm for feature importance evaluation. Energies. 2017;**10**(8):1168

[35] Bai S, Li M, Kong R, Han S, Li H, Qin L. Data mining approach to construction productivity prediction for cutter suction dredgers. Automation in Construction. 2019;**105**:102833

Chapter 3

Voice Identification Using Classification Algorithms

Orken Mamyrbayev, Nurbapa Mekebayev, Mussa Turdalyuly, Nurzhamal Oshanova, Tolga Ihsan Medeni and Aigerim Yessentay

Abstract

This article discusses the classification algorithms for the problem of personality identification by voice using machine learning methods. We used the MFCC algorithm in the speech preprocessing process. To solve the problem, a comparative analysis of five classification algorithms was carried out. In the first experiment, the support vector method was determined—0.90 and multilayer perceptron—0.83, that showed the best results. In the second experiment, a multilayer perceptron with an accuracy of 0.93 was proposed using the Robust scaler method for personal identification. Therefore, to solve this problem, it is possible to use a multi-layer perceptron, taking into account the specifics of the speech signal.

Keywords: speaker identification, classification, speech recognition, MFCC

1. Introduction

In the era of informatization, many high-tech products gradually entered our daily life and significantly changed our life habits. On the other hand, information technologies continue to evolve towards a more human-centered approach. Biometric identification technology, which provides us with simpler and more convenient methods for identifying specific people, has gradually replaced some of the existing authentication methods that should be explored before people will be able to manage them properly Face recognition systems used in public places, law enforcement organizations [1], and Siri voice mobile assistant on iPhone, Bixby Voice on Galaxy [2], are examples of biometric identification results.

The recognition of a person by his voice is one of the forms of biometric authentication, which makes it possible to identify a person by a combination of unique voice characteristics and refers to dynamic methods of biometrics. Speaker recognition is a technology that can automatically identify the speaker based on the speech waveform, that reflects the physiological and behavioral characteristics of speech parameters from the speaker. Like traditional speaker recognition systems, there are two stages, namely, training and testing. These are the main stages of speaker recognition. Learning is the process of extracting phonetic characteristics from a speaker that has already been recorded or saved as a sample, storing them in a database, and familiarizing the system with the characteristics of the speaker's voice. Testing is the process of comparing questionable sound and phonetic characteristics from a speaker recognition database. Two popular sets of features, often used in the analysis of the speech signal are the Mel frequency cepstral coefficients (MFCC) and the linear prediction cepstral coefficients (LPCC). The most popular recognition models are vector quantization (VQ), dynamic time warping (DTW), and artificial neural network (ANN) [3].

The study of speech technologies for the Kazakh language is conducted in Kazakhstan. Kazakh language belongs to agglutinative languages. Agglutinative languages that have a system in which the dominant type of inflection is the "gluing" of various formants of suffixes or prefixes, each of which has only one meaning. The Turkic, Mongolian, Korean languages are agglutinative. In our country, the personal identification system in the Kazakh language has not been developed yet and research in this area is relevant.

This article deals with the problem of identification of a person using the classification model through the use of artificial neural networks. The paper is organized as follows. Section 2 describes the work on the relevant scientific research area. Section 3 discusses data preprocessing methods. Section 4 describes the methodology of automatic identification. Sections 5 and 6 discuss the results of the experiment and the conclusion.

2. Related works

A common feature of agglutinative languages, such as Finnish, Kazakh, and Turkish, is that, until now, attempts at personal identification and speech recognition have not led to comparable performance with English systems [4]. The reason for this is not only the difficulty of modeling the language, but also the lack of suitable resources for speech and text learning. In [5, 6] the systems aim to reduce active vocabulary and language models to a possible size by clustering and focusing.

Recently, neural networks have become dominant in various machine learning fields. One of them is natural language processing (the sequence of characters/words can be considered as another type of signals) in which the multilayer perceptron (MLP) and long-short-term memory (LSTM) network, two standard classifiers are widely used for the tasks of disambiguating morphological forms or identifying the boundary of sentence/tokens.

In many papers [7, 8], it was shown that the use of ANNs in conjunction with the HMM can improve the accuracy of speech recognition. Acoustic models are usually based on deep neural networks, which are artificial neural networks of direct propagation, containing more than one hidden layer between the input and output layers. For training, the backpropagation method is used.

The review article considers that feature extraction is one of the most important tasks in the identification system, which significantly affects the process and performance of the system. In the review analysis, the existing proposals and implementations of methods for identifying the features of the identification system were considered. Analysis of the results shows that MFCC-based approaches have been used more than any other approach and, moreover, it was revealed that the current trend of the identification system research is to solve important identification system problems, such as adaptability, complexity, multilingual recognition and noise resistance [9].

In one of the works [10], speech pre-processing method was considered using the VAD algorithm, which proves that this algorithm improves the performance of speech recognition. The study presents the principles of operation and the block diagram of the VAD algorithm in recognition of Kazakh speech.

Toleu et al. [11] proposed character-based MLP and LSTM models that can jointly identify the boundary of sentences and tokens. In order to extract the high-level abstract features, the proposed models project the characters embedding into low-dimensional space which could allow us to judge the different variety of signals. The models were tested for three languages: English, Italian and Kazakh. The experimental results show that character-based MLP and LSTM models for sentence and token segmentation have positive effects in terms of F-measure and the error rates compared to existing models.

Clustering algorithms are used to partition an existing set of speech segments into groups according to similarity of their attributes. Parametric algorithms for determining initial points (centroids) and subsequent cluster propagation are proposed in [12] and can be applied for speech classification task solving.

The work [13] considers the question of using a throat microphone (laryngophone) as an additional modality for phonetic segmentation of the speech signal into acoustic sub-word units. The new algorithm is proposed for the automatic speech signal segmentation based on the use of changing dynamics analysis of the throat-acoustic correlation (TAC) coefficients, which can be used for subsequent speech segment classification.

In the works of Russian scientists can be found a study on the recognition of continuous Russian speech, using deep belief networks (DBN), described in [14]. A method using finite state transducers was used for speech recognition and it was shown that the proposed method allows to increase the accuracy of speech recognition in comparison with hidden Markov models.

3. Feature extraction and configuration parameters

In identification tasks, the main process is speech pre-processing. In this study, we select MFCC [15] as a tool for extracting voice dynamics functions. The speech pre-processing process is described in **Figure 1**.

Voice signals in the time domain change very quickly and dramatically, but if we convert speech signals from the time domain to the frequency domain [16], then the corresponding spectrum can be clearly defined. Our system separates the signals into frames and calls the window function to increase the continuity of voice signals in the frame. DCT is being used for quantitative evaluation of spectral energy data into data units that can be analyzed by MFCC [17]. The MFCC parameters are in the range of analyzed frequencies 300–8000 Hz, as well as 16 cepstral.

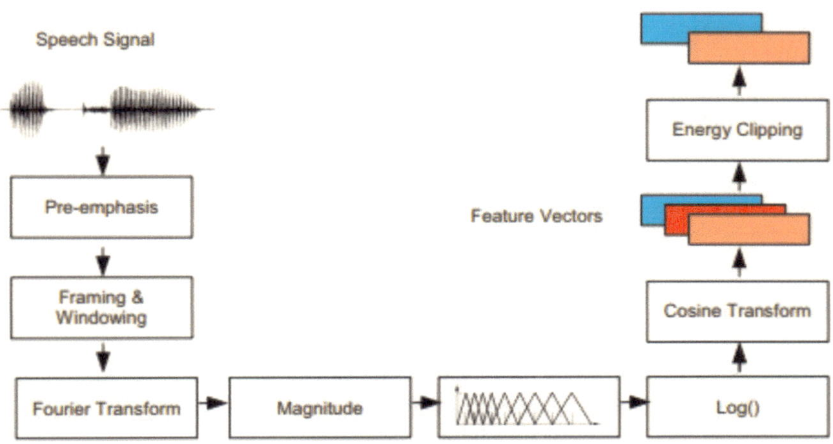

Figure 1.
Steps involved in extracting MFCC feature vectors.

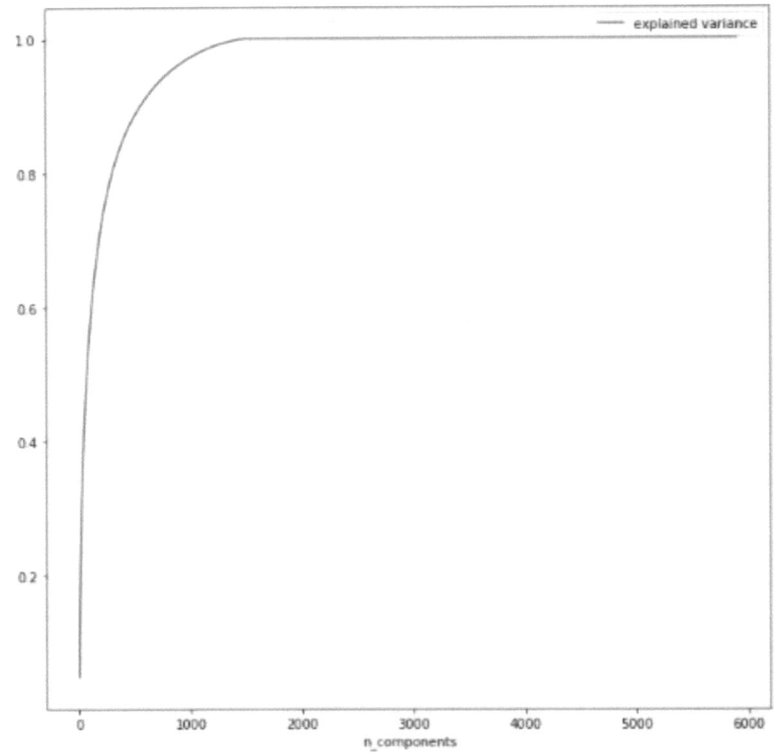

Figure 2.
Preservation of dispersion with decreasing dimension by the method of principal components.

As a result, 5904 features were obtained for each audio file. Each audio file was marked with the initials of the speaker whose voice was recorded in it. The resulting dataset had a dimension of 1480 × 5904.

To visualize the data, the principal components method was used to reduce the dimension of the vector space from 5904 features to two- and three-dimensional space [18]. Maintaining the dispersion in the reduction of the dimensionality by the principal component analysis shown in **Figure 2**.

As can be seen from the above graph, 100% of the variance is preserved when the data dimension is reduced to 1479 features. However, as experiments with classification models and data standardizers have shown, such a reduction in dimension critically affects the accuracy of the classification.

4. The proposed speech identification system

The methodology of our work is as follows:

4.1 The design of the experimental data

Data for analysis were provided by the laboratory of "computer engineering of intelligent systems." The data set consists of 1480 audio recordings from 20 speakers with 74–75 recordings. Each audio recording consists of phrases in Kazakh with an average length of 6 seconds. To identify the speaker, we collected the following data: name, gender, place of birth, year of birth (**Table 1**).

Label	Origin	Name	Middle name	Gender	Birthplace	Year of birth
MZA	Masimkanova	Zhazira	Auezbekkyzy	Female	Almaty	20.03.1982
IMT	Iskakova	Moldir	Tasbolatkyzy	Female	Almaty	01.01.1994
DAZ	Duisenbaeva	Aigerim	Zhanbolatovna	Female	Almaty	15.05.1995
ZEA	Zhetpisbaev	Erlan	Alibekovich	Man	Almaty	23.05.1995
SSM	Samrat	Sanjar	Muhametkaliuly	Man	Almaty	12.07.1996
...	...	...	...	...	...	...

Table 1.
Information about the speakers.

To increase the accuracy of recording audio materials, a soundproofing, professional recording studio from Vocalbooth.com was used.

All audio materials have the same characteristics:

- file extension: .wav;
- digital conversion method: PCM;
- discrete frequency: 44.1 kHz;
- digit capacity: 16 bits;
- number of audio channels: one (mono).

The sound and recording of one speaker took an average of 40–50 minutes of time including the time required to prepare the speaker, equipment and doubles, which corresponds to the 74–75 files received, a total length of 7–8 minutes for each speaker.

4.2 Classification algorithms

For the speaker identification problem we took the following known classification algorithms.

4.2.1 Extra-Trees algorithm

The Extra-Trees algorithm builds an ensemble of unpruned decision or regression trees. The algorithm's two main differences with other tree-based ensemble methods are that it splits nodes by choosing cut-points fully at random and that it uses the whole learning sample to grow the trees. The Extra-Trees algorithm is given in **Table 2**.

It has two parameters: K, the number of attributes randomly selected at each node and n_{min}, the minimum sample size for splitting a node. It is used several times with the original learning sample to generate an ensemble model.

4.2.2 KNN algorithm

The K-nearest-neighbor (KNN) algorithm measures [19] the distance between a query scenario and a set of scenarios in the data set.

Trees_node(M)
Input signal: the local learning subset M corresponding to the node
Output signal: a tree $[a < a_c]$ or nothing

- If **Tree(S)** is TRUE then return nothing;
- Otherwise select K attributes $\{a_1,\ldots,a_K\}$ among all non constant (in S) candidate attributes;
- Draw K trees $\{s_1,\ldots,s_K\}$, where s_i = Random_split(S, a_i), $\forall i = 1,\ldots, K$;
- Return a tree s_* such that Score(s_*, S) = $\max_{i=1,\ldots,K}$ Score(s_i, S).

Random_split(S,*a*)
Inputs: a subset S and an attribute *a*
Output: a split

- Let a^S_{max} and a^S_{min} denote the maximal and minimal value of a in S;
- Draw a random cut-point a_c uniformly in $[a^S_{min}, a^S_{max}]$;
- Return the tree $[a < a_c]$.

Tree(S)
Input: a subset S
Output: a Boolean

- If $|S| < n_{min}$, then return TRUE;
- If all attributes are constant in S, then return TRUE;
- If the output is constant in S, then return TRUE;
- Otherwise, return FALSE.

Table 2.
Extra-Trees algorithm.

To classify each of the test sample objects, the following operations should be performed sequentially:

- To calculate the distance to each training sample feature.
- Select k objects of the training sample, the distance to which is minimal.
- The class of the object being classified is the class most often found among the k nearest neighbors.

4.2.3 SVC algorithm

In order to use an SVC to solve a linearly separable, binary classification problem we need to:

- create **H**, where $H_{ij} = y_i y_j x_i \cdot x_j$
- find α so that
- $\sum_{i=1}^{L} \alpha_i - \frac{1}{2}\alpha^T H \alpha$
- is maximized, subject to the constraints
- $\alpha_i \geq 0\ \forall_i\ and\ \sum_{i=1}^{L} \alpha_i y_i = 0$
- this is using a QP solver.
- calculate $w = \sum_{i=1}^{L} \alpha_i y_i x_i$

- determine the of Support Vectors S by finding the indices such that $\alpha_i > 0$
- calculate $b = \frac{1}{N_s} \sum_{s \in S} \left(\alpha_m y_m x_m \cdot x_s \right.$
- each new point x' is classified by evaluating $y' = \text{sgn}\,(w \cdot x' + b)$.

4.2.4 MLPClassifier algorithm

Multi-layer perceptron is a supervised learning algorithm that learns [20] a function $f(X) = R_n : R_n \to R^0$ by training on a speech dataset, where n is the number of dimensions for input and 0 is the number of dimensions for output. Given a set of features $X = x_1, x_2, \dots, x_n$, it can learn a non-linear function approximator for either classification (**Figure 3**).

The input layer consists of $x_1, x_2, \dots, x_n$ representing the input features. The output layer receives the values from the last hidden layer and transforms them into output values.

4.2.5 Gaussian NB algorithm

Naive Bayes gives the probability of a data point $X = x_1, x_2, \dots, x_n$ belonging to class C_k as proportional to a simple product of $n + 1$ factors. The class prior $p(C_k)$ plus n conditional feature probabilities $p(x_i|C_k)$. Specifically,

$$p(C_a) \prod_{i=1}^{n} p(x_i|C_a) > p(C_b) \prod_{i=1}^{n} p(x_i|C_b)$$

$$p(C_a|x_1, \dots, x_n) > p(C_b|x_1, \dots, x_n)$$

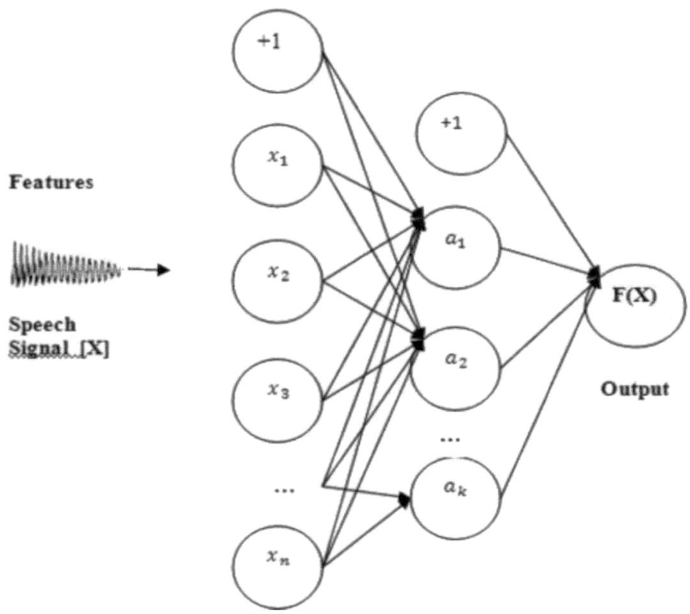

Figure 3.
MLP architecture.

Thus, the most likely class assignment for a data point $x_1, x_2, \ldots, x_n$ can be found by calculating $p(C_a)\prod_{i=1}^{n} p(x_i|C_k)$ for $k = 1, \ldots, K$ and assigning $x_1, x_2, \ldots, x_n$ the class C_k for which this value is largest.

5. Results and discussion

In Section 4.2 we applied the algorithms considered for the problem of personality identification and made a comparative analysis. Comparative analysis and

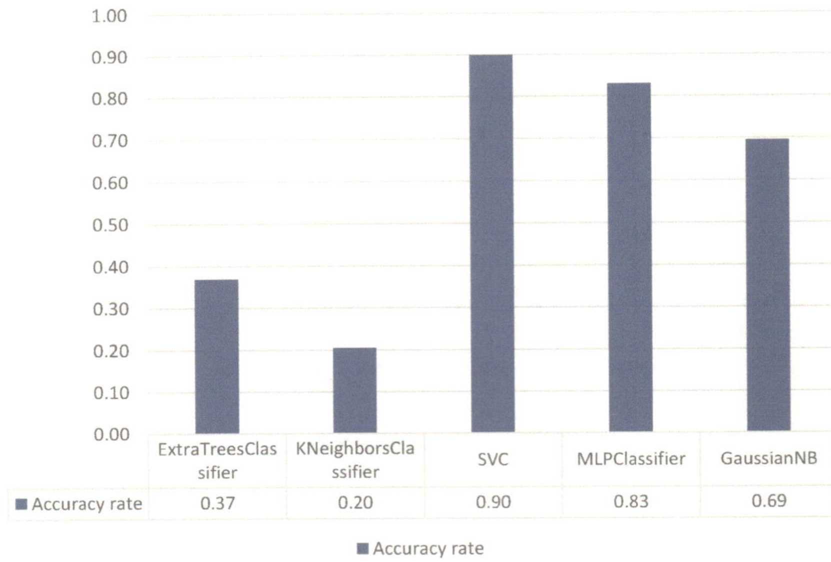

Figure 4.
Classification accuracy on a data set.

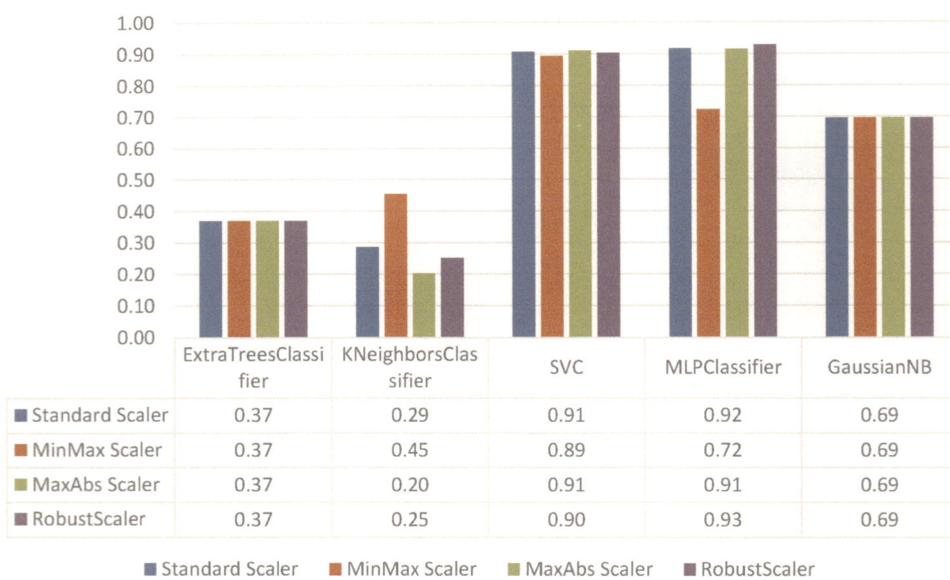

Figure 5.
Accuracy of classification when scaling data by various methods.

experiments have shown that the best results were obtained using the support vector machine and multilayer perceptron (**Figure 4**).

As can be seen from the diagram, the support vector machine and the multilayer perceptron showed the best results, 0.90 and 0.83, respectively.

To improve the results by different methods, the scaling was carried out and the results were slightly changed (**Figure 5**).

Now the highest accuracy was shown by the multilayer perceptron—0.93 when scaled by the Robust scaler method, and the support vector machine was relegated to the background, although it improved its result in accuracy from 0.90 to 0.91 when scaled by the Standard scaler and MaxAb scaler methods.

If we reduce the dimension of speech features to 1479 using the principal components method, the classification accuracy will change as in **Table 3**.

The purpose of the comparative analysis was to determine the degree of influence of classification algorithms for the problem of personality identification, as well as a comparative evaluation of the SVC and MLPClassifier Algorithms. Experiments conducted on the training set of speech data showed results that allow to

Algorithm name	Standard scaler	MinMaxScaler	MaxAbsScaler	RobustScaler
ExtraTreesClassifier	0.128125	0.128125	0.128125	0.128125
KNeighborsClassifier	0.043571	0.052143	0.050089	0.060982
SVC	0.051875	0.134732	0.097500	0.157679
MLPClassifier	0.002589	0.051875	0.082054	0.098393
GaussianNB	0.324286	0.324286	0.324286	0.324286

Table 3.
The accuracy of classification on the data with a decrease in dimension.

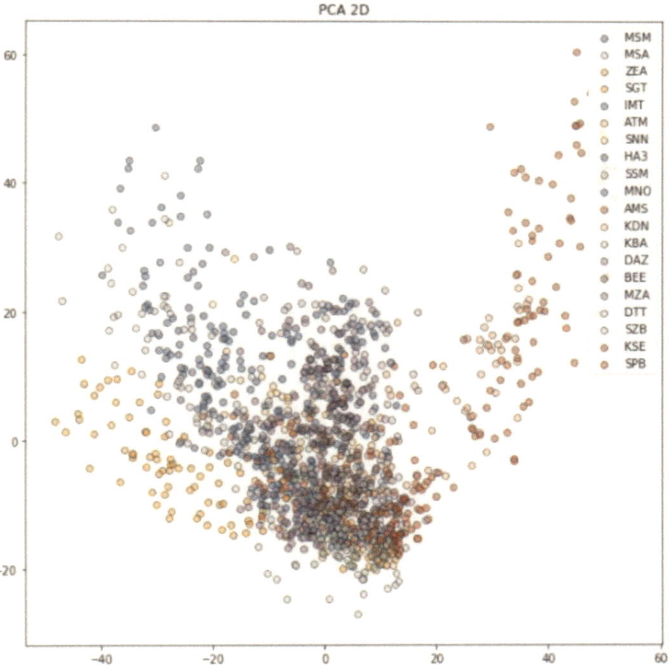

Figure 6.
Two-dimensional representation of speech data.

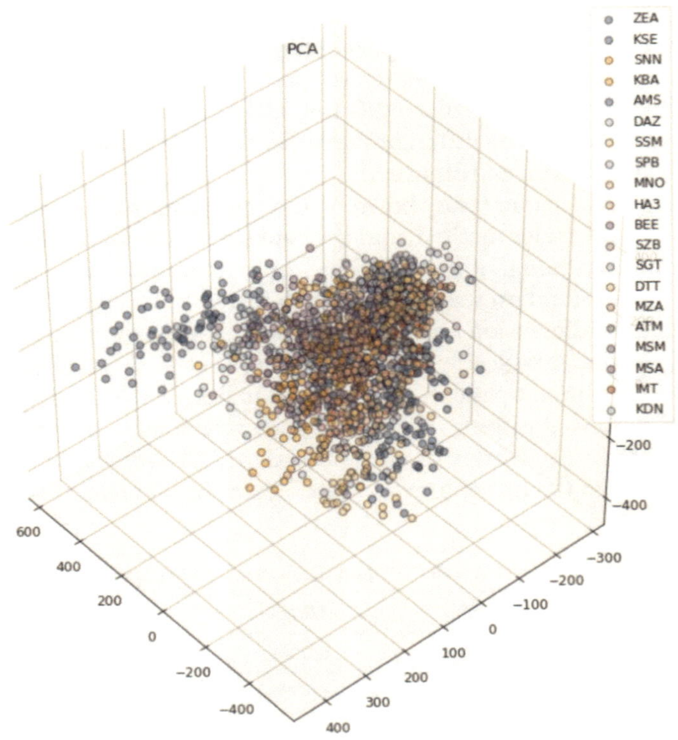

Figure 7.
Three-dimensional representation of speech data.

speak about the prospects of these algorithms. The data obtained were presented in **Figures 6** and 7.

The results of the classification when scaling data by various methods turned out to be significantly different from the results obtained during preliminary experiments.

6. Conclusions

In this paper, a number of classification algorithms and speech preprocessing issues were considered. Based on the analysis of the experimental results, a multilayer perceptron with an accuracy of 0.93 was proposed for scaling by the Robust scaler method and we will be able to classify the speech signal with the help of a multilayer perceptron. Further from the data we identified the personality.

In our further studies, we would like to solve the problem of verifying the identity of the data obtained.

Acknowledgements

This work was supported by the Ministry of Education and Science of the Republic of Kazakhstan. IRN AP05131207 Development of technologies for multilingual automatic speech recognition using deep neural networks.

Author details

Orken Mamyrbayev*, Nurbapa Mekebayev, Mussa Turdalyuly,
Nurzhamal Oshanova, Tolga Ihsan Medeni and Aigerim Yessentay
Institute of Information and Computational Technologies, Almaty, Kazakhstan

*Address all correspondence to: morkenj@mail.ru

References

[1] Zhan C, Li W, Ogunbona P. Face recognition from single sample based on human face perception. In: International Conference Image and Vision Computing New Zealand; 2009. pp. 56-61

[2] Beigi H. Fundamentals of Speaker Recognition. Springer Science & Business Media; 2011

[3] Yella S, Gupta N, Dougherty M. Comparison of pattern recognition techniques for the classification of impact acoustic emissions. Transportation Research Part C: Emerging Technologies. 2007;**15**(6): 345-360

[4] Aida-zade K, Xocayev A, Rustamov S. Speech recognition using support vector machines. In: AICT'16. 10th IEEE International Conference on Application of Information and Communication Technologies; 2016

[5] Serizel R, Giuliani D. Vocal tract length normalization approaches to DNN-based children's and adults' speech recognition. In: IEEE Workshop on Spoken Language Technology; 2014. pp. 135-140

[6] Popović B, Ostrogonac S, Pakoci E, Jakovljević N, Delić V. Deep neural network based continuous speech recognition for Serbian using the Kaldi toolkit. In: Ronzhin A, Potapova R, Fakotakis N, editors; SPECOM 2015; LNCS; Heidelberg: Springer; Vol. 9319; 2015. pp. 186-192

[7] Psutka J, Ircing P, Psutka JV, Hajič J, Byrne WJ, Mirovsky J. Automatic Transcription of Czech, Russian, and Slovak Spontaneous Speech in the MALACH Project. In: Proceedings of Eurospeech. Lisboa; Portugal; 4-8 September 2005. pp. 1349-1352

[8] Hinton G, Deng L, Yu D, Dahl G, Mohamed A, Jaitly N, et al. Deep neural networks for acoustic modeling in speech recognition: The shared views of four research groups. IEEE Signal Processing Magazine. 2012;**29**(6):82-97

[9] Tirumala SS, Shahamiri SR, Garhwal AS, Wang R. Speaker identification features extraction methods: A systematic review. Expert Systems with Applications. 2017;**90**: 250-271

[10] Kalimoldayev MN, Alimkhan K, Mamyrbayev OJ. Methods for applying VAD in Kazakh speech recognition systems. International Journal of Speech Technology. 2014b;**17**(2):199-204

[11] Toleu A, Tolegen G, Makazhanov A. Character-aware neural morphological disambiguation. In: Proceedings of the 55th Annual Meeting of the Association for Computational Linguistics; ACL; 2017. pp. 666-671

[12] Krassovitskiy A, Mussabayev R. Energy-based centroid identification and cluster propagation with noise detection. In: Nguyen N, Pimenidis E, Khan Z, Trawiński B, editors. Computational Collective Intelligence. Lecture Notes in Computer Science. Vol. 11055. Cham: Springer; 2018. pp. 523-533. DOI: 10.1007/978-3-319-98443-8_48. ICCCI 2018

[13] Mussabayev RR, Kalimoldayev MN, Amirgaliyev Ye N, Mussabayev TR. Automatic speech segmentation using throat-acoustic correlation coefficients. Open Engineering. 2016;**6**:335-346

[14] Karpov A, Kipyatkova I, Ronzhin A. Very large vocabulary ASR for spoken Russian with syntactic and morphemic analysis. In: Proc. INTERSPEECH-2011; Florence, Italy; 2011. pp. 3161-3164

[15] Mamyrbayev O, Turdalyuly M, Mekebayev N, Alimhan K,

Kydyrbekova A, Turdalykyzy T. Automatic recognition of Kazakh speech using deep neural networks. In: Asian Conference on Intelligent Information and Database Systems; 07 March 2019. pp. 465-474

[16] Mamyrbayev OZ, Kunanbayeva MM, Sadybekov KS, Kalyzhanova AU, Mamyrbayeva AZ. One of the methods of segmentation of speech signal on syllables. Bulletin of the National Academy of Sciences of the Republic of Kazakhstan. 2015:286-290

[17] Aida-Zade K, Ardil C, Rustamov S. Investigation of combined use of MFCC and LPC features in speech recognition systems. World Academy of Science, Engineering and Technology International Journal of Computer and Information Engineering. 2007: 2647-2653

[18] Voitko VV, Bevz SV, Burbelo SM, et al. Automated system of audio components analysis and synthesis In: Proceedings of SPIE; 2019. p. 110450V. DOI: 10.1117/12.2522313

[19] Hazmoune S, Bougamouza F, Mazouzi S. A new hybrid framework based on hidden Markov models and K-nearest neighbors for speech recognition. International Journal of Speech Technology. 2018;**21**(3):689-704

[20] Ribeiro FC, Santos CRT, Cortez PC, et al. Binary neural networks for classification of voice commands from throat microphone. IEEE Access. 2018; **6**:70130-70144

Chapter 4

Parallel Processing for Range Assignment Problem in Wireless Sensor Networks

M. Prasanna Lakshmi and D. Pushparaj Shetty

Abstract

Wireless sensor network is a collection of autonomous devices called sensor nodes which sense the environmental factors such as temperature, pressure, humidity, moisture, etc. The nodes sense the data, process it and transmit to the other nodes within their transmission range through radio propagation. Energy minimization in wireless sensor networks is a significant problem since the nodes are powered by a small battery of limited capacity. In case of networks with several thousand nodes, the simulation of algorithms can be very slow. The parallel computing model provides significantly faster simulation time for larger networks. Parallel processing involves executing the program instructions by dividing them among multiple processors with the objective of reducing the running time. So, we propose algorithms for the range assignment problem in wireless sensor networks using the parallel processing techniques. We also discuss the complexity of the proposed algorithms and significance of the parallel processing techniques in detail. The proposed techniques will be useful for implementing the distributed algorithms in WSNs.

Keywords: range assignment, parallel processing, minimum spanning tree, wireless sensor networks, algorithm, complexity

1. Introduction

In this section, preliminaries of Wireless Sensor Networks (WSNs) and parallel processing are discussed in detail.

1.1 Wireless sensor networks

Wireless Sensor Network (WSN) is a group of spatially dispersed autonomous devices called sensor nodes equipped with a radio transceiver along with an antenna. The devices are responsible for sending and receiving radio signals. Transmission range is assigned to the nodes to facilitate the communication between the nodes of a network. The sensor nodes sense the physical conditions of the environment such as temperature, pressure, sound, heat, light and humidity, etc., at various locations, process the data and transmit to the other nodes that lie within their transmission range through radio propagation. WSN monitors a physical system in real world and has applications in several fields such as environmental monitoring, biological detection, military and health care, etc. [1]. Because of the power constraints for the nodes using batteries with limited capacity, minimization of energy consumption is a significant problem in WSNs [2].

Constructing an efficient topology by assigning transmission range to the nodes of a network to minimize the total energy consumption has become a major challenge in WSNs. In general, the transmission range assigned to a node can be tuned so that the required connectivity constraint for the resultant network is satisfied and the total energy is minimized; this class of problems is categorized as topology control problems [3]. Some of the specified constraints include simple connectivity, bi-connectivity, k-connectivity, etc. [4]. By using an appropriate topology, the energy utilization of the network can be minimized which results in an increased lifetime of a WSN.

Transmission range of a node is the range within which the data sent by other node is received properly [5]. Two nodes in a WSN can communicate if and only if one node is in the transmission range of the other. Transmission range of a node u with the range r in three different dimensions is shown in the **Figure 1**. In general, the communication is multi hop in nature, and a node transmits the data to the destination node in its range using the relay nodes. In order to minimize the energy consumption, it is desirable to use short edges rather than the long energy-inefficient edges.

Let P_s be the range of source node u to transmit the signal to the destination node v. Then a signal transmitted by the source node u is attenuated by a factor: $P_r \propto \frac{P_s}{dist(u,v)^{\alpha}}$, where $dist(u,v)$ is the Euclidean distance between u and v, and α is the distance range gradient or path loss coefficient that depends on various environmental factors and generally lies between 2 and 6 [6]. In this chapter, the terms range and power are used interchangeably.

In reality, some problems use mathematical models that describe the problem in a systematic way and enable us to solve the problem efficiently. Graphs can be used to model many relations and represent many physical problems in the real world. A WSN is modeled as an undirected graph to solve the energy minimization problems. In this model, a vertex of the undirected graph represents a node and the edge joining two vertices represents the communication link between the nodes. The nodes are deployed on a Euclidean plane and each pair of two nodes is associated to the Euclidean distance between the nodes. Each node is located by its x and y coordinates and the distance function is used to find the Euclidean distance between any two nodes x and y, and is denoted by $w(xy)$ which is given by $w(xy) = \sqrt{(x_1 - x_2)^2 + (y_1 - y_2)^2}$.

Figure 2 illustrates the graph theoretic modeling of a WSN with four sensor nodes $v_1, v_2, v_3,$ and v_4. In this, **Figure 2(a)** shows a network of four nodes with

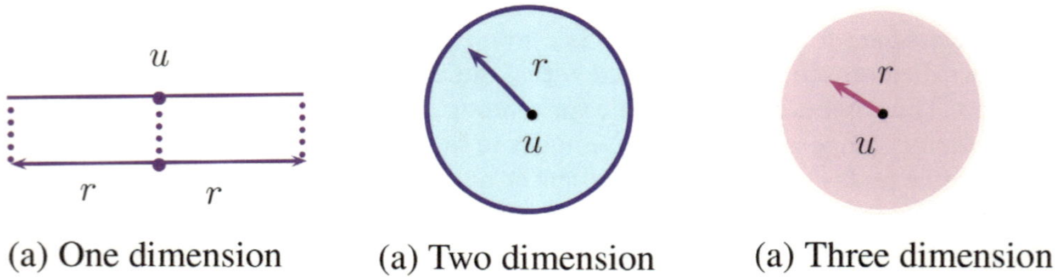

Figure 1.
Radio coverage in different dimensional networks.

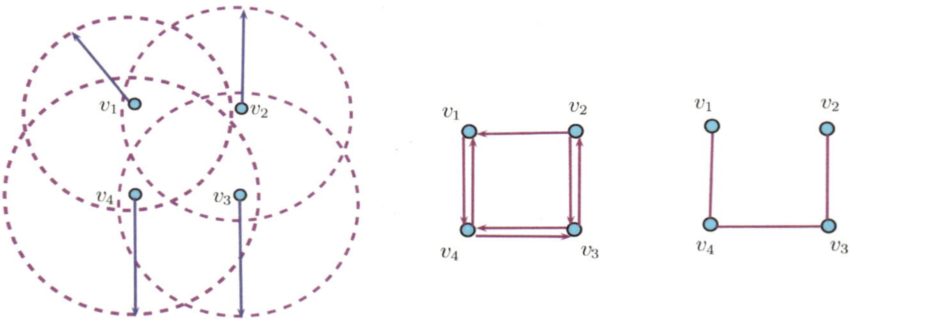

(a) The nodes and their transmission range (b) Asymmetric topology (c) Symmetric topology

Figure 2.
Illustration of graph theoretic modeling.

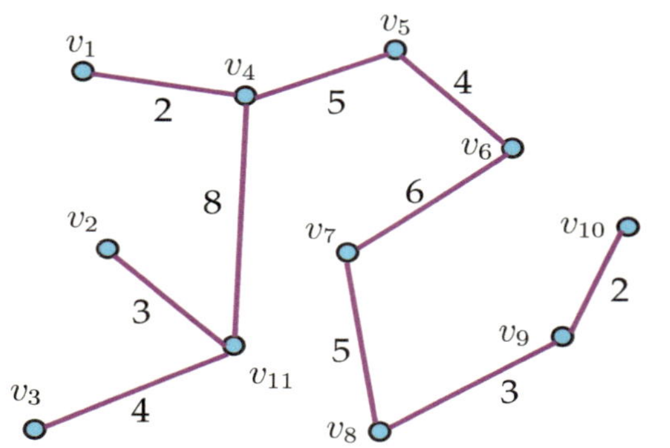

(a) Reduced Topology of eleven vertices

Vertex	Range
v_1	2
v_2	3
v_3	4
v_4	8
v_5	5
v_6	6
v_7	6
v_8	5
v_9	3
v_{10}	2
v_{11}	8

(b) Range assignment

Figure 3.
Illustration of range assignment.

their respective transmission ranges and its asymmetric and symmetric versions are shown in **Figure 2(b)** and **(c)**, respectively [7]. There is an edge between two vertices in symmetric graph if and only if it is bidirectional. A network is said to be bidirectional if and only if each edge is bidirectional and there exists a bidirectional path between every pair of two nodes of the network. Since the signal transmitted by a sensor node must be acknowledged by the receiver node, the bidirectional links are given preference.

Throughout this chapter, the considered topology is bidirectional. Therefore, we use an undirected graph along with the weight function w to represent a WSN. Once a WSN is formulated as an undirected graph $G = (V, E, w)$, the transmission range $R(v)$ assigned to each vertex $v \in G$ is the maximum of all its adjacent edge weights and is mathematically given as follows:

$$R(v) = max\,\{w(vu)|uv \in E(G)\}. \quad (1)$$

The total range of the graph which is the sum of the ranges of all the vertices of G is given by $R(G) = \Sigma_{i=1}^{n} R(v_i)$. **Figure 3** is presented to illustrate the range assignment of a WSN. In this, **Figure 3(a)** shows a spanning tree of 11 vertices along with weights assigned to the edges and **Figure 3(b)** shows the range values assigned to the vertices as the maximum of all its adjacent edge weights in the spanning tree, as given in Eq. (1).

Minimum range assignment problem in a WSN is to assign transmission range to the nodes of the network such that the resultant subgraph H satisfies the specified constraints and the total range of the network, i.e., $R(H)$ is minimized [8]. Let H be a spanning subgraph of the given graph $G = (V, E, w)$, $R_H(v)$ be the range assigned to vertex $v \in V(H)$ as given in Eq. (1) and $R(H)$ be total range of the subgraph H. For various problems considered in this chapter, our interest is to find a spanning tree with minimum range.

Chen et al. [9] studied the problem of strong connectivity in multi hop packet radio networks and proposed a 2-approximation algorithm which initially considers an undirected Minimum Spanning Tree (MST) and later establishes the bidirectional connectivity. Strong Minimum Energy Topology problem (SMET) in a WSN is studied by Cheng et al. [10], in which the objective is to assign transmission range to the nodes of the network such that the total range of the network is minimized and the reduced topology is connected. The authors have proved that the SMET problem is NP-complete and proposed two algorithms namely: Minimum Spanning Tree (MST) based Heuristic which is of 2-approximation ratio and an Incremental Power Greedy Heuristic and performed the simulation to show that the Incremental Power Greedy Heuristic performs better than the MST Heuristic. Formulation of the SMET problem is as follows:

Problem: Strong Minimum Energy Topology problem (SMET).

Instance: A complete graph $G = (V, E, w)$, where $w : V \times V \rightarrow \mathbb{R}^+$ is a weight function.

Question: A range assignment such that the induced subgraph H is connected and the and the total range of H, i.e., $R(H)$ is minimized.

1.2 Parallel processing

Large scale WSNs such as environmental monitoring systems produce large amount of data in the order of Peta bytes. The challenges in this case are storage and computation as the sensors are constrained by their resources which are scarcely available [11]. If the large amount of data is processed centrally, all data needs to be transmitted to a central server using multi hop transmissions which leads to high computational cost. One of the best ways to avoid such problem is to exploit the advantages of the distributed storage and parallel processing. Instead of transmitting data to a central server, data is stored and processed in network which reduces the computational cost. In this process, the computational task is decomposed into many small tasks, and the computation is executed in parallel by the distributed sensors.

Sullivan et al. [12] proposed a set of new parallel algorithms for a tree decomposition-based approach to solve the maximum weighted independent set problem. Independent set problem for a given graph is to find a subset of vertices with maximum cardinality such that no two vertices in the graph are adjacent to each other. Maximum Weighted independent set problem in which, the vertices are assigned weights, is to find an independent set with maximum weight [13]. Zhu et al. [14] proposed parallel algorithm for the hierarchical-based protocol for WSNs based on the Low-Energy Adaptive Clustering Hierarchy (LEACH) algorithm in order to improve the routing efficiency of the networks. Authors have implemented the algorithms using the parallel processing technique and showed that this technique has increased the overall performance. Lounis et al. [15] presented a new model for parallel computing of energy consumption in WSNs, and implemented on GPU architecture. Through simulation, authors showed that the proposed model provides simulation times significantly faster than that of the sequential model for large networks and for long simulations.

Andoni et al. [16] gave algorithms for geometric graph problems in the modern parallel models. The authors found a $(1 + \epsilon)$ approximation algorithm for a

Geometric Minimum Spanning tree (MST) problem. In geometric MST, set of n points are plotted on a low dimension space such as R^d and other bounded metric spaces, in which weights of the edges are the distances between their endpoints. Authors also gave a general algorithmic framework that also applies to Earth-Mover Distance and the transportation cost problem. Katsigiannis et al. [17] presented a Helper Threading scheme for parallelizing the Kruskal's algorithm [18] which is used for finding a Minimum Spanning Forest (MSF). The implementation employs one main thread, which executes the serial algorithm, and several helper threads, that run in parallel and reduce the work of the main thread.

In this chapter, we discuss the parallel processing techniques for the range assignment problem for simple connectivity. We propose algorithms for this problem using parallel processing techniques and determine the complexity. The proposed algorithms initially find an MST and assign range to each vertex of the network based on its range in the Euclidean MST. We find the Euclidean MST using well known algorithms: Prim's and Kruskal's and later we assign the range using parallel processing technique for both the cases. We also discuss the complexity of the proposed algorithms in detail for both the algorithms.

Simulation is an act of imitating operation of real world process or system over time [19]. The simulation is performed on a model which represents the system. It is known that simulations closely reflect accurate behavior of the system. Simulations help in evaluating the algorithm before investing on the physical hardware.

2. Related work

Finding a Euclidean MST is an important problem in Graph Theory and there are several algorithms in the literature [20]. MST problems have significant applications in computer and communication network design, as well as indirect applications in fields such as computer vision and cluster analysis [21]. Parallelization of Prim's algorithm is studied by Deo and Yoo [22] and the proposed algorithms used shared memory computers. Gonina et al. [23] studied the parallel implementation of the Prims's algorithm for finding the MST of dense graphs. Authors presented the simulation results which demonstrate the advantage of proposed algorithms.

Vladimir et al. [24] studied serial variants of Prim's and Kruskal's algorithm and presented their parallelization, targeting message passing parallel machine with distributed memory. Authors have considered large graphs and presented experimental results which show that Prim's algorithm is a good choice for dense graphs while Kruskal's algorithm is better for sparse graphs. Authors have proposed algorithms based on the sequential algorithms of Prim and Kruskal, targeting message passing parallel machine with distributed memory. The proposed algorithm which uses the Prim's algorithm has a running time of $O\left(\frac{n^2}{k}\right) + O(n \log k)$ and the one which uses the Kruskal's algorithm has a running time of $O\left(\frac{n^2}{k}\right) + O(n^2 \log k)$ where n is the number of vertices and k is the number of partitioned subsets of V such that each subset has $\frac{n}{k}$ consecutive vertices and their edges. In this section, we discuss the parallel processing algorithms for computing Euclidean MST using both Prim's and Kruskal's algorithms separately. In both the algorithms, each processor gets a subset of vertices from the given set of vertices to process.

The considered input for the parallelization of both Prim's and Kruskal's algorithm by Vladimir et al. [24] is an undirected simple graph with weights assigned to the edges. For our research the input considered is a complete graph along with edge weights, where the edge weights are the Euclidean distance between the nodes.

2.1 Parallelization of Prim's algorithm by Loncar et al.

Prim's algorithm starts from an arbitrary vertex and then grows the subtree by choosing a new vertex from the set of unvisited vertices and adding it to the subtree in each iteration. The Prim's algorithm runs until the number of edges becomes $n - 1$ or all the vertices are added to the subtree. The complexity of the Prim's algorithm is $O(n^2)$ where n is the number of vertices of the given graph. In this algorithm, the input considered is an undirected graph with n vertices and m edges along with the edge weights in which weight of edge uv is denoted by $w(uv)$ [24].

The weights are represented using the adjacency matrix which is defined as follows:

$$a_{ij} = \begin{pmatrix} w(v_iv_j) & \text{if } (v_iv_j \in E) \\ 0 & \text{otherwise.} \end{pmatrix} \tag{2}$$

Prim's algorithm is implemented using the auxiliary array d of length n which stores the weights from each vertex to the MST. The lightest weight edge is chosen from d and is added to MST in each iteration and the array d is updated accordingly. The main loop of the Prim's algorithm cannot be parallelized since the minimum edge weights incident to MST change in each iteration. So, only two steps are parallelized: finding a minimum weight edge that connects a new vertex not in MST to a vertex in MST and updating the array d. The complexity of the proposed algorithm is $O\left(\frac{n^2}{k}\right) + O(n \log k)$ where n is the number of vertices and k is the number of processors.

2.2 Parallelization of Kruskal's algorithm by Loncar et al.

Kruskal's algorithm finds a minimum weighted edge which connects two components in the forest in each iteration, i.e., it grows multiple trees in parallel. Initially, Kruskal's algorithm creates a forest considering each vertex as a tree and next it sorts all the edges. Each time a minimum weighted edge is chosen and added to the forest if it connects two different trees else it is discarded. This process is repeated until the forest becomes a spanning tree. The complexity of the proposed algorithm is $O(m \log n)$ where m is the number of edges and n is the number of vertices [24].

Similar to the Prim's algorithm, we have adjacency matrix which stores the weights between the vertices, and each processor is assigned a subset of vertices. In parallelization of Kruskal's algorithm, each processor sorts the edges of its vertices according to their weights. At each process, a local MST is found using Kruskal's algorithm and finally, all the local MST's are merged. The complexity of the proposed algorithm is $O\left(\frac{n^2}{k}\right) + O(n^2 \log k)$ where n is the number of vertices and k is the number of processors.

3. Parallelization of range assignment problem

In this section, we present the proposed algorithms for range assignment problem using parallel processing techniques. The proposed algorithms adopt the algorithms for parallelization of both Prim's and Kruskal's algorithms by Loncar et al. [24]. In this environment, the nodes are deployed on a Euclidean plane and each pair is associated with the Euclidean distances between those nodes. Since a WSN is modeled as an undirected graph along with a weight function, the input for these algorithms is a complete graph $G = (V, E, w)$ where the weight function is as explained in Section 1. The range assigned to a vertex v is denoted by $R(v)$.

Objective of this problem is to find a spanning tree such that the total range of the spanning tree is minimized. In the proposed algorithms, we initially find the spanning tree of the given complete graph and assign the range to the vertices based on their range in the Euclidean MST using parallel processing. We propose two algorithms: one using the Prim's algorithm and the other using the Kruskal's algorithm. We also compute the complexity of the proposed algorithm in both the cases. The list of notations used in this chapter is given in **Table 1**.

3.1 Range assignment using Prim's algorithm

Let $G = (V, E, w)$ be the given complete graph. In this algorithm initially, we find a spanning tree by parallelizing the Prim's algorithm and range is assigned to the vertices based on their range in the MST by using parallel processing. The formulation of the problem is as follows:

Problem*:* Parallel Algorithm for Range assignment using Prim's algorithm (PARA-Prim's).

Input*:* A complete graph $G = (V, E, w)$, where w is a weight function and P the set of processors.

Objective*:* To find a spanning tree such that the total energy is minimized, using parallel processing.

In the proposed algorithm, we consider a complete graph and the number of processors as input and compute a spanning tree which minimizes the total range. In this algorithm, we partition the vertices into k groups and each processor gets $\frac{n}{k}$ consecutive vertices and their edges. Each processor also contains auxiliary array d of the vertices which maintains the distances of the vertices to the MST. Let V_i and d_i be the set of vertices and the auxiliary array of the processor p_i, respectively.

Each process finds a minimum weighted edge which connects the MST with vertices of the set V_i and sends to the leader processor using all-to-one reduction. The leader processor selects one with the minimum weight among all the received edges, adds it to MST and broadcasts to all the processors. Processors mark the vertices connected to MST and update their auxiliary array d and this process is repeated until all the vertices are added to the subtree.

Next, we assign range to the vertices in the following way: for a vertex, each process finds the farthest vertex to it in the MST and sends the distance between

Notation	Explanation
V	Set of vertices
E	Set of edges
n	Number of nodes
$w(uv)$	Euclidean distance between u and v
$R(v)$	Range of vertex v
d	Auxiliary array
P	The set of processors
p_i	i^{th} processor
k	Number of processors
T	Resultant spanning tree

Table 1.
List of notations.

them to the leader processor using all-to-one reduction. Later, leader processor selects one with maximum edge weight among all the received edge weights and that weight is assigned as range to the vertex v. This process of range assignment is repeated for all the vertices. The sequence of steps is presented in Algorithm 1.

Theorem 1.1 [24] The parallelization of Prim's algorithm takes $O\left(\frac{n^2}{k}\right) + O(n \log k)$ running time where n is the number of nodes and k is the number of processors.

Algorithm 1: PARA-Prim's

```
begin
    Let |P| = k, |V| = n, T = φ, j = 1, Visited[i] = 0 for i = 1 to n
    for i = 1 to k do
        p_i = {v_j, v_{j+1}, ..., v_{j+n/k}}
        j = n/k + 1
    end
    while |V(T)| < n − 1 do
        for i = 1 to k do
            for j = 1 to n/k do
                min = ∞
                for k = 1 to n/k do
                    if p_i[j] ≠ p_i[k] && visited[p_i[j]] ≠ 0 && w(p_i[j] p_i[k]) < min
                      then
                        e[i] = p_i[j] p_i[k]
                    end
                end
            end
        end
        for i = 1 to k do
            if w(e[i]) < w(e_min) then
                e_min = uv = e_i
            end
        end
        if Visited[u] = 0 || Visited[v] = 0 then
            T = T ∪ {e_min}
            Visited[v] = 1
            update the corresponding d[v]
        end
        for i = 1 to n do
            for j = 1 to k do
                max[i] = 0
                for k = 1 to n/k do
                    if v_i ≠ p_j[k] && v_i p_j[k] ∈ T && w(v_i p_j[k]) > max then
                        max[i] = w(v_i p_j[k])
                    end
                end
            end
            max = max[1]
            for i = 1 to k do
                if max[i] > max then
                    max = max[i]
                end
            end
            R[v_i] = max
        end
    end
    return T, R
end
```

Theorem 1.2 The algorithm PARA-Prim's takes $O\left(\frac{n^2}{k}\right) + O(n \log k)$ running time, where n is the number of nodes and k is the number of processors.

Proof: From Theorem 1.1, it is clear that finding the MST by parallelizing the Prim's algorithm takes $O\left(\frac{n^2}{k}\right) + O(n \log k)$ running time for n number of nodes and k number of processors. In PARA-Prim's algorithm the range assignment is done based on the MST obtained by parallelizing the Prim's algorithm. For each vertex, we find the farthest vertex in each processor which takes the running time of $O\left(\frac{n}{k}\right)$ and for all such vertices it takes a running time of $O\left(\frac{n^2}{k}\right)$. Next each processor sends the farthest vertex to the leader process and leader processor finds the global farthest vertex which takes a running time of $O\left(\frac{n}{k}\right)$ and for all such n vertices it takes a running time of $O\left(\frac{n^2}{k}\right)$. So the running time of PARA-Prim's algorithm is $O\left(\frac{n^2}{k}\right) + O(n \log k)$.

3.2 Range assignment using Kruskal's algorithm

Similar to the Prim's algorithm in this algorithm, we have k number of processors assigned with $\frac{n}{k}$ set of consecutive vertices. Each processor sorts the edges of its vertices and finds a local MST using the Kruskal's algorithm. Finally all the local MSTs are merged in the following way: first processor sends its set of edges of the local MST to the second processor and forms a new local MST by using the set of edges from both the processors. Now, the first process remains no longer and terminates, and this process of merging continues until single process remains. Range assignment is done to the vertices in parallel way similar to the PARA-Prim's algorithm as explained in the previous subsection. The formulation of the problem is as follows:

Problem*:* Parallel Algorithm for Range assignment using Kruskal's algorithm (PARA-Kruskal's).

Input*:* A complete graph $G = (V, E, w)$, where w is a weight function and P the set of processors.

Objective*:* To find a spanning tree such that the total energy is minimized, using parallel processing.

The sequence of steps for the above explained procedure is presented in Algorithm 2 named PARA-Kruskal's.

Theorem 1.3 [24] The parallelization of Kruskal's algorithm takes $O\left(\frac{n^2}{k}\right) + O(n^2 \log k)$ where n is the number of vertices and k is the number of processors.

Theorem 1.4 The algorithm PARA-Kruskal's takes $O\left(\frac{n^2}{k}\right) + O(n^2 \log k)$ running time, where n is the number of nodes and k is the number of processors.

Proof: From Theorem 1.3, it is clear that, the running time of the parallelization of Kruskal's algorithm is $O\left(\frac{n^2}{k}\right) + O(n^2 \log k)$. Next, we do the range assignment to the vertices based on its range in the MST formed by parallelizing the Kruskal's algorithm. From Theorem 1.3, it is clear that the running time of the range assignment for all the vertices is $O\left(\frac{n^2}{k}\right)$. So, the running time of the PARA-Kruskal's algorithm is $O\left(\frac{n^2}{k}\right) + O(n^2 \log k)$.

Algorithm 2: PARA-Kruskal's

```
begin
    Let |P| = k, |V| = n
    Follow the steps from 3 to 5 of Algorithm 1
    for i = 1 to k do
        T_i = Φ
        Let E_i be the set of all edges of the vertices of p_i
        Sort the edges of E_i according to their edge weights
        while |V(T_i) < n/k − 1 do
            for j = 1 to n/k do
                min = ∞
                for k = 1 to n/k do
                    if w(p_i[j] p_i[k]) < min then
                        min = w(p_i[j] p_i[k])
                        a = p_i[j]
                        b = p_i[k]
                    end
                end
            end
            if T_i ∪ {ab} is acyclic then
                T_i = T_i ∪ {ab}
            end
        end
    end
    for i = 1 to k − 1 do
        Let A be the set of all edges of T_i
        Let B be the set of all edges of T_{i+1}
        p_i sends A to p_{i+1}
        p_{i+1} finds T as a new local MST from A ∪ B
    end
    Repeat the steps from 28 to 44 of Algorithm 1
    return T, R
end
```

4. Comparison with the state of arts

Solving the range assignment problem by computing Euclidean MST using Prim's algorithm and Kruskal's algorithm takes running time of $O(n^2)$ and $O(n^2 \log n)$, respectively. For WSNs with large number of sensor nodes, this complexity is quite high which can be improved by using parallel processing techniques. Simulation can be performed on hardware that supports multiple cores in order to realize the improvements over serial programming environments. The complexities of the range assignment problem using Prim's algorithm and Kruskal's algorithm by employing parallel processing techniques are $O\left(\frac{n^2}{k}\right) + O(n \log k)$ and $O\left(\frac{n^2}{k}\right) + O(n^2 \log k)$, respectively. So, theoretically there is a significant improvement in the complexity of the range assignment problem by applying parallel processing techniques. Simulation can be performed to study the stability of the proposed algorithms and to compare the proposed algorithms with the existing algorithms for the range assignment algorithms.

5. Conclusions

In this research, we have studied the range assignment problem by employing parallel processing techniques of both Prim's and Kruskal's algorithms.

The complexity of the proposed two algorithms is discussed in detail and it is shown that complexity can be improved by using parallel processing techniques for larger networks. It is an interesting problem to study variations of range assignment problems using parallel processing techniques. The implementation of the proposed algorithms can be realized using CUDA programming which supports programming GPU with multiple cores.

Abbreviations

WSN	wireless sensor network
MST	minimum spanning tree
MSF	minimum spanning forest
PARA	parallel algorithm for range assignment

Author details

M. Prasanna Lakshmi[1*†] and D. Pushparaj Shetty[2†]

1 Indian Institute of Information Technology, Sri City, Chittor, India

2 National Institute of Technology Karnataka, Mangalore, India

*Address all correspondence to: prasannasainitw@gmail.com

† These authors are contributed equally.

References

[1] Akyildiz IF, Su W, Sankarasubramaniam Y, Cayirci E. Wireless sensor networks: A survey. Computer Networks. 2002;**38**(4):393-422

[2] Clementi, AE, Penna P, Silvestri R. The power range assignment problem in radio networks on the plane. In: Annual Symposium on Theoretical Aspects of Computer Science. Berlin, Heidelberg: Springer; February 2000. pp. 651-660

[3] Lloyd EL, Liu R, Marathe MV, Ramanathan R, Ravi SS. Algorithmic aspects of topology control problems for ad hoc networks. Mobile Networks and Applications. 2005;**10**(1–2):19-34

[4] West DB. Introduction to Graph Theory. Vol. 2. Upper Saddle River, NJ: Prentice hall; 1996

[5] Santi P. Topology control in wireless ad hoc and sensor networks. ACM Computing Surveys (CSUR). 2005; **37**(2):164-194

[6] Carmi P, Chaitman-Yerushalmi L. On the minimum cost range assignment problem. In: International Symposium on Algorithms and Computation. Berlin, Heidelberg: Springer; 2015. pp. 95-105

[7] Calinescu G, Wan PJ. Range assignment for high connectivity in wireless ad hoc networks. In: International Conference on Ad-Hoc Networks and Wireless. Berlin, Heidelberg: Springer; 2003. pp. 235-246

[8] Fuchs B. On the hardness of range assignment problems. Networks: An International Journal. 2008;**52**(4):183-195

[9] Chen WT, Huang NF. The strongly connecting problem on multihop packet radio networks. IEEE Transactions on Communications. 1989;**37**(3):293-295

[10] Cheng X, Narahari B, Simha R, Cheng MX, Liu D. Strong minimum energy topology in wireless sensor networks: NP-completeness and heuristics. IEEE Transactions on Mobile Computing. 2003;**2**(3):248-256

[11] Wang Y, Wang Y. Distributed storage and parallel processing in large-scale wireless sensor networks. In: High Performance Computing Workshop. Cetraro, Italy: IOS Press; 2010. pp. 288-305

[12] Sullivan BD, Weerapurage D, GroÃńr C. Parallel algorithms for graph optimization using tree decompositions. In: 2013 IEEE International Symposium on Parallel & Distributed Processing, Workshops and Phd Forum. IEEE; 2013. pp. 1838-1847

[13] Minty GJ. On maximal independent sets of vertices in claw-free graphs. Journal of Combinatorial Theory, Series B. 1980;**28**(3):284-304

[14] Zhu Y, Yao Q, George G, Wu S, Zhang C. Parallel LEACH algorithm for wireless sensor networks. In: Proceedings of the International Conference on Parallel and Distributed Processing Techniques and Applications (PDPTA). The Steering Committee of The World Congress in Computer Science, Computer Engineering and Applied Computing (WorldComp); 2012. p. 1

[15] Lounis M, Bounceur A, Laga A, Pottier B. GPU-based parallel computing of energy consumption in wireless sensor networks. In: 2015 European Conference on Networks and Communications (EuCNC). IEEE; 2015, June. pp. 290-295

[16] Andoni A, Nikolov A, Onak K, Yaroslavtsev G. Parallel algorithms for geometric graph problems. In: Proceedings of the Forty-Sixth Annual ACM Symposium on Theory of Computing. ACM; 2014. pp. 574-583

[17] Katsigiannis A, Anastopoulos N, Nikas K, Koziris N. An approach to parallelize Kruskal's algorithm using helper threads. In: 2012 IEEE 26th International Parallel and Distributed Processing Symposium Workshops & PhD Forum. IEEE; 2012. pp. 1601-1610

[18] Cormen TH, Leiserson CE, Rivest RL, Stein C. Introduction to Algorithms. Cambridge, Massachusetts, London, England: McGraw-Hill Book Company; 2009

[19] Sobeih A, Hou JC, Kung LC, Li N, Zhang H, Chen WP, et al. J-Sim: A simulation and emulation environment for wireless sensor networks. IEEE Wireless Communications. 2006;**13**(4): 104-119

[20] Rosen KH. Handbook of Discrete and Combinatorial Mathematics. Boca Raton, London, New York, Washington, DC: CRC Press; 2017

[21] Graham RL, Hell P. On the history of the minimum spanning tree problem. Annals of the History of Computing. 1985;**7**(1):43-57

[22] Deo N, Yoo YB. Parallel Algorithms for the Minimun Spanning Tree Problem. Computer Science Department: Washington State University; 1981

[23] Gonina E, KalÃľ LV. Parallel PrimâĂŹs Algorithm on Dense Graphs with a Novel Extension. 2007

[24] LonÄŅar V, Åăkrbic S. Parallel implementation of minimum spanning tree algorithms using MPI. In: 2012 IEEE 13th International Symposium on Computational Intelligence and Informatics (CINTI). IEEE; 2012. pp. 35-38

Section 2

Intelligent Systems

Chapter 5

Vehicle Secrecy Parameters for V2V Communications

Na-Young Ahn and Dong Hoon Lee

Abstract

This paper studies the parameters affecting secrecy capacity in vehicle communication. The vehicle secrecy parameters largely include vehicle driving-related parameters, antenna-related parameters for transmitting and receiving signals, path-related parameters for indirect communication, and noise-related parameters using a fading channel. Although many researches have been conducted on antenna-related parameters and noise-related parameters considered in general wireless communication, relatively little research has been made on parameters caused by the vehicle itself. These vehicle secrecy parameters also imply that secrecy capacity can be varied by the user. In the future, this study will be a very informative topic when trying to perform vehicle communication while maintaining a certain level of security capacity. In the coming autonomous driving era, this research is very necessary and will help to carry out vehicle communications more safely.

Keywords: secrecy capacity, vehicle secrecy parameter, physical layer security, vehicle speed, antenna, relay, fading

1. Introduction

Vehicle-to-vehicle communication functionality is essential for general 5G communications [1, 2]. This functionality aims to achieve the safe operation of autonomous vehicles by sharing vehicle driving-related information, such as basic security messages (BSMs). To guarantee vehicle-to-vehicle functionality, security must be the foundation of design and privacy must be a priority. Previous studies have already made significant progress on security beyond the vehicular network layer [3, 4]. Yet, despite this, existing vehicle security schemes demonstrate insufficient computing power and large power consumption with respect to processing received or transmitted data from a large number of vehicles. To overcome these difficulties, researches on physical layer security [5–7] have attempted to develop secure data communication methods based on the physical properties of the wireless channel.

In information theory, channel (or Shannon) capacity is known as a maximal amount of information that can be transmitted through a wireless channel [8, 9]. In general, channel capacity is given as

$$C = W\log(1 + SNR), \tag{1}$$

where W is the channel bandwidth, and SNR is the signal-to-noise ratio. Secrecy capacity denotes the channel capacity of a legitimate channel less the channel

IntechOpen

capacity of a wiretap channel. That is, secrecy capacity is a maximum data rate that is achievable between the legitimate TX-RX pair, subject to the constraints on information attainable by an unauthorized receiver [10]. For a Gaussian wiretap channel, secrecy capacity C_s is:

$$C_s = \frac{1}{2}\log\left(1+\frac{P}{N_m}\right) - \frac{1}{2}\log\left(1+\frac{P}{N_w}\right), \tag{2}$$

where P is the transmitter's power, N_m is the receiver's noise, and N_w is the eavesdropper's noise.

Secrecy capacity means an entropy that can conceptually transmit secrets securely without taps. It is a value obtained by subtracting the channel capacity for performing illegal communication from the channel capacity for performing mathematically legitimate communications? In information theory, channel capacity is the maximum rate that can be transmitted without error. Thus, the unit of secrecy capacity is bits/sec/Hz. Could this concept be meaningful in vehicle communication? We are asking questions, and so on. Nobody knows the existence of wiretapping. In vehicle communication, valid data are transmitted over the air in four directions, and anyone can obtain it if desired. Even if it is encrypted, the existence of a quantum computer makes it possible for this threat to cause serious problems. We naturally cannot but consider wireless channels, but protected channels, in vehicular communications. So how can we protect wireless channels, but secure channels from eavesdroppers? The answer is physical layer security. Therefore, research on modeling for vehicle communications is inevitable.

Secrecy capacity can be controlled in real time. The parameters affecting these security capacities are classified as follows: vehicle-related parameters, antenna-related parameters, communication path-related parameters, and noise-related parameters.

2. Vehicle-related parameters

Vehicle-related parameters are very important for vehicle accidents and physical safety as parameters related to the operation of the vehicle while driving. The vehicle-related parameters include speed of the autonomous vehicle, the response time, etc., as mentioned by Ahn et al. [11].

2.1 Vehicle speed

It is generally assumed that secrecy capacity will be increased according to the vehicle speed. For example, it is assumed that there will be a difference in eavesdropping data transmitted from a vehicle running at low speeds and eavesdropping data transmitted from vehicles running at high speeds. Considering the Doppler effect, there is a theoretical study on this [12]. Chopra et al. suggested that as the vehicle speed increases, the eavesdropper is less likely to succeed in eavesdropping [12].

Chopra has announced that as speed of the vehicle increases, the probability of an attacker's success gradually decreases, referring to **Figure 1**. A common guess is that you will get your hands on these results. SNR value for speed is drastically reduced according to the effects of beam merge and Doppler shift, and the probability of success of the eavesdropper is expected to sharply decrease (**Figure 1**).

However, this is related to naive wireless communication, and it is difficult to apply it to a vehicle in operation. As mentioned in the previous section, the autonomous vehicle speed is closely related to the safety distance, and the safety distance eventually affects the channel capacity between vehicles. According to my modeling and simulation results, as the vehicle speed increases, secrecy capacity becomes rather small.

Ahn et al. concluded that vehicle speed is closely related to the safety distance to prevent impulsiveness, and this safety distance ultimately affects secrecy capacity, resulting in a close relationship between vehicle speed and secrecy capacity, referring to **Figure 2**. The result was, surprisingly, that the faster the vehicle speed was, the less secrecy capacity was increased, rather than increasing (**Figure 2**) [11].

Some say that secrecy capacity will increase with speed of the vehicle, and others say that secrecy capacity will be reduced depending on speed of the vehicle. Who is wrong and who is right? I think that these results are derived from the fact that the definition of secrecy capacity for vehicles has not been established. In addition, it is presumed that there is a totally different answer depending on the setting, whether or not the eavesdropper is present, whether the eavesdropper is moving, and the information of the eavesdropper. It means that this lack of research on secrecy capacity in vehicle communication.

So, what information do these facts give us? The relationship between vehicle speed and secrecy capacity is simply not defined. Nevertheless, there is a connection between speed of the vehicle and secrecy capacity. The important thing is that you can control secrecy capacity at the speed of the vehicle.

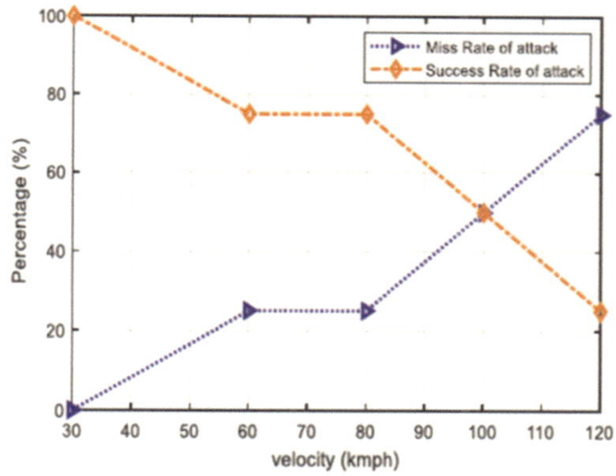

Figure 1.
Percentage of miss/success rate of attack for eavesdroppers according to different velocities of user [12].

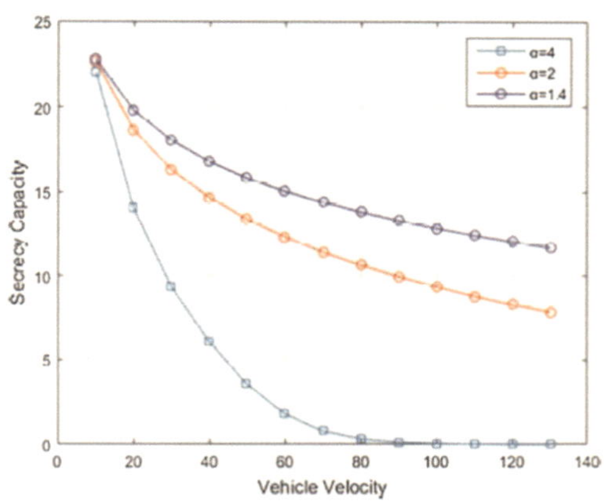

Figure 2.
The relationship of secrecy capacity according to speed of vehicle [11].

2.2 Vehicle system parameter

What do the system parameters of the car mean? System parameters are an important factor in determining the collision distance between vehicles. For example, response time has a significant impact on break distance. The break distance is proportional to the safety distance. The response time will eventually affect the safety distance, and this safety distance will eventually affect secrecy capacity.

The faster the response time from one vehicle to the reception and processing of signals from other vehicles, the sooner it is expected to help secure safety. If you think about it, it would be better to respond quickly to ensure safety. In fact, in my simulation results, it was confirmed that the shorter the response time, the greater secrecy capacity. However, this cannot be the correct answer in a real environment. Why? What is the purpose of using secure capacity? Its purpose is to ensure the safety of communication from any eavesdropper.

In order to secure the communication, the encryption method is basically used in vehicle communication. There is a difference between the response time using an encryption scheme and the response time without an encryption scheme. For security, encryption will be used, and if encryption is used, response time will naturally be longer. This long response time will eventually lead to a reduction in secrecy capacity.

To summarize, in order to achieve security based on physical layer security, a short response time is advantageous. On the other hand, to achieve application layer security, response time is inevitably long. The trade-off between physical layer security and application security is inevitable.

In any case, it is important to note that secrecy capacity is closely related to the response time of the system. In addition, it is generally possible to select the relevant mode whether or not to apply application security in the vehicle system. This means that the response time associated with secrecy capacity is not fixed but selectable by the user. The conclusion is that secrecy capacity is controllable according to the selected response time.

2.3 Speed limit

Does the law govern secrecy capacity? That is right. All vehicles that drive on the road must be moving in compliance with speed limit. I have conceptually presented and calculated secrecy capacity of a rolling car at the intersection. The conclusion was not beyond the expected range. It has been confirmed that as speed limit increases, secrecy capacity decreases. If you comply with vehicle regulations, your secrecy capacity will remain constant. In the real world, however, there is no vehicle that keeps the law. Are these only vehicle-related parameters?

3. Antenna-related parameters

Antenna beam radiation technology has been studied variously for physical layer security in wireless communication. Examples include transmission power allocation, artificial noise generation, jamming, and beam direction determination.

3.1 Transmission power

According to my confirmation, secrecy capacity changes according to the transmission power. For example, it has been confirmed that secrecy capacity is increased so that the transmission power can be increased. Then, in order to vary secrecy capacity, it is necessary to check whether the intensity of the transmission power can be controlled in real time. However, there have been many researches on adaptive transmission power control in view of power consumption [13, 14]. The study of the majority of adaptive transmit power control aims at minimizing power consumption and storing energy. It is known that transmission power control is possible although the purpose is different. This means that secrecy capacity can be controlled in real time through transmission power control.

3.2 Beamforming

Beamforming technology is a very old technology of physical layer security. By forming a beam to be transmitted to the target device, the attack opportunity itself is deprived of an attacker having malicious purpose in the other direction. Recently, a technique for maximizing secrecy capacity using beamforming has been disclosed [15]. 3D beamforming technology is introduced beyond 2D to increase secrecy capacity [16, 17].

3.3 Artificial noise/jamming

A method for increasing secrecy capacity using artificial noise or jamming is disclosed [18, 19]. The jamming strategy is to send an artificial noise signal to the eavesdropper to effectively reduce their channel quality from correct reception. If an artificial noise signal is aimed at an eavesdropper, the quality of the eavesdropper's channel may be degraded. According to the eavesdropping channel model, a complete secret can be achieved if the eavesdropper's channel state is worse than a certain level, i.e., the legitimate receiver's channel state. Thus, radio interference can be a practical physical layer-based security measure, especially if the transmitted information needs to be protected from unintentional manual eavesdroppers.

3.4 Antenna gain

The size and shape of the small antenna will change the effective area of the antenna and the power output of the receiver antenna will be changed to affect SNRs. To obtain an antenna gain G, a relatively small antenna is needed at high frequencies [20, 21],

$$G = \frac{4\pi A_e}{\lambda^2} = \frac{4\pi f^2 A_e}{C^2}, \tag{3}$$

where A_e is the effective area, f is the carrier frequency, C is the speed of light, and λ is the carrier wavelength. As can be seen from the equation, if the effective area of the receiving antenna is increased, SNR value can be improved. Increasing the carrier frequency of the transmitter can improve SNR value.

3.5 Array antenna

The array antenna disperses the incident laser beam into the respective antenna elements through a plurality of directional couplers, modulates the phase or frequency of the dispersed laser beam, and adjusts the traveling direction of the output laser beam [22]. In general, the array antenna can vary the elevation angle and the azimuth angle of the beam. What this means is that the host vehicle can concentrate and transmit the beam to a specific target vehicle [23, 24]. Naturally, secrecy capacity can be improved as compared with not. In order to keep secrecy capacity

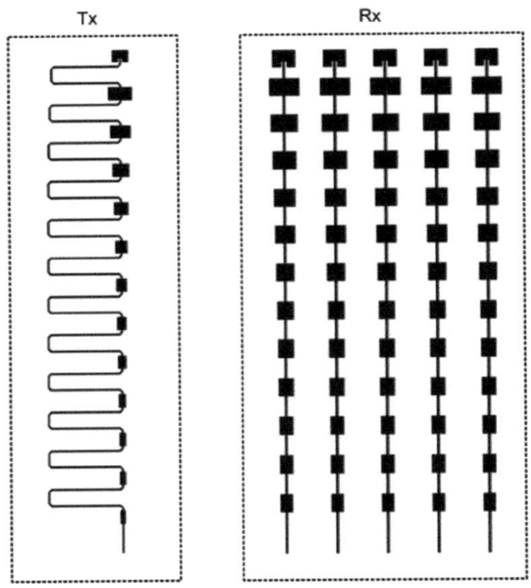

Figure 3.
Array antenna for vehicles.

of the vehicle above a certain value, this characteristic of the array antenna can be used. For example, when secrecy capacity of the vehicle is below a certain value, secrecy capacity can be increased by controlling the phase/frequency of the array antenna.

Referring to **Figure 3**, the array antenna includes a transmitter Tx having a snake feed line structure composed of a plurality of antennas and a receiver Rx having a structure of a plurality of lines composed of a plurality of antennas. Each antenna of Tx and Rx may be implemented in a microstrip structure.

4. Path-related parameters

4.1 Relay communication

As described above, secrecy capacity decreases as the vehicle speed increases. Decreased secrecy capacity as the vehicle speed increases can be compensated by a cooperative relay communication. In V2V communication, secrecy capacity can be improved by adoption at one relay R between the host vehicle A and the target vehicle B. In a general, secrecy capacity of a cooperative relay communication is higher than that of the direct communication without a relay [25–28]. For simplicity of the analysis of secrecy capacity, we assume that the system model comprises one relay R between the host vehicle A and the target vehicle B, as shown in **Figure 4**.

Channel capacity of the legitimated channel is expressed as:

$$C_1(A, B) = W \log_2\left(1 + \left(\frac{P_A h_{AB}}{P_R h_{RB} + \sigma_B^2}\right)\right), \qquad (4)$$

where P_A and P_R are the transmission powers of the host vehicle A and the relay R, respectively; h_{AB} is the channel gain between the host vehicle A and the target vehicle B; h_{RB} is the channel gain between the relay R and the target vehicle B; σ_B^2 is an additive white Gaussian noise at the target vehicle B; and W is a bandwidth.

And channel capacity of the wiretap channel is given by:

$$C_2(A,E) = W\log_2\left(1+\left(\frac{P_A h_{AE}}{P_R h_{RE}+\sigma_E^2}\right)\right), \tag{5}$$

where h_{AE} is the channel gain between vehicle A and eavesdropper E, and σ_E^2 is the additive white Gaussian noise at vehicle B. Then, secrecy capacity with the cooperative relay communication is denoted by

$$C_R = W\left[\log_2\left(1+\left(\frac{P_A h_{AB}}{P_R h_{RB}+\sigma_B^2}\right)\right) - \log_2\left(1+\left(\frac{P_A h_{AE}}{P_R h_{RE}+\sigma_E^2}\right)\right)\right]. \tag{6}$$

Figure 5 shows secrecy capacity with and without the relay R. Referring to **Figure 5**, we can see that the relay R helps to improve secrecy capacity in total. We confirmed that V2V communication using the relay may enhance secrecy capacity. The relationship between the existence of the relay and secrecy capacity can be summarized as shown in **the following table, Table 1**.

As described above, when relay communication is performed, an improvement in secrecy capacity is basically expected as compared with the case where relay communication is not performed. For this reason, relay communication is basically installed in vehicle communication. However, relay communication still has the issue of relay selection and communication rejection of relay object. The important

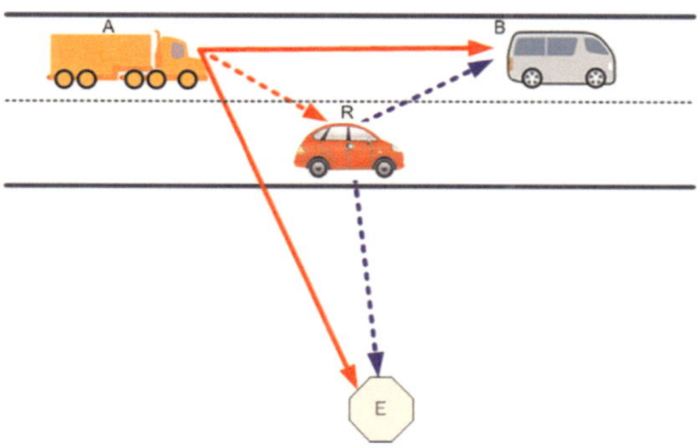

Figure 4.
System model with a relay between vehicle A and vehicle B.

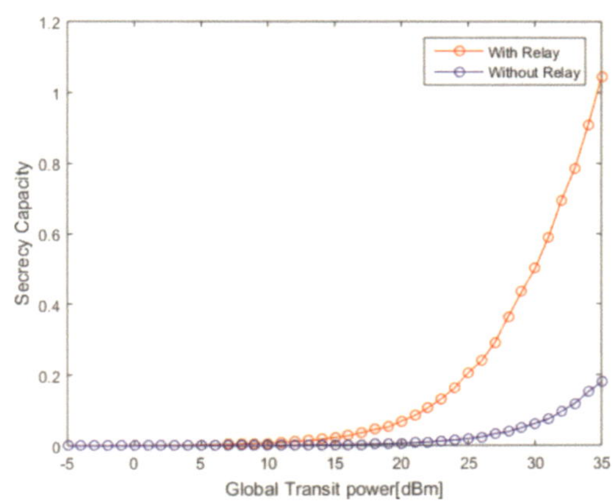

Figure 5.
Secrecy capacity variance according to relay existence [11].

Relay	Off	Down	Secrecy capacity
	On	Up	

Table 1.
Relationship between the relay mode and secrecy capacity [11].

thing is that the choice of relay communication can be a factor enough to be a variable factor of secrecy capacity.

4.2 RSU-assisted communication

When direct communication is not possible, Road Side Unit (RSU) can be used for vehicle communication [29–31]. RSU can be used to deliver a message from one vehicle to another. Here, the other vehicle may be an unspecified number, usually according to the broadcast message. Therefore, since the target vehicle cannot be specified, it is logically difficult to define secrecy capacity. At the same time, however, the presence of eavesdroppers is also unspecific, so secrecy capacity is expected to increase. When direct vehicle communication is not possible, indirect vehicle communication via RSU can be achieved. For example, when serious defects in secrecy capacity are found or expected, indirect vehicle communication via RSU may be required. When secrecy capacity control is not possible, that is, vehicle communication using RSU can be used last.

5. Noise-related parameters

Will noise-related parameters affect secrecy capacity? According to the concept of secrecy capacity, noise is of course a big influence. However, the problem is whether such noise is controllable in communication. Although it is a technology related to noise removal in other fields, wavelet transform is used to reduce power analysis attacks by removing noise [32–34]. Preventing DPA through wavelet transform is basically data modulation to ensure the randomness of transmitted data. However, the concern of noise related to secrecy capacity is not data modulation, but an issue related to the environment of the channel being transmitted. Is there a technique to eliminate or reduce channel noise, even if there is a technique to increase the noise of the channel? Although there is a technique of relatively increasing or decreasing the size of a signal, there is no technique for reducing or eliminating the absolute amount of noise.

However, it cannot be said that at some point the noise of the channel becomes the noise of the channel at another point. This suggests that the channel noise at another point in time may be changed depending on the channel noise at a certain point in time. Recently, research has been introduced to improve the characteristics of a wireless signal by reducing channel noise using adaptive equalization/filter algorithms [35, 36].

5.1 Additive white Gaussian noise channel

The channels used in communication are classified into various channels according to the noise component existing in the medium in which the signal is moving, not the medium itself. There are a lot of kinds of noise in the air, and because these noises occur at all frequencies, this noise is called white noise.
The visible light, which is a type of electromagnetic wave with high frequency, has

a different color depending on the frequency, but it is attached as if noise of various frequencies is gathered like a white color when a plurality of colored light is overlapped.

Additive White Gaussian Noise (AWGN) channel is the most common type of channel, and is a channel that produces even noise across the entire frequency band. AWGN is a random noise without any special peripheral elements. This channel only contains the sum of the white noise. Here white noise follows Gaussian density. The corresponding probability density function p(x) is expressed by the following equation [37, 38]:

$$p(x) = \frac{1}{\sigma\sqrt{2\pi}} \exp\left(\frac{(x-\mu)^2}{2\sigma^2}\right), \tag{7}$$

where μ is the mean, σ is the standard deviation, and σ^2 is the variance.

5.2 Rician fading channel

In a wireless channel environment, when the direct wave (line-of-sight) mainly appears in an environment such as a room dominated by a reflected wave, the probability distribution characteristic of the radio reception signal shows a Rician distribution [39, 40]. Rician probability density function p(x) is expressed by the following equation:

$$p(x) = \frac{x}{\sigma^2} \exp\left(-\frac{(x^2+A)^2}{2\sigma^2}\right) I_0\left(\frac{A_x}{\sigma^2}\right), \quad A \geq 0, \tag{8}$$

where σ^2 is the variance of the received total reflected power, that is, the multipath signal, A is the magnitude of line of sight (LOS) or dominant component, and I_o is the zero-order transformed Bessel function.

5.3 Rayleigh fading channel

Rayleigh distribution is a stochastic model that shows the statistical time-varying characteristics of the envelope of random amplitudes of the received signal by independent quadrature components (such as in-phase and quadrature) or multiple multipath signals when the indirect wave (reflected wave, etc.) predominates over the direct channel in the wireless channel environment [41–43].

Rician factor K is dominant for multipath power (σ^2), and the ratio of the minute power ($A^2/2$) is represented by $K = A^2/2\sigma^2$. When K = 0, the Rician distribution is a Rayleigh distribution. Rayleigh probability density function p(x) is expressed by the following equation:

$$p(x) = \frac{x}{\sigma^2} \exp\left(-\frac{x^2}{2\sigma^2}\right) I_0\left(\frac{A_x}{\sigma^2}\right), A \geq 0. \tag{9}$$

5.4 Nakagami fading channel

The multipath fading phenomenon of signals in wireless mobile communication is a very important problem. In modeling such a fading channel, a Nakagami fading model is known to be fit and theoretically suitable. The Nakagami fading model has the following probability density function for the envelope of the signal [44–47]:

$$p(x) = \frac{2m^m x^{2m-1}}{\Gamma(m)\Omega^m} \exp\left(-\frac{mx^2}{\Omega}\right), x \geq 0, \tag{10}$$

where x is the size of the Nakagami fading signal, Ω is the mean square value of x, Γ(m) is the gamma function, and m is the rate of the modality, which determines the shape of the distribution and is a fading index that indicates the degree of fading. When m = 1, Rayleigh distribution is obtained. When m = 0.5, Gaussian distribution is gained.

Generally, selective diversity is used to reduce the influence of fading. The selection diversity is a method of selecting the largest signal among the signals received through the multipath. If the signal received at the k-th receiving end is the largest, the received signal can be represented by a sinusoidal wave:

$$y_k(t) = u_k \exp(j2\pi f_0 t), \tag{11}$$

where u_k is the magnitude of the k-th received signal envelope and f_0 is the carrier frequency. Therefore, the average power of the received signal is $u_k^2/2$. Assuming that the average noise power is N, the SNR value of the signal received at the k-th receiving end is as follows.

$$\gamma_k = \frac{u_k^2}{2N}, \tag{12}$$

Therefore, the probability density function for SNR of the k-th receiver experiencing Nakagami fading is expressed by the following equation.

$$p_\gamma(\gamma) = \frac{m_k^{m_k} \gamma_k^{m_k - 1}}{\Gamma(m_k)\gamma_0^{m_k}} \exp\left(-\frac{m_k \gamma_k}{\gamma_0}\right), \ \gamma \geq 0, \tag{13}$$

where γ_k is the SNR value of the k-th receiver, γ_0 is the average SNR value in the presence of fading, and m_k is the SNR value of the k-th receiver. The probability distribution function is the integral of the probability density function:

$$p_\gamma(\gamma) = \int_0^\gamma \left(\frac{m_k^{m_k} \gamma_k^{m_k - 1}}{\Gamma(m_k)\gamma_0^{m_k}} \exp\left(-\frac{m_k \gamma_k}{\gamma_0}\right)\right) dx, \ \gamma \geq 0, \tag{14}$$

Assuming that all receivers are independent of each other, the probability that SNR value of the signal received after the selective combining is less than x is expressed by the following equation:

$$P_r(\gamma_{sc} \leq x) = \prod_{k=1}^{D} P_r(\gamma_k \leq x), \tag{15}$$

where γ_{sc} is the SNR value after selective combining and D is the number of receivers. Assuming that the average SNR value and the fading parameter m_k are the same at each receiving end, the SNR value of the received signal has the same probability distribution. Therefore, the SNR value of the signal received is expressed by:

$$P_{\gamma,sc}(x) = [P_r(\gamma_k \leq x)]^D = [P_r(x)]^D, \tag{16}$$

As a result, the probability density function for SNR of the received signal after selective combining is expressed by the following equation as a differential value of Eq. (16):

$$P_{\gamma,sc}(\gamma) = D\left[P_{\gamma}(\gamma)\right]^{D-1}P_{\gamma}(\gamma). \tag{17}$$

If the number of antennas of the receiver is larger than that of the transmitter, the signal with the highest SNR among the signals of the multipath can be used or the signal can be combined. If the number of antennas of the transmitter is larger than that of the receiver, beamforming can increase the SNR value obtained by the receiver. The number of antennas D of the receiving end may affect the SNR value.

6. Conclusions

We studied vehicle-related parameters that affect secrecy capacity of vehicle communication, such as vehicle speed, response time, and speed limit. When considering the Doppler effect according to the speed of the vehicle, secrecy capacity is studied to be proportional to the speed, but when considering the safety distance of autonomous driving, secrecy capacity is inversely proportional to the speed of the vehicle. In general, the relationship between secrecy capacity according to the antenna-related parameters, beamforming, jamming, the size of the antenna, and the number of antennas was also discussed. We also looked at increasing secrecy capacity in indirect communication through relays rather than direct communication, and improving security through assistive devices. Finally, we examined how the noise-related parameters according to the fading model influence secrecy capacity. In conclusion, it was confirmed that various vehicle secrecy parameters exist in vehicle communication, and these parameters can be changed by a user. This makes it possible for us to be able to communicate with the vehicle later, while maintaining a certain level of security. It is expected that the road to secure the security of the vehicle radio channel used for these vehicle secrecy parameters will begin not long.

Acknowledgements

We sincerely appreciate Professor S.J. Oh for teaching physical layer security.

Author details

Na-Young Ahn and Dong Hoon Lee*
The Graduate School of Information Security at Korea University, Seoul, Korea

*Address all correspondence to: donghlee@korea.ac.kr

References

[1] Sun L, Du Q. Secure data dissemination for intelligent transportation systems. In: Secure and Trustworthy Transportation Cyber-Physical Systems, Springer Briefs in Computer Science. 2017. pp. 99-140

[2] Camacho F, Cárdenas C, Muñoz D. Emerging technologies and research challenges for intelligent transportation systems: 5G, HetNets, and SDN. International Journal on Interactive Design and Manufacturing. 2018;**12**(1): 327-335

[3] Whyte W, Weimerskirch A, Kumar V, Hehn T. A security credential management system for V2V communications. In: 2013 IEEE Vehicular Networking Conference (VNC). 2013

[4] Lei A, Cruickshank H, Cao Y, Asuquo P, Ogah CPA, Sun Z. Blockchain-based dynamic key management for heterogeneous intelligent transportation systems. IEEE Internet of Things Journal. 2017;**4**(6):1832-1843

[5] Liang L, Peng H, Li GY, Shen XS. Vehicular communications: A physical layer perspective. IEEE Transactions on Vehicular Technology. 2017;**66**(12): 10647-10659

[6] Han D, Bai B, Chen W. Secure V2V communications via relays: Resource allocation and performance analysis. IEEE Wireless Communications Letters. 2017;**6**(3):342-345

[7] Eltayeb ME, Choi J, Al-Naffouri TY, Heath RW Jr. Enhancing secrecy with multi-antenna transmission in millimeter wave vehicular communication systems. IEEE Transactions on Vehicular Technology. 2017;**66**(9):8139-8151

[8] Available from: https://en.wikipedia.org/wiki/Information_theory

[9] Bloch M, Barros J, Rodrigues MRD, McLaughlin SW. Wireless information-theoretic security. IEEE Transactions on Information Theory. 2008;**54**(6): 2515-2534

[10] Zou Y, Zhu J, Wang X, Leung VCM. Improving physical-layer security in wireless communications using diversity techniques. IEEE Network. 2015;**29**(1):42-48

[11] Ahn NY, Lee DH, Oh S-J. Vehicle communication using secrecy capacity. In: Kapoor S, Arai K, Bhatia R, editors. Proceedings of the Future Technologies Conference (FTC) 2018 - Volume 2, Advances in Intelligent Systems and Computing. Vol. 881. Switzerland: Springer Verlag. 2019. pp. 158-172. Available at: https://doi.org/10.1007/978-3-030-02683-7_13

[12] Chopra G, Jha RK, Jain S. Novel beamforming approach for secure communication in UDN to maximize secrecy rate and fairness security assessment. IEEE Internet of Things Journal. 2019;**6**(4):5935-5947

[13] Lin S, Zhang J, Zhou G, Gu L, Stankovic JA, He T. Adaptive transmission power control for wireless sensor networks. In: Proc. 4th Int. Conf. Embedded Netw. Sensor Syst. 2006. pp. 223-236

[14] Zareei M, Vargas-Rosales C, Villalpando-Hernandez R, Azpilicueta L, Anisi MH, Rehmani MH. The effects of an adaptive and distributed transmission power control on the performance of energy harvesting sensor networks. Computer Networks. 2018;**137**:69-82

[15] Nandan N, Majhi S, Wu H. Maximizing secrecy capacity of underlay MIMO-CRN through bi-directional zero-forcing beamforming.

IEEE Transactions on Wireless Communications. 2018;**17**(8):5327-5337

[16] Yaacoub E, Al-Husseini M, Chehab A, Abualsaud K, Khattab T, Guizani M. 3D beamforming with massive cylindrical arrays for physical layer secure data transmission. IEEE Communications Letters. 2019;**23**(5): 830-833

[17] Cho S, Chen G, Coon JP. Enhancement of physical layer security with simultaneous beamforming and jamming for visible light communication systems. IEEE Transactions on Information Forensics and Security. 2019;**14**(10):2633-2648

[18] Wang S, Jiang X-Q, Wang P, Zhu Y. An artificial noise assisted secrecy-enhancing scheme for space–time shift keying systems. Physical Communication. 2019;**35**:100693

[19] Huo Y, Tian Y, Ma L, Cheng X, Jing T. Jamming strategies for physical layer security. IEEE Wireless Communications. 2018;**25**(1):148-153

[20] Compston AJ, Fluhler JD, Schantz HG. A fundamental limit on antenna gain for electrically small antennas. In: 2008 IEEE Sarnoff Symposium; Princeton, NJ. 2008. pp. 1-5

[21] Hong T, Liu C, Kadoch M. Machine learning based antenna design for physical layer security in ambient backscatter communications. Wireless Communications and Mobile Computing. 2019:4870656

[22] Trichili A, Park K, Zghal M, Ooi BS, Alouini M. Communicating using spatial mode multiplexing: Potentials, challenges and perspectives. IEEE Communication Surveys and Tutorials; 2019:1

[23] Inomata M, Imai T, Kitao K, Okumura Y, Motoharu S, Takatori Y. Radio propagation prediction for high frequency bands using hybrid method of ray-tracing and ER model with point cloud of urban environments. In: IET Conference Proceedings. 2018

[24] Guntupalli AB, Wu K. 60 GHz circularly-polarized smart antenna system for high throughput two-dimensional scan cognitive radio. In: 2013 IEEE MTT-S International Microwave Symposium Digest (MTT); Seattle, WA. 2013. pp. 1-3

[25] Sun L, Ren P, Du Q. Distributed source-relay selection scheme for vehicular relaying networks under eavesdropping attacks. EURASIP Journal on Wireless Communications and Networking. 2014;**1**:1-11

[26] Zheng T, Wang H, Huang R, Mu P. Adaptive DF relaying transmission for security. In: 2015 IEEE Globecom Workshops (GC Wkshps); San Diego, CA. 2015. pp. 1-6

[27] Han D, Bai B, Chen W. Secure V2V communications via relays: Resource allocation and performance analysis. IEEE Wireless Communications Letters. 2017;**6**(3):342-345

[28] Chen JS, Yang CY, Hwang MS. The capacity analysis in the secure cooperative communication system. International Journal of Network Security. 2017;**19**(6):863-869

[29] Wu Y et al. Secrecy-driven resource management for vehicular computation offloading networks. IEEE Network. 2018;**32**(3):84-91

[30] Wang J, Liu J, Kato N. Networking and communications in autonomous driving: A survey. IEEE Communication Surveys and Tutorials. 2019;**21**(2): 1243-1274

[31] Wang L, Liu X. NOTSA: Novel OBU with three-level security architecture

for internet of vehicles. IEEE Internet of Things Journal. 2018;**5**(5):3548-3558

[32] Pelletier H, Charvet X. Improving the DPA Attack Using Wavelet Transform; 26-29 September 2005. Honolulu, Hawaii, USA: NIST's Physical Security Testing Workshop; 2005

[33] Ai J, Wang Z, Zhou X, Ou C. Improved wavelet transform for noise reduction in power analysis attacks. In: 2016 IEEE International Conference on Signal and Image Processing (ICSIP); Beijing. 2016. pp. 602-606

[34] Dong X et al. A wavelet-based power analysis attack against random delay countermeasure. In: 2018 Asian Hardware Oriented Security and Trust Symposium (AsianHOST); Hong Kong. 2018. pp. 19-24

[35] Martinek R, Vanus J, Bilik P, Al-Wohaishi M, Zidek J, Wen H. The implementation of equalization algorithms for real transmission channels. In: 2016 IEEE International Instrumentation and Measurement Technology Conference Proceedings; Taipei. 2016. pp. 1-6

[36] Martinek R et al. Modelling of wireless fading channels with RF impairments using virtual instruments. In: 2016 IEEE 17th Annual Wireless and Microwave Technology Conference (WAMICON); Clearwater, FL. 2016. pp. 1-6

[37] Shu Z, Yang Y, Qian Y, Hu RQ. Impact of interference on secrecy capacity in a cognitive radio network. In: 2011 IEEE Global Telecommunications Conference—GLOBECOM 2011; Kathmandu. 2011. pp. 1-6

[38] Yacoub MD. The α-μ distribution: A physical fading model for the Stacy distribution. IEEE Transactions on Vehicular Technology. 2007;**56**(1):27-34

[39] Qu S, Fleisher SM. Double differential MPSK on the fast Rician fading channel. IEEE Transactions on Vehicular Technology. 1992;**41**(3): 278-295

[40] Hua Y, Wang Y. On the saturate throughput of IEEE 802.11 DCF with capture effect in Rician fading channel. In: 2008 4th International Conference on Wireless Communications, Networking and Mobile Computing; Dalian. 2008. pp. 1-4

[41] Donald AMM, Olivier JC. A comparative study of deterministic and stochastic sum-of-sinusoids models of Rayleigh-fading wireless channels. In: 2007 IEEE Wireless Communications and Networking Conference, Kowloon. 2007. pp. 2027-2031

[42] Afzal A, Hassan SA. A stochastic geometry approach for outage analysis of ad hoc SISO networks in Rayleigh fading. In: 2013 IEEE Global Communications Conference (GLOBECOM); Atlanta, GA. 2013. pp. 336-341

[43] Feng Q, Li W, Huang L. Analysis of spontaneous Raman and Rayleigh scatterings in distributed Fiber Raman amplification systems based on a random distribution model. IEEE Photonics Journal. 2017;**9**(6):1-8. Art No. 7205008

[44] Abbas SA, Sheikh AU. A geometric theory of Nakagami fading multipath mobile radio channel with physical interpretations. In: Proceedings of Vehicular Technology Conference. Vol. 2. Atlanta, GA, USA: VTC; 1996. pp. 637-641

[45] Zhang QT. A generic correlated Nakagami fading model for wireless communications. IEEE Transactions on Communications. 2003;**51**(11): 1745-1748

[46] Lu J, Han Y. Application of multipath shape factors in Nakagami-m fading channel. In: 2009 International Conference on Wireless Communications & Signal Processing; Nanjing. 2009. pp. 1-4

[47] Nguyen T, Tran X. Performance of cooperative NOMA system with a full-duplex relay over Nakagami-m fading channels. In: 2019 3rd International Conference on Recent Advances in Signal Processing, Telecommunications & Computing (SigTelCom); Hanoi, Vietnam. 2019. pp. 130-134

Chapter 6

The Novel Applications of Deep Reservoir Computing in Cyber-Security and Wireless Communication

Kian Hamedani, Zhou Zhou, Kangjun Bai and Lingjia Liu

Abstract

This chapter introduces the novel applications of deep reservoir computing (RC) systems in cyber-security and wireless communication. The RC systems are a new class of recurrent neural networks (RNNs). Traditional RNNs are very challenging to train due to vanishing/exploding gradients. However, the RC systems are easier to train and have shown similar or even better performances compared with traditional RNNs. It is very essential to study the spatio-temporal correlations in cyber-security and wireless communication domains. Therefore, RC models are good choices to explore the spatio-temporal correlations. In this chapter, we explore the applications and performance of delayed feedback reservoirs (DFRs), and echo state networks (ESNs) in the cyber-security of smart grids and symbol detection in MIMO-OFDM systems, respectively. DFRs and ESNs are two different types of RC models. We also introduce the spiking structure of DFRs as spiking artificial neural networks are more energy efficient and biologically plausible as well.

Keywords: recurrent neural networks, reservoir computing, delayed feedback reservoir, echo state networks, cyber-security, smart grids, MIMO-OFDM

1. Introduction

Smart grids are a new generation of power grids, which provide more intelligent and efficient power transmission and distribution. However, the smart grids are vulnerable to security challenges unless properly protected. False data injection (FDI) attacks are the first and most common type of attacks in smart grids. Two major types of FDI attacks are known in smart grids. These two major types are single-period or opportunistic and multi-period or dynamic attack, respectively. In single-period attack, the adversary waits until it finds the opportunity to launch the attack instantaneously. On the other hand, in dynamic attacks, the adversary launches the attack gradually and through time toward its desired state. The single-period attacks are widely studied in the literature and they are more easily detected by the supervisory control and data acquisition (SCADA). In this chapter, we focus to study the multi-period or dynamic attacks [1–5].

State vector estimation (SVE) is the first technique to tackle the FDI detection in smart grids. However, SVE fails to detect stealth FDI attacks with low magnitudes.

In recent years, both supervised and unsupervised machine learning (ML) approaches have been proposed to study FDI detection in smart grids. Generally, ML-based techniques have shown better performances than SVE. However, the ML techniques that have been proposed so far are not capable to capture the rich spatio-temporal correlations that exist between different components of smart grids. Therefore, in this chapter, we introduce spiking delayed feedback reservoirs (DFRs) to tackle the FDI detection problem in smart grids as they are very energy efficient and also can capture the spatio-temporal correlations between different components of smart grids. DFRs are an energy efficient class of reservoir computing systems [6–8].

Figure 1 demonstrates the structure of a reservoir computing (RC) system. As it can be seen, there are three layers in RC systems. They are the input, reservoir, and output layer, respectively. The architecture of RC systems is based on recurrent neural networks (RNNs). However, unlike the RNNs, the weights of the hidden (reservoir) layer are fixed and do not go through a training. The reservoir weights have to be initialized such that the echo state property is satisfied. Echo state property implies that in order to form a memory, the largest eigenvalue of the reservoir weights has to be less than 1. The largest eigenvalue of the reservoir layer's weights is a design parameter and plays an important role in the performance of the RC systems. DFRs, echo state networks (ESNs), and liquid state machines (LSMs) are three different categories of RC systems. The strength of RNNs is employed as the reservoir or liquid states. In the reservoirs or liquid states, the weights of synaptic connections are fixed and do not require any training. The output weights are the only sets of weights that require training in RC models. This results in reducing the computational complexity of RC models compared to traditional RNNs [9–12].

Equation (1) expresses the states of reservoir nodes,

$$s(t) = f\left[W_{res}^{res}.s(t-1) + W_{in}^{res}.x(t-1)\right], \tag{1}$$

where $s(t)$ is the state of reservoir node at time t; $x(t-1)$ corresponds to the input signal at time $t-1$; W_{res}^{res} and W_{in}^{res} correspond to the weights of randomly generated reservoir and input connection, respectively; and $\hat{y}$ represents the estimated output that can be expressed in terms of input and weight connections,

$$\hat{y} = W_{res}^{out}s(t) + W_{in}^{out}.x(t-1) + W_{bias}^{out}, \tag{2}$$

where W_{res}^{out} are the output weights of the neurons that form the reservoir layer; W_{in}^{out} correspond to the feedback weights from output layer to reservoir layer; and

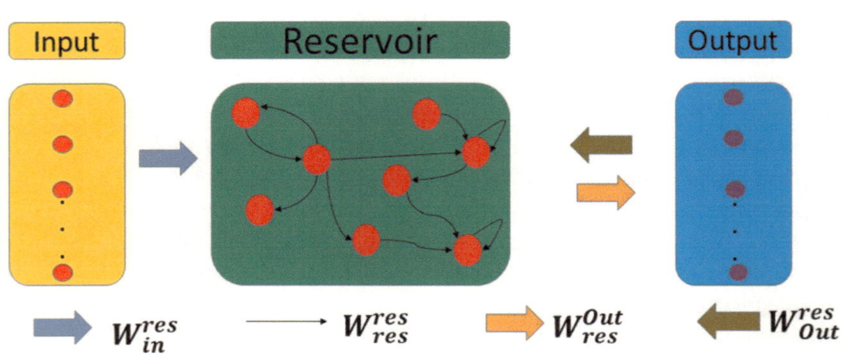

Figure 1.
Structure of reservoir computing.

W_{bias}^{out} is the set of weights for bias values training. The process of nonlinear mapping is accomplished by the neurons in the reservoir layer. The neurons in the reservoir layer own two major properties: (1) high dimensionality and (2) forming a short term memory that spatio-temporal patterns can be memorized. Several studies have shown that these two properties are satisfied only if the neurons at the reservoir layer operate at the edge of chaos. Satisfying the echo states property, is the key to make the reservoir neurons work at the edge of chaos. The lower computational complexity and the flexible reservoir implementation of RC models make them very suitable for unconventional computing paradigms applications.

The DFR is a ring topology of RC systems, where a single artificial neuron and a delay loop together form the reservoir layer. There are multiple choices available for the single artificial neuron of the DFR. In this chapter, we introduce spiking neurons as the nonlinear single neuron of the DFR. Spiking neurons are one of the several mathematical models that are introduced to model the biological neurons. Spikes are the main signals that the neurons of the brain use for communication. Hence, the mathematical representation of the biological neurons as spikes tends to be more biological plausible. Energy efficiency is another motivation to use the spiking neurons. TrueNorth chip consumes only 70 milliWatts (mW) to run 1 million spiking neurons with 256 million synapses [13–15]. The energy efficiency of spiking neural networks (SNNs) makes them a suitable choice for hardware implementations of artificial neurons as well [16, 17].

So far, several models for spiking neurons including leaky-integrate-and-fire (LIF) and the Hodgkin-Huxley have been proposed to mimic the behavior of our brains' neurons [18]. The LIF models of spiking neurons have been used more commonly than other spiking artificial models of neurons due to their simplicity and ease of hardware implementation [19, 20]. The spiking neurons fire a spike as soon as a stimulating current is applied on their membrane, which makes the voltage of the membrane exceeds a certain threshold value. The relationship between the stimulating current and the voltage of membrane is expressed as follows:

$$\tau_m \frac{dv_m}{dt} = -(V_m - E) + (I_{noise} + I_s)R_m, \tag{3}$$

where V_m is the membrane voltage; $\tau_m = R_m C_m$ corresponds to the neuron's time constant; C_m and R_m are the capacitance and the resistance of the membrane, respectively; E represents the resting voltage; I_{noise} is noise current; and I_s is stimulus current [21]. We set R_m to 1 mega *ohms* and C_m = 10 nano Farads (nF).

In **Figure 2**, the topology of our proposed spiking DFR is demonstrated. There are multiple blocks in this structure. The input block is where the smart grids' measurements are received. These measurements have to be first encoded before getting processed by DFR. There are two major types of encoding schemes for spiking neurons, namely rate encoding and temporal encoding [22]. Rate encoding has been vastly studied in the literature. However, recent studies have shown that temporal encoding schemes are more efficient and are superior to rate encoding schemes. The exact time that spike fires is used for temporal encoding of spikes. However, in rate encoding schemes, the number of the spikes that are fired by the neuron is used to encode the stimulus.

It has been shown in several experiments that temporal encoding is more likely to be the encoding scheme, which is leveraged by biological neurons. The neurons in the lateral geniculate nucleus, retina, and the visual cortex respond to the stimuli with milliseconds (ms) precision. The computational complexity of temporal

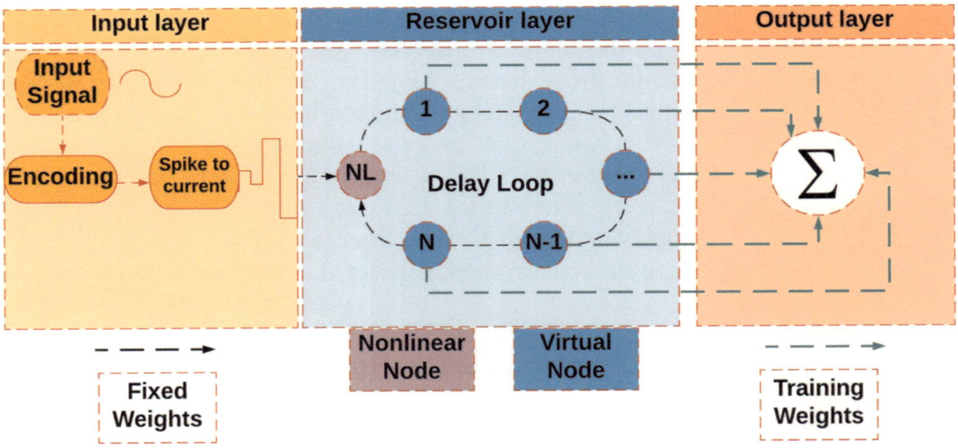

Figure 2.
Spiking delayed feedback reservoir computing.

encoding schemes has also made them superior to rate encoding approaches [23]. Therefore, in this chapter, we focus on temporal encoding schemes.

After the smart grids' measurements are encoded, the encoded data is then converted to the analog current. This current is next fed in to the nonlinear node, which in our case, is a LIF neuron. For each current signal, its corresponding spike train is generated by the LIF neuron, and this spike train goes through a delay loop. The delay loop along with the LIF neuron forms the reservoir layer of DFR. We repeat this process as long as the corresponding reservoir states of each smart grid's measurements are generated. The interspike intervals (ISI) of each spike trains are used as the training feature of the readout layer [24]. In this chapter, a multi-layer perceptron (MLP) is used as the readout layer. The features extracted in the reservoir layer are used for training the MLP layer. For each class of data, i.e., compromised and uncompromised, a proper label is assigned. We consider 1 as the label of compromised samples, and 0 for uncompromised samples.

Equation (4) expresses the governing equation for DFR,

$$\dot{x} = -x(t) + F(x(t-\tau), I(t), \theta), \tag{4}$$

where F is a differentiable nonlinear function; τ is the delay loop, which is a hyperparameter that requires tuning; $x(t)$ corresponds to the reservoirs states of DFR; and $I(t)$ is the input stimulus current signal along with a masking scheme. The total delay time, τ, is divided into N equidistant delay units within the delay loop. Dividing the total delay into N equidistant delay units is expressed as follows:

$$\tau = N\theta, \tag{5}$$

where θ represents the time interval between reservoir virtual nodes. Unlike the conventional RC model, the number of nonlinear nodes of DFR is drastically reduced, due to the ring topology of DFR. The weights of the output MLP layer are the only weights that undergo the training process [16].

DFRs have drawn a lot of attentions due to their capability to map the data from low dimensional space to high dimensional space. As it can be seen in **Figure 3**, by mapping the data from low dimensional space to high dimensional space, the non-linearly separable data becomes linearly separable. The chaos theory through Lyapunov analysis has shown that delay systems can show high dimensional behavior if the delay value is tuned properly such that the delay system operates at the edge of chaos. The Lyapunov dimension of a delay chaos system directly is

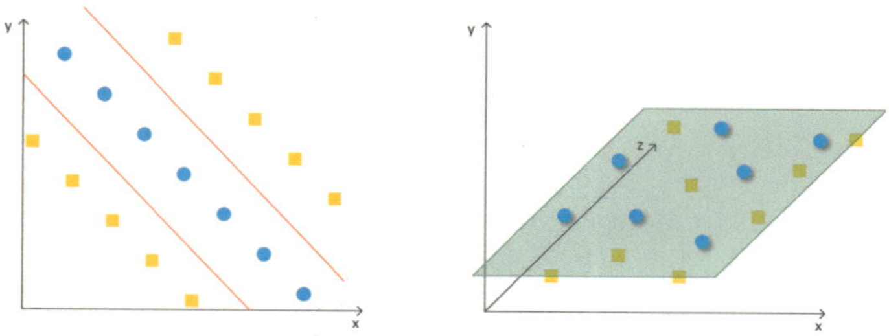

Figure 3.
High dimensional mapping of data using DFR.

determined by to the delay value [25]. In this chapter, we will examine the effect of delay value on the performance of DFR while detecting the dynamic hidden attacks in smart grids.

In this chapter, we will also look at symbol detection in multiple-input multiple-output orthogonal frequency division multiplexing (MIMO-OFDM) systems. In wireless communication systems, multicarrier access techniques are realized through OFDM. In fact, frequency-selective fading channels are converted to multiple flat-fading subchannels [26–28]. Spectral efficiency, transceiver structure, channel capacity, and robustness against interference are all improved as a result of applying OFDM in wireless communication systems [29–33]. MIMO systems are also extensively leveraged in different wireless communication systems including HSPA+(3G), WiMAx(4G), and long term evolution (4G LTE). By using MIMO systems, the capacity of wireless link is improved through the transmission of symbols on multiple paths. The system which is realized through the combination of MIMO and OFDM systems is called a MIMO-OFDM system [34–38]. A MIMO-OFDM system has shown to be very effective in utilizing the benefits of both MIMO and OFDM systems.

In order to detect the transmitted symbols accurately at the receiver (Rx), it is very essential to estimate the wireless channel state information (CSI) precisely [39–41]. CSI estimation is one of the major challenges of MIMO-OFDM systems. There are generally two major approaches for CSI estimation. The first approach leverages blind channel estimation to obtain the statistical properties of the channel [42]. The second category of CSI estimation techniques is based on training the symbols sent by transmitter (Tx) and received by (Rx) [29, 43, 44]. Training-based CSI estimation techniques have been adopted in many advanced communication systems including 3GPP LTE/LTE-Advanced. In the former category of CSI estimation techniques, no computational overhead is inferred, but they are good only for the channels that are varying very slowly with respect to time [45]. The latter category, i.e., training-based category can be applied for any channel regardless of their statistical properties. Therefore, the learning-based techniques including artificial neural networks have been vastly studied in literature [46–48] as the wireless channel estimation mechanism. RNNs have also been studied in [49–52] for CSI estimation and symbol detection. Due to the difficulties of training, the conventional RNNs, we introduce echo state networks (ESN) for symbol detection and CSI estimation in MIMO-OFDM wireless communication systems.

2. Problem formulation of smart grids attack detection

The state and topology of smart grids are the two major targets that are manipulated by the adversaries [53]. The state of the smart grids is the key factor in determining the measurements values. A linear function H and the environment noise are the other two factors that determine the measurements values.

$$z = Hx + n, \tag{6}$$

where z is the measurement vector that represents the real parts of the line flows and bus injections; H is a linear function; x is the state vector, and n is the environment noise [53]. Equation (6) can be written as follows in case the meters are compromised by an adversary,

$$\begin{aligned} \check{z} &= z + a, \\ \check{z} &= Hx + n + a, \end{aligned} \tag{7}$$

where a is the attack vector. The attack represented in Eq. (7) is an observable attack. The attack can also be hidden by the attacker. In this chapter, we consider the attacks as hidden dynamic attacks. The hidden attack is defined as $a = Hc$, and Eq. (6) is reformulated as follows,

$$\begin{aligned} \check{z} &= Hx + n + Hc \\ \check{z} &= H(x + c) + n, \end{aligned} \tag{8}$$

where c is the desired state of the adversary, where the attacker wants to drift the normal state of the smart grid toward its desired state by hiding it in the H matrix. Hidden attacks are more challenging to be detected. The adversaries launch dynamic attacks such that the state of the smart grid system is drifted toward their desired state gradually. Dynamic attacks are defined as a function of time as the adversary achieves its desired state gradually and through time. In single-period attacks, the variations of the attacks magnitude are sudden and abrupt, and are more easily detected. The formulation of dynamic attack used in this chapter is as follows:

$$\check{z}(t) = Hx(t) + n + a(t). \tag{9}$$

The dynamic attack $a(t)$ is time dependent, and we also assume that the adversary has access to H matrix. Thus, the attack can be performed as hidden or unobservable. In hidden attacks, the attack $a(t)$ can be expressed as $a(t) = Hc(t)$, and $c(t)$ is defined as follows:

$$c(t) = Acos\left(2\pi f_c t\right) \times N(0,1), \tag{10}$$

where A is the magnitude of attack; cos is cosine function; f_c corresponds to the frequency of attack and we set that equal to 1 in this chapter, and $N(0,1)$ is a normally distributed vector in which its mean is zero and its variance is 1.

$$\check{z}(t) = H\left(x + Acos\left(2\pi f_c t\right) \times N(0,1)\right) + n. \tag{11}$$

MATPOWER is a publicly available toolbox [54] that can be used to simulate the smart grids. In this chapter, we use MATPOWER to simulate the meters of a smart grid with 14 buses. There are totally 34 different meters in an IEEE-14 bus smart grid. We assume that the level of the access that the adversary can have to the meters of the system can range from 0 to 34. The level of access is defined as the number of meters that can be compromised by the attacker. In this chapter, the dataset that we use for train, test, and validation is assumed to be unbalanced.

A dataset is called unbalanced when the ratio of compromised and uncompromised samples is not equal. In this chapter, it is assumed that 80% of the samples are uncompromised and 20% are compromised. Totally, 10,000 samples for training and 10,000 samples for test and validation are generated using MATPOWER.

3. Attack detection performance of DFR

The performance metrics for evaluation are ***accuracy*** and ***F1***. ***Accuracy*** and ***F1*** are defined as:

$$\text{Accuracy} = (\text{TP} + \text{TN})/(\text{TP} + \text{TN} + \text{FP} + \text{FN}), \tag{12}$$

$$\text{F1} = 2 \times \frac{Precision \times Recall}{Precision + Recall}, \tag{13}$$

where $Precision = \frac{TP}{TP+FP}$; and $Recall = \frac{TP}{TP+FN}$ and TP, TN, FP, and FN correspond to the number of true positive, true negative, false positive, and false negative samples, respectively.

Accuracy of attack detection for three different methods and magnitude of attacks, A = 0.1, 1, and 10.

In order to evaluate the performance of our proposed spiking DFR model, we compare our results with a MLP and a SNN. The MLP is trained using backpropagation algorithm and SNN is trained using precise spike driven (PSD) algorithm. In PSD, temporal encoding is leveraged as the encoding scheme. PSD is used to learn the hetero-associations that exist in spatio-temporal spike patterns and is introduced in [21]. As it can be seen in **Figures 4** and **5**, spiking DFR + MLP outperforms both MLP and SNN in terms of ***accuracy*** and ***F1***. That is due to the fact that the spiking DFR + MLP is capable to map the data from low dimensional space to high dimensional space, and also captures the spatio-temporal correlation that exists between different components of smart grids. Based on our simulation results, the average **accuracy** of attack detection is increased up to **94.6%** when the combination of spiking neurons, DFR, and MLP is realized in a single platform. This improvement is observed for all different magnitude of attacks and number of compromised measurements. In our baseline model where only SNNs are used, the average **accuracy** is **77.92%**. This improvement implies that the average **accuracy** is improved about **17%** through our introduced hybrid spiking DFR and MLP model. ***F1*** measure shows even more significant improvement brought about. ***F1*** that is achieved through combination of spiking neurons, DFR, and MLP is **78%**. However, the ***F1*** which is achieved by SNN and PSD algorithm for dynamic attack detection is about **25%**, which means that our introduced model increases the ***F1*** for **53%**.

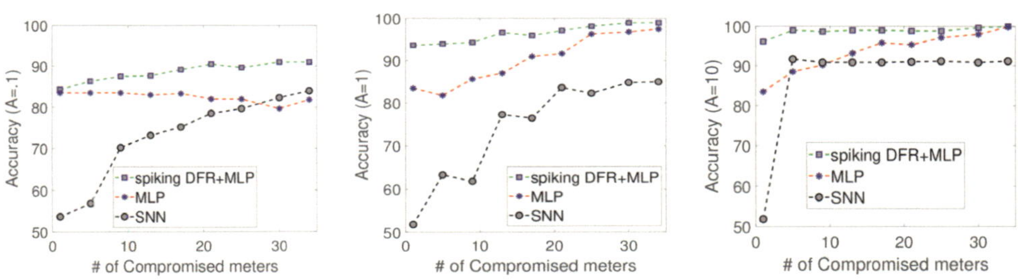

Figure 4.
Accuracy of attack detection for three different methods and magnitude of attacks, A = 0.1, 1, and 10.

3.1 Delay effect on the performance of DFR

As it was mentioned in Section 1, the DFRs cannot show high dimensional behavior unless the delay value is tuned properly that the DFR operates at the edge of chaos. At this part, we show that delay value can significantly affect the performance of DFR for hidden dynamic attack detection on smart grids. **Figure 6** demonstrates the performance of DFR for different values of delay. As it can be seen in **Figure 6**, for delay equal to 40 milliseconds (ms), the performance of spiking DFR + MLP achieves the highest value in terms of ***F1*** and ***accuracy***. However, for delay value equal to 10 ms, the lowest performances are obtained. This observation implies that only for a proper delay value, the spiking DFR + MLP can operate at the edge of chaos and show high dimensional behavior. The phase portrait behavior of DFR with respect to varying the delay time is shown in **Figure 7**. The dynamic behavior of the delay systems can be tracked through phase portraits and chaotic or periodic behavior of the system can be demonstrated. It is suggested in [25] that if the delay of dynamic system is tuned properly, it can show high dimensional behavior. We also investigate the solution of the delay differential equation (DDE) to further explore the dynamic behaviors of our introduced model. As demonstrated in **Figure 7**, DDE is leveraged to model the dynamic behavior of nonlinear function while the delay is varying.

Figure 7 shows that varying the delay value can shift the behavior of delay system from periodic to edge of chaos region and completely chaotic.

3.2 Complexity analysis

In this section, the complexity of our approach in terms of training time is analyzed. The computational complexity of the introduced spiking DFR + MLP

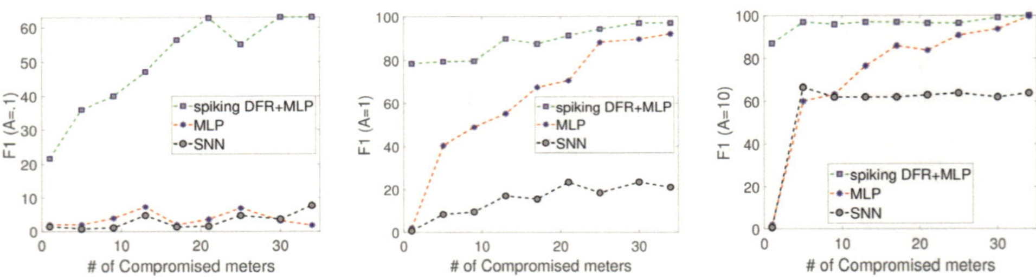

Figure 5.
F1 of attack detection for three different methods and magnitude of attacks, A = 0.1, 1, and 10.

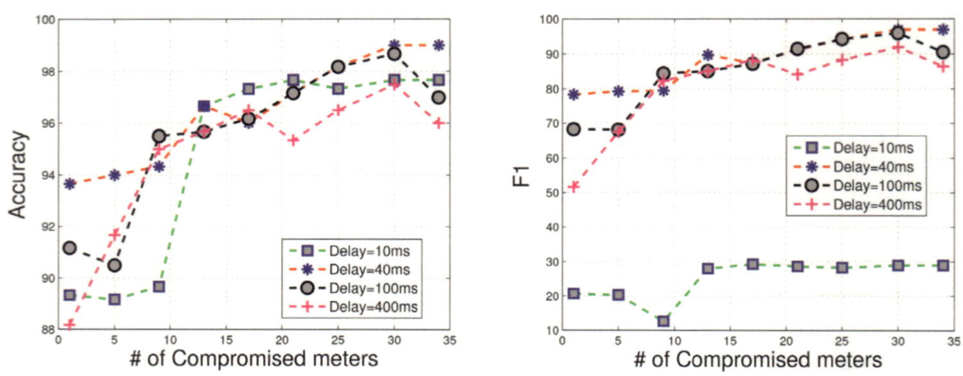

Figure 6.
Effects of different values of the delay on the performance when the A = 1.

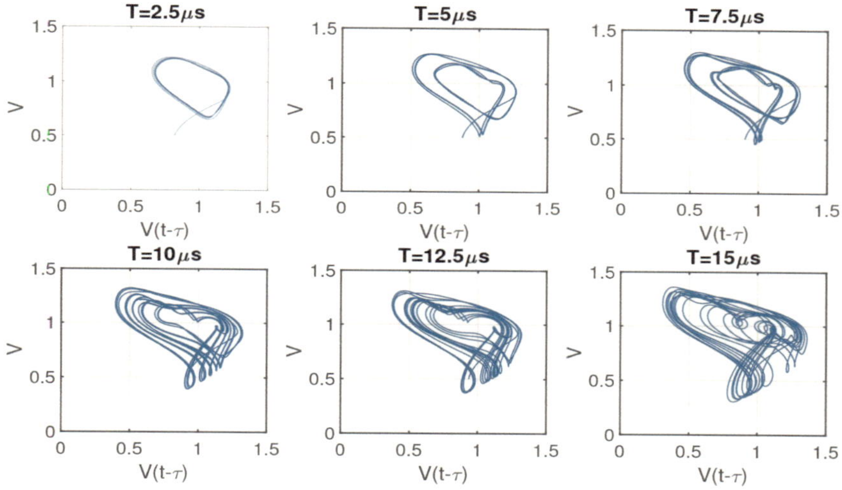

Figure 7.
Phase portraits of DFR.

Algorithm	Training time
Spiking DFR + MLP	16.69 s
MLP	3.2 s
SNN	90 s

Table 1.
Computational complexity analysis.

is associated with calculating the state of the reservoir layer, and updating the weights of readout layer during training. In the introduced spiking DFR model, the weights of input and reservoir layers are fixed and do not undergo any training. That is the fact that makes DFRs significantly computationally efficient compared to other types of RNNs. In traditional RNNs, all the hidden layers require to be trained. Due to the training of all hidden layers, the RNNs are very difficult to train. The measure of complexity is equivalent to the total number of floating-point operations (FLOPs). The training time of RC-based learning techniques correspond to the complexity of model as well [55]. In order to evaluate the computational complexity of our proposed model, the training time of our model is compared with the baseline approaches, i.e., MLP and SNN. **Table 1** presents the training times (complexity) of spiking DFR + MLP, MLP, and SNN.

The SNN which is trained by PSD algorithm shows the highest computational complexity, as it can be seen in **Table 1**. The spiking DFR + MLP and MLP rank as the second and third computationally complex algorithms, respectively. As it can be seen in **Figure 2**, there are some building blocks in the spiking DFR + MLP. Therefore, the computational complexity of spiking DFR + MLP is higher than a simple MLP. Temporal encoding, spike to current, and reservoir blocks are the blocks that exist in our introduced model. However, the superiority of our model in terms of performance makes it justified for us to use this model as the attack detection platform in smart grids.

4. Reservoir computing-based symbol detection

4.1 Received signal

We assume there are N_r antennas at Rx; and N_t antennas at Tx. The received signal can be expressed as:

$$\boldsymbol{y}(t) = \sum_i \boldsymbol{h}_i(t) \circledast x_i(t) + \boldsymbol{n}(t) \tag{14}$$

where $\boldsymbol{n}(t)$ is the additive noise; $\circledast$ stands for the convolution operation; $\boldsymbol{h}_i(t) \in \mathbb{C}^{N_r \times 1}$ is the channel from the ith Tx antenna to the Rx; and $x_i(t)$ is the associated transmitted signal, which is defined as:

$$x_i(t) = \sum_{p=0}^{\infty} \sum_{n=0}^{N_c} g(t - pT_s) s_i[n,p] e^{j\pi(nf_0 + f_c)t} \tag{15}$$

where n is the index of subcarrier; p is the index of time instance; f_c is the carrier frequency; $s[n,p]$ is modulation symbols; f_0 is the frequency space between each subcarrier component; N_c is the number of subcarriers; and $g(t)$ is the waveform function with finite time support which is usually selected as:

$$g(t) = \begin{cases} 1 & \mathrm{t} \in (0, T_s] \\ 0 & \text{otherwise} \end{cases}$$

The channel model is defined according to the ray-tracing principle

$$\boldsymbol{h}(t) := \sum_k \alpha_k \boldsymbol{a}(\theta_k) \delta(t - \tau_k) \tag{16}$$

where k is the index of channel taps; θ_k stands for the angle of arrival (DoA); α_k is the associated path gain; and τ_k is the delay parameter.

4.2 Symbol detection framework

In symbol detection, we aim to estimate $s[n,p]$ belonging to all transmission antennas and time channel use, where the general framework is shown in **Figure 8**. For this problem, the interference from different antennas and OFDM symbols need to be canceled out. Rather than estimating the underlying channel information, in our approach, the reservoir computing network $\mathcal{RC}$ is applied to $\boldsymbol{y}(t)$ to retrieve the transmitted waveform. At the learning stage, the objective is written as:

$$\min_{\mathbf{W}_{out}} L\left(\mathcal{RC}(\boldsymbol{y}(t)), \{x_i(t)\}_i\right) \tag{17}$$

where L is the loss function. Through learning the output weight of RC, it yields an interference cancellation manner, which can recover the transmitted signals. Meanwhile, this relies on a symbol level synchronization among multiple antennas. Alternatively, the symbol detection can be learned through a decomposed manner.

Following this way, we can rewrite the received signal model (14) as:

$$\boldsymbol{y}(t) = \boldsymbol{h}_k(t) \circledast x_k(t) + \sum_{j \neq k} \boldsymbol{h}_j(t) \circledast x_j(t) + \boldsymbol{n}(t) \tag{18}$$

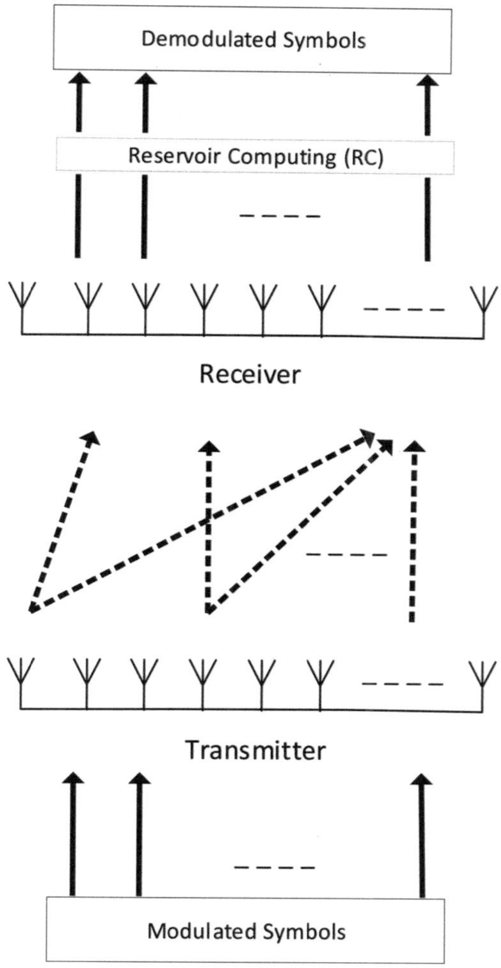

Figure 8.
Symbol detection framework.

where k is the index of interested user; and the remained terms are treated as the interference to the k th user. Given a user index k, the symbol detection is conducted by learn a RC by solving

$$\min_{\mathcal{RC}_k} L\left(\mathcal{RC}_k\left(\mathbf{y}(t)\right), x_k(t)\right). \tag{19}$$

The symbol detection requires learning k RCs, correspondingly. The trained RCs generate estimated symbols for each stream independently.

Moreover, an input buffer can be incorporated to further improve the symbol detection performance as proposed in [31]. To this end, the input of RC at time t_0 is a batch $\{\mathbf{y}(t)\}_{t=t_0}^{t_0+T}$, where T is the length of the buffer.

4.3 One layer learning

We consider the special case when the output is only with one layer. According to the dynamic equation of inner states, denoted as $\{\mathbf{s}(t)\}_{t=0}^{T_a-1}$, where T_a is the length of the training data [56], by stacking the states into a matrix $S{:=}[s(0), s(1), \cdots, s(T_a - 1)]$. The output weights can be updated according to

$$\min_{W} \|WS - \overline{X}\|_2 \tag{20}$$

where $\overline{X} \in \mathbb{C}^{N \times T_a}$ is the target waveform at transmitter side, in which N denotes the number of streams; and W is the output layer to be learned. Accordingly, the target waveform $\overline{X}$ can be chosen as the time domain presentation of scattered pilots or comb pilots. For the target of scattered pilots, the (i, t) th entry of $\overline{X}$ is defined as

$$x_i(t) = \sum_{p=0}^{\infty} \sum_{n \in \Omega_p} g(t - pT_s) s_i[n, p] e^{j\pi(nf_0 + f_c)t} \tag{21}$$

where Ω_p stands for the index of the sub-carriers selected as pilots in the pth OFDM symbol. Specially, for the comb pilots, Ω_P is defined as all the subcarriers at a certain OFDM symbol or several subarriers across all OFDM symbols.

For solving the problem (20), W can be calculated once whole batch of training data are collected, which is through the following pseudo-inverse operation

$$\hat{W} = \overline{X} S^{+} \tag{22}$$

or thorough an online version, such as gradient descent or recursive least squares [57]. For multiple output layers, it follows the same method as multiple layers feed-forward neural networks via the forward backward propagation procedure [58].

4.4 Simulation results

In **Figure 9**, it demonstrates the BER performance of reservoir computing-based symbol detection methods: simple echo state networks (ESN) and echo state networks with windows (WESN) to the conventional methods: linear minimum mean

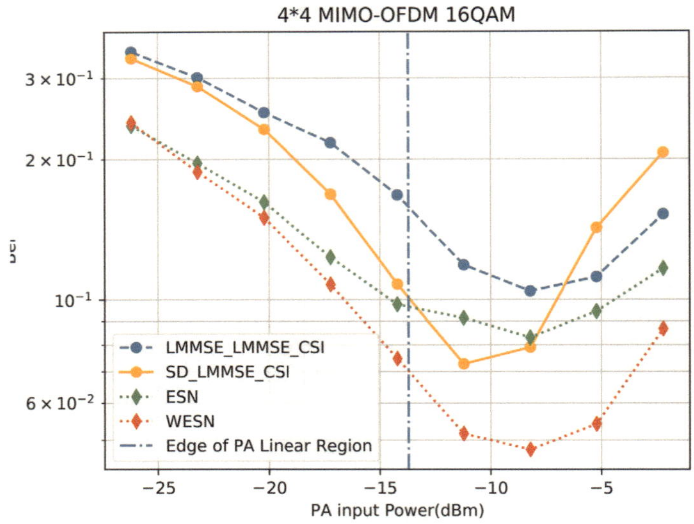

Figure 9.
BER comparison of reservoir computing-based symbol detection methods (ESN and ESN) to conventional methods (LMMSE and sphere decoding).

squared error (LMMSE) and sphere decoding (SD). For the conventional methods, the CSI is obtained by LMMSE channel estimation [59, 60]. Here, we also consider the impact by PA non-linearity at the transmitter side. When the transmitted signal goes throughout the nonlinear region of PA, the signal suffers strong distortion, which can lead to a poor BER performance. Meanwhile, from this figure, we can observe the learning-based methods perform the best at low SNR regime and nonlinear region. This is because conventional methods rely on accurate CSI, which cannot be obtained in these two cases, while learning-based methods are robust against the model-based methods.

5. Conclusion

In this chapter, the emerging applications of spiking DFRs and ESNs were explored. We introduced the combination of spiking neurons, DFRs, and MLPs as the main platform to detect FDI attacks in smart grids. Our simulation results showed that spiking DFR + MLP outperforms SNN, and MLP in terms of ***accuracy*** and ***F1***, respectively. The combination of DFRs and spiking neurons is capable of mapping the data to high dimensional space and capturing the spatio-temporal correlations, which exist between different components of smart grids. The effect of delay value on the performance of DFR was also studied in this chapter. We showed that DFRs can show high dimensional behaviors only for the delay values that make them operate at the edge of chaos. The computational complexity of our introduced model was also studied. In the use case of ESN for MIMO-OFDM symbol detection, we see this learning-based framework can perform better than conventional channel model-based methods when the obtained channel information is imperfect or model mismatch exists. The cost of learning is very few, i.e., it does not require a large size of pilots, which permits the application of this technique in practical system.

Acknowledgements

The work of K. Hamedani, L. Liu and Z. Zhou are supported in part by the U.S. National Science Foundation under grants ECCS-1802710, ECCS-1811497, CNS-1811720, and CCF-1937487.

Author details

Kian Hamedani*, Zhou Zhou, Kangjun Bai and Lingjia Liu
Electrical and Computer Engineering Department, Virginia Tech, Blacksburg, USA

*Address all correspondence to: hkian@vt.edu

References

[1] Li J, Liu L, Zhao C, Hamedani K, Atat R, Yi Y. Enabling sustainable cyber physical security systems through neuromorphic computing. IEEE Transactions on Sustainable Computing. 2017;**3**(2):112-125

[2] Atat R, Liu L, Chen H, Wu J, Li H, Yi Y. Enabling cyber-physical communication in 5g cellular networks: Challenges, spatial spectrum sensing, and cyber-security. IET Cyber-Physical Systems: Theory and Applications. 2017; **2**(1):49-54

[3] Atat R, Liu L, Ashdown J, Medley MJ, Matyjas JD, Yi Y. A physical layer security scheme for mobile health cyber-physical systems. IEEE Internet of Things Journal. 2017;**5**(1):295-309

[4] Li Y, Ng BL, Trayer M, Liu L. Automated residential demand response: Algorithmic implications of pricing models. IEEE Transactions on Smart Grid. 2012;**3**(4):1712-1721

[5] Atat R, Liu L, Wu J, Ashdown J, Yi Y. Green massive traffic offloading for cyber-physical systems over heterogeneous cellular networks. ACM/ Springer Journal of Mobile Networks and Applications. 2018;**24**(4):1-9

[6] Atat R, Liu L, Wu J, Li G, Ye C, Yang Y. Big data meet cyber-physical systems: A panoramic survey. IEEE Access. 2018;**6**:73603-73636

[7] Atat R, Liu L, Yi Y. Privacy protection scheme for ehealth systems: A stochastic geometry approach. In: 2016 IEEE Global Communications Conference (GLOBECOM); IEEE; 2016. pp. 1-6

[8] Wang X, Liu L, Zhu L, Tang T. Joint security and QoS provisioning in train-centric CBTC systems under sybil attacks. IEEE Access. 2019;**7**: 91169-91182

[9] Mosleh S, Sahin C, Liu L, Zheng R, Yi Y. An energy efficient decoding scheme for nonlinear MIMO-OFDM network using reservoir computing. In: 2016 International Joint Conference on Neural Networks (IJCNN); IEEE; 2016. pp. 1166-1173

[10] Zhao C, Danesh W, Wysocki BT, Yi Y. Neuromorphic encoding system design with chaos based CMOS analog neuron. In: 2015 IEEE Symposium on Computational Intelligence for Security and Defense Applications (CISDA); IEEE; 2015. pp. 1-6

[11] Zhao C, Li J, Liu L, Koutha LS, Liu J, Yi Y. Novel spike based reservoir node design with high performance spike delay loop. In: Proceedings of the 3rd ACM International Conference on Nanoscale Computing and Communication; ACM; 2016. p. 14

[12] Bay K, An Q, Yi Y. Deep-DFR: A memristive deep delayed feedback reservoir computing system with hybrid neural network topology. In: Proceedings of the 56th Annual Design Automation Conference 2019; ACM; 2019. p. 54

[13] Esser SK, Merolla PA, Arthur JV, Cassidy AS, Appuswamy R, Andreopoulos A, et al. Convolutional networks for fast, energy-efficient neuromorphic computing. Proceedings of the National Academy of Sciences. 2016;**113**(41):11441-11446

[14] Li J, Zhao C, Hamedani K, Yi Y. Analog hardware implementation of spike-based delayed feedback reservoir computing system. In: 2017 International Joint Conference on Neural Networks (IJCNN); IEEE; 2017. pp. 3439-3446

[15] Li J, Bay K, Liu L, Yi Y. A deep learning based approach for analog hardware implementation of delayed

feedback reservoir computing system. In: 2018 19th International Symposium on Quality Electronic Design (ISQED); IEEE; 2018. pp. 308-313

[16] Bay K, Li J, Hamedani K, Yi Y. Enabling an new era of brain-inspired computing: Energy-efficient spiking neural network with ring topology. In: 2018 55th ACM/ESDA/IEEE Design Automation Conf. (DAC); 2018. pp. 1-6

[17] Zhao C, Hamedani K, Li J, Yi Y. Analog spike-timing-dependent resistive crossbar design for brain inspired computing. IEEE Journal on Emerging and Selected Topics in Circuits and Systems. 2017;**8**(1):38-50

[18] Hu J, Tang H, Tan KC, Li H, Shi L. A spike-timing-based integrated model for pattern recognition. Neural Computation. 2013;**25**(2):450-472

[19] Zhao C, Wysocki BT, Thiem CD, McDonald NR, Li J, Liu L, et al. Energy efficient spiking temporal encoder design for neuromorphic computing systems. IEEE Transactions on Multi-Scale Computing Systems. 2016;**2**(4): 265-276

[20] Hamedani K, Liu L, Atat R, Wu J, Yi Y. Reservoir computing meets smart grids: Attack detection using delayed feedback networks. IEEE Transactions on Industrial Informatics. 2017;**14**(2): 734-743

[21] Yu Q, Tang H, Tan KC, Li H. Precise-spike-driven synaptic plasticity: Learning hetero-association of spatiotemporal spike patterns. PLoS ONE. 2013;**8**(11):e78318

[22] Zhao C, Li J, An H, Yi Y. Energy efficient analog spiking temporal encoder with verification and recovery scheme for neuromorphic computing systems. In: 2017 18th International Symposium on Quality Electronic Design (ISQED); IEEE; 2017. pp. 138-143

[23] C. Zhao, B. T. Wysocki, Y. Liu, C. D. Thiem, N. R. McDonald, and Y. Yi, "Spike-time-dependent encoding for neuromorphic processors," ACM Journal on Emerging Technologies in Computing Systems, vol. 12, no. 3, pp. 23:1-23:21, Sep. 2015

[24] Zhao C, Yi Y, Li J, Fu X, Liu L. Interspike-interval-based analog spike-time-dependent encoder for neuromorphic processors. IEEE Transactions on Very Large Scale Integration (VLSI) Systems. 2017;**25**(8): 2193-2205

[25] Hegger R, Bünner MJ, Kantz H, Giaquinta A. Identifying and modeling delay feedback systems. Physical Review Letters. 1998;**81**(3):558

[26] She C, Yang C, Liu L. Energy-efficient resource allocation for MIMO-OFDM systems serving random sources with statistical QoS requirement. IEEE Transactions on Communications. 2015; **63**(11):4125-4141

[27] Almosa H, Mosleh S, Perrins E, Liu L. Downlink channel estimation with limited feedback for FDD multi-user massive MIMO with spatial channel correlation. In: 2018 IEEE International Conference on Communications (ICC); IEEE; 2018. pp. 1-6

[28] Mosleh S, Liu L, Ashdown JD, Perrins E, Turck K. Content-based user association and MIMO operation over cached Cloud-RAN networks. arXiv preprint arXiv:1906.11318; 2019

[29] Tse D, Viswanath P. Fundamentals of Wireless Communication. Cambridge University Press; 2005

[30] Shafin R, Liu L, Zhang J, Wu Y-C. DoA estimation and capacity analysis for 3-D millimeter wave massive-MIMO/FD-MIMO OFDM systems. IEEE Transactions on Wireless Communications. 2016;**15**(10): 6963-6978

[31] Zhou Z, Liu L, Chang H-H. Learn to demodulate: MIMO-OFDM symbol detection through downlink pilots. arXiv preprint arXiv:1907.01516; 2019

[32] Atat R, Ma J, Chen H, Lee U, Ashdown J, Liu L. Cognitive relay networks with energy and mutual-information accumulation. In: IEEE INFOCOM 2018-IEEE Conference on Computer Communications Workshops (INFOCOM WKSHPS); IEEE; 2018. pp. 640-644

[33] Mahmood FE, Perrins ES, Liu L. Energy consumption vs. bit rate analysis toward massive MIMO systems. In: 2018 IEEE International Smart Cities Conference (ISC2); IEEE; 2018. pp. 1-7

[34] Porwal R, Agrawal H, Vyas R. MIMO OFDM space time coding-spatial multiplexing increasing performance and spectral efficiency in wireless systems. International Journal for Scientific Research and Development. 2014;**2**(06):2321-0613

[35] Shafin R, Liu L, Li Y, Wang A, Zhang J. Angle and delay estimation for 3-D massive MIMO/FD-MIMO systems based on parametric channel modeling. IEEE Transactions on Wireless Communications. 2017;**16**(8):5370-5383

[36] Shafin R, Liu L, Zhang J. DoA estimation and RMSE characterization for 3D massive-MIMO/FD-MIMO OFDM system. In: 2015 IEEE Global Communications Conference (GLOBECOM); IEEE; 2015. pp. 1-6

[37] Shafin R, Jiang M, Ma S, Piazzi L, Liu L. Joint parametric channel estimation and performance characterization for 3D massive MIMO OFDM systems. In: 2018 IEEE International Conference on Communications (ICC); IEEE; 2018. pp. 1-6

[38] Shafin R, Liu L. DoA estimation and performance analysis for multi-cell multi-user 3D mmwave massive-MIMO OFDM system. In: 2017 IEEE Wireless Communications and Networking Conference (WCNC); IEEE; 2017. pp. 1-6

[39] Liu L, Chen R, Geirhofer S, Sayana K, Shi Z, Zhou Y. Downlink MIMO in LTE-advanced: SU-MIMO vs. MU-MIMO. IEEE Communications Magazine. 2012;**50**(2):140-147

[40] Mahmood FE, Perrins ES, Liu L. Modeling and analysis of energy consumption for MIMO systems. In: 2017 IEEE Wireless Communications and Networking Conference (WCNC); IEEE; 2017. pp. 1-6

[41] Shafin R, Liu L, Zhang JC. On the Channel Estimation for 3D Massive MIMO Systems. E-LETTER; 2014

[42] Ozdemir MK, Arslan H. Channel estimation for wireless OFDM systems. IEEE Communication Surveys and Tutorials. 2007;**9**(2):18-48

[43] Shafin R, Liu L, Ashdown J, Matyjas J, Zhang J. On the channel estimation of multi-cell massive FD-MIMO systems. In: 2018 IEEE International Conference on Communications (ICC); IEEE; 2018. pp. 1-6

[44] Shafin R, Chen H, Nam YH, Hur S, Park J, Reed J, et al. Self-tuning sectorization: Deep reinforcement learning meets broadcast beam optimization. arXiv preprint arXiv: 1906.06021; 2019

[45] Shafin R, Liu L. Multi-cell multi-user massive FD-MIMO: Downlink precoding and throughput analysis. IEEE Transactions on Wireless Communications. 2018;**18**(1):487-502

[46] Liodakis G, Arvanitis D, Vardiambasis I. Neural network-based digital receiver for radio

communications. WSEAS Transactions on Systems. 2004;**3**(10):3308-3313

[47] Cai H, Zhao X-h. MIMO-OFDM channel estimation based on neural network. In: 2010 6th International Conference on Wireless Communications Networking and Mobile Computing (WiCOM); IEEE; 2010. pp. 1-4

[48] Shafin R, Liu L, Chandrasekhar V, Chen H, Reed J, et al. Artificial intelligence-enabled cellular networks: A critical path to beyond-5g and 6g. arXiv preprint arXiv:1907.07862; 2019

[49] Sarma KK, Mitra A. Modeling MIMO channels using a class of complex recurrent neural network architectures. AEU International Journal of Electronics and Communications. 2012;**66**(4):322-331

[50] Routray G, Kanungo P. Rayleigh fading MIMO channel prediction using RNN with genetic algorithm. In: International Conference on Computational Intelligence and Information Technology; Springer; 2011. pp. 21-29

[51] Chang H-H, Song H, Yi Y, Zhang J, He H, Liu L. Distributive dynamic spectrum access through deep reinforcement learning: A reservoir computing-based approach. IEEE Internet of Things Journal. 2018;**6**(2): 1938-1948

[52] Mahmood F, Perrins E, Liu L. Energy-efficient wireless communications: From energy modeling to performance evaluation. IEEE Transactions on Vehicular Technology. 2019;**68**(8):7643-7654

[53] Kim J, Tong L, Thomas RJ. Dynamic attacks on power systems economic dispatch. In: 2014 48th Asilomar Conference on Signals, Systems and Computers; IEEE; 2014. pp. 345-349

[54] Zimmerman RD, Murillo-Sánchez CE, Thomas RJ, et al. Matpower: Steady-state operations, planning, and analysis tools for power systems research and education. IEEE Transactions on Power Apparatus and Systems. 2011;**26**(1):12-19

[55] Mosleh S, Liu L, Sahin C, Zheng YR, Yi Y. Brain-inspired wireless communications: Where reservoir computing meets MIMO-OFDM. IEEE Transactions on Neural Networks and Learning Systems. 2017;**29**(10): 4694-4708

[56] Shafin R, Liu L, Ashdown J, Matyjas J, Medley M, Wysocki B, et al. Realizing green symbol detection via reservoir computing: An energy-efficiency perspective. In: 2018 IEEE International Conference on Communications (ICC); IEEE; 2018. pp. 1-6

[57] Jaeger H. Adaptive nonlinear system identification with echo state networks. In: Advances in Neural Information Processing Systems; 2003. pp. 609-616

[58] Hecht-Nielsen R. Theory of the backpropagation neural network. In: Neural Networks for Perception. Elsevier; 1992. pp. 65-93

[59] Cheng L, Wu Y-C, Ma S, Zhang J, Liu L. Channel estimation in full-dimensional massive MIMO system using one training symbol. In: 2017 IEEE 18th International Workshop on Signal Processing Advances in Wireless Communications (SPAWC); IEEE; 2017. pp. 1-5

[60] Danesh W, Zhao C, Wysocki BT, Medley MJ, Thawdar NN, Yi Y. Channel estimation in wireless OFDM systems using reservoir computing. In: 2015 IEEE Symposium on Computational Intelligence for Security and Defense Applications (CISDA); IEEE; 2015. pp. 1-5

Chapter 7

SMS Security on Android Using RC4 Algorithm

Kaung Htet Myint

Abstract

SMS plays an important role in mobile communication systems. The sending side is acting like as a server for the receiving side that receives short message service at the receiving side. SMS does not incorporate a procedure to accord security for the text sent as data. A majority of the applications for mobile devices are designed and implemented without taking security into account. SMS messages are not normally encrypted by default. Confidentiality is the notion of making sure that data is not made accessible or exposed to unauthorized people. Encryption is the main approach to confidentiality. Both symmetric and asymmetric encryption can be employed. As confidentiality was the original purpose of cryptology, this chapter is introduced as a data confidentiality approach to SMS on Android. It encompasses SMS network architecture as well as cryptographic protocols as theory background and it also deals with design, implementation, and confidentiality assessment of RC4 stream cipher for SMS data confidentiality on mobile networks.

Keywords: SMS, security, encryption, confidentiality, cryptography

1. Introduction

Data confidentiality is the notion of making sure that data is not made accessible or exposed to unauthorized people and is approximately comparable with secrecy. Measurements approximated to confirm confidential information is intended to avoid useful data from unauthorized users' getting them, creating certain hurdles which authorized users can surmount. Access will be limited to intended persons to interpret the facts in query. The facts to be classified in accordance with the quantity and damaged type are public. This type drops into unauthorized users. Strict procedures can be fulfilled in accordance with those classifications. An encipher security system can be used for the security of data confidentiality.

A text messaging service component of most telephones, Internet, and mobile-device systems is known as short message service (SMS). Standardized communication protocols are used to permit smart phones to transfer short text messages. Short message service is also commonly referred to as a "text message." The user can conduct a message of up to 160 characters to another device with a SMS. In SMS, longer messages will automatically be fragmented into several parts. This type of text messaging is supported by most cell phones.

The formal name for text messaging is SMS. Short message service is a way to conduct short, text-only messages from one phone to another. These messages are usually conducted over a cellular data network.

The procedure for conducting SMS is launching the Messages application on the phone. Tap on the Compose Message button. Enter the phone number or name of the contact you want to text. Type your message and finally hit Send. These days, there exist a number of security issues and vulnerabilities related to SMS [1, 2].

Cryptography is related with the procedure of changing ordinary plain text into unintelligible text and vice versa [3]. SMS sent for data confidentiality over mobile networks can be protected by RC4 stream cipher [4]. The objective of this chapter is to offer data secrecy during the SMS messages transmission to prevent them from being received by illegal parties and to ensure the authenticity of the message from the genuine sender.

2. Related works

Phyo Su Khin proposed a short message service (SMS) security for mobile devices with AES algorithm, which focused on the security of short message service (SMS) based on advanced encryption standard (AES) with 128 bits which allows user to encrypt messages before it is transmitted over the network with the use of encryption to protect SMS messages. This application can run on Android devices. The sender and the receiver use the same key to encrypt and decrypt the message as per user requirement in order to improve security and to get high confidentiality.

Aye Mya MoMo proposed image encryption based on XTS-AES MODE where a secure image encryption using XTS-AES and WHIRLPOOL Hash function was implemented. This system improves integrity and confidentiality and is suitable for parallel operation.

Myo Thinzar Aung proposed a secure video streaming system using SRTP and RC4 algorithm where Ronald Rivest symmetric key algorithm (RC4) is used for data encryption and then the encrypted data is embedded into secure real-time transport protocol (SRTP) header. Data acknowledgement is generated to the sender and receiver by using secure real-time transport control protocol (SRTCP).

Yu Loon Ng proposed short message service (SMS) security solution for mobile devices, where the focus is on the security of short message service (SMS) and the Global System for Mobile communication (GSM) network, and the use of encryption to protect SMS messages and encryption schemes was conducted to understand the properties of different encryption schemes and their applicability to SMS messages. The selected scheme was implemented in the form of a Secure SMS Chat application to validate the viability of the selected encryption scheme.

3. Basic concepts of SMS technology

By cooperating with the cellular network, short message service transmits text messages from one phone to other phones. These devices require short messaging entities (SMEs). These are starting points (sender) and endpoints (receiver) for SMS messages. They never connect directly with each other [5]. They always connect with a short message service center (SMSC). A mobile telephone can be an SME. Computer containing a messaging software [6], which can connect directly with the SMSC of the service source, can be an SME. Two types of SMS messages conditional on the character of the device in the network are mobile-originated (MO) messages and mobile-terminated (MT) messages. The mobile phone sends MO messages to the SMSC and receives MT messages. These MO and MT messages are encrypted in a different way during conduction [7].

The Common Channel Signaling System 7 (SS7) conveys SMS messages. A worldwide standard that describes the processes and procedures for exchanging

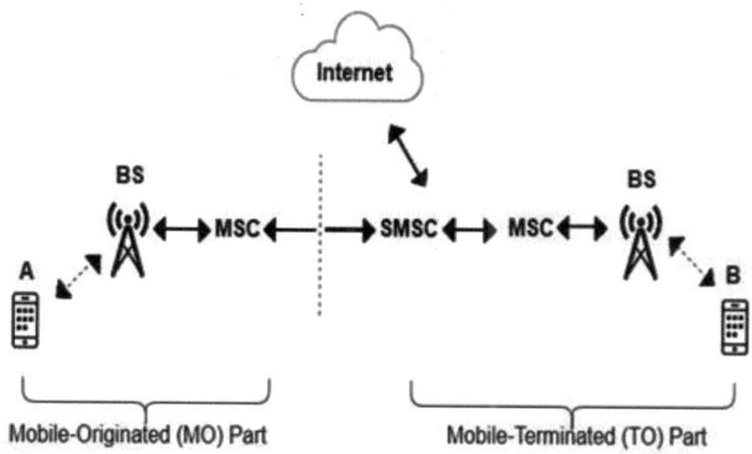

Figure 1.
Mobile network architecture.

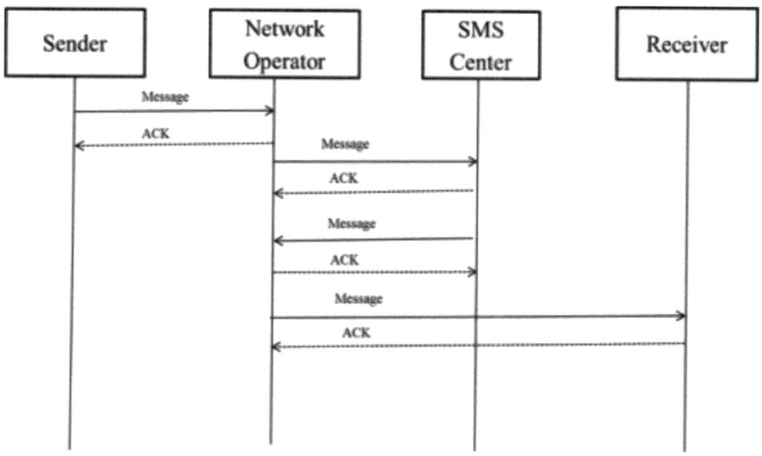

Figure 2.
Message flow of SMS network.

data among network components of wire line and wireless phone carriers is known as SS7 [5]. These components use the SS7 procedure to give-and-take control data for call system, movement control, etc. Theoretically, the common SMS mobile network architecture contains two parts known as mobile originating (MO) part and mobile terminating (MT) part (**Figure 1**). The wireless structure for network part of the sending mobile switching center changes all circulation into and out of the structure in spite of the source are known as MO. The other part contains an improper location and the termination of MSC for the phone, as well as a central stock and onward server is called SMS Centre. It is accountable for receiving information and keeping information (**Figure 2**).

4. Cryptography

Cryptographic algorithms can be separated into: symmetric key algorithms and asymmetric key algorithms. The general concept of RC4 is it uses a symmetric-keystream cipher as shown in **Figure 3**. A stream cipher stands for a symmetric key cipher where plaintext digits are merged with a keystream.

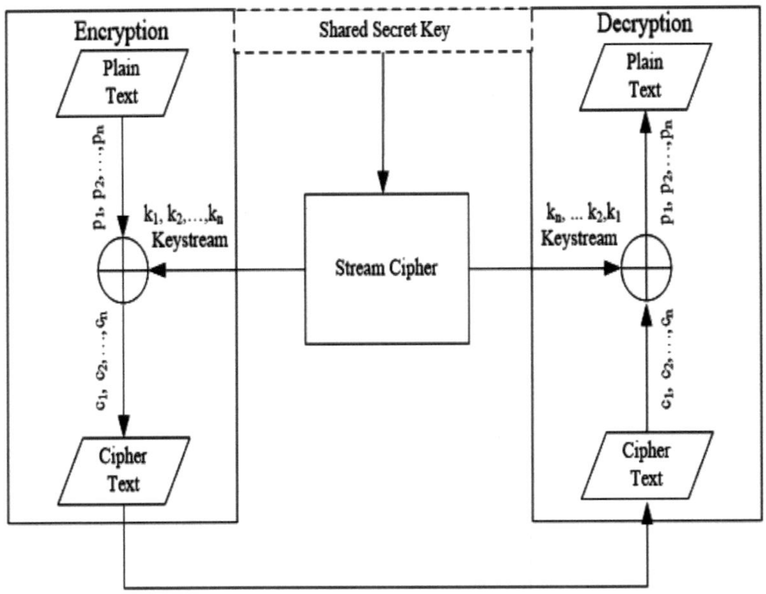

Figure 3.
Stream cipher.

5. RC4 stream cipher

Rivest Cipher 4 (RC4) is very popular because it is simple and can be very fast. It is an adjustable stream size key cipher that included bytes focused on processes. It is founded on the practice of unplanned arrangement. RC4 makes bits of a pseudo-random stream (a keystream). As with any stream cipher, these can be used for the procedure of hiding a data in disguising its material (encryption) by merging it with the message to be sent securely from the source to the intended endpoint of the message (plaintext) using bit-wise exclusive OR. A procedure to revert cipher text into plain text (decryption) is executed in the similar way. This stream cipher includes two parts.

5.1 Key-scheduling algorithm (KSA)

The key-scheduling algorithm is used to start up the arrangement in the range "S." The number of bytes in the key is called "keylength" and can be in the array $1 \leq \text{keylength} \leq 256$. It is used to start up the arrangement in the "S" box. Keylength stands for number of bytes in key and ranges from 1 to 256. The key-scheduling algorithm (KSA) [3] is as follows:

```
Begin
for i from 0 to 255
    S[ i] :=i
endfor
j:=0
for i from 0 to 255
    j:=(j+S[ i] +key[ i mod keylength] )
mod 256
    swap values of S[ i] and S[ j]
endfor
end
```

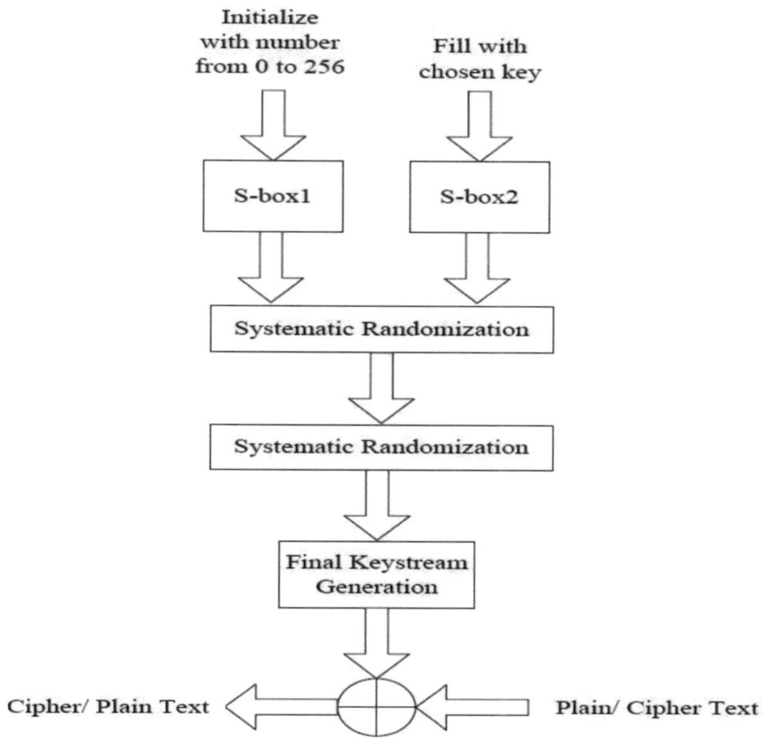

Figure 4.
General RC4 stream cipher.

5.2 Pseudorandom generation algorithm (PRGA)

The arrangement is started with a variable length key, characteristically between 40 and 2048 bits, via the key-scheduling algorithm (KSA). The stream of bits is created using the **pseudorandom generation algorithm (PRGA)**. RC4 creates a keystream. After that, the stream of bits is created by a PRGA. It amends the condition and outputs a byte of the keystream.

```
Begin
    i:=0
    j:=0
    while GeneratingOutput:
    i=(i+1) mod 256
    j=(j+S[ i] ) mod 256
    swap values of S[ i] and S[ j]
    K:= S[ (S[ i] +S[ j] ) mod 256]
    output K
    endwhile
end
```

First, we implement RC4 stream cipher by using key-scheduling algorithm (KSA) and pseudorandom generation algorithm (PRGA) in Java programming language (**Figure 4**).

6. Design and implementation

The SendSMS mobile application receives SMS plain text, password, and phone number of the receiver as inputs and comes out as a cipher text. The cipher text is passed through mobile network communication channel. The ReceiveSMS mobile application receives the cipher text that is passed through the mobile network communication channel, the password, and the phone number of the sender as inputs and comes out as a SMS plain text. The implementation of two smart phone applications is displayed in **Figure 5**.

The *SendSMS* mobile application is used by the creator and the *ReceiveSMS* mobile application is used by the intended person (**Figure 6**). The creator must input the phone number of the intended person, password, and SMS message to *SendSMS* smart phone application and press *Send Message* button. The intended person must input the phone number of the creator and the same password used by the creator to *ReceiveSMS* mobile application and press *ReceiveMessage* button. Then,

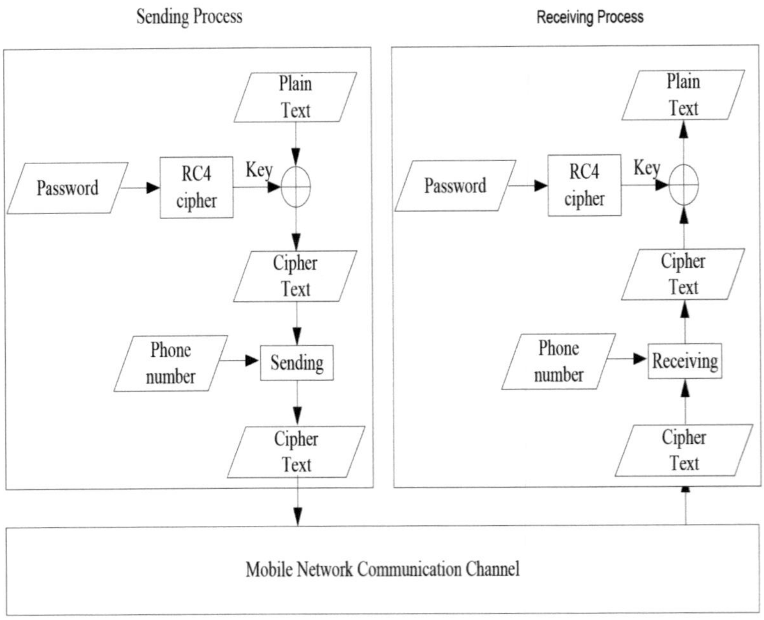

Figure 5.
Design for implementation.

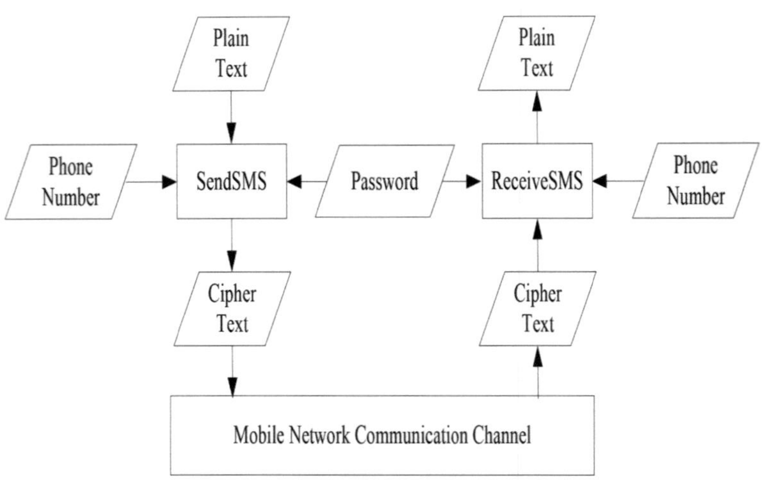

Figure 6.
Data flow diagram.

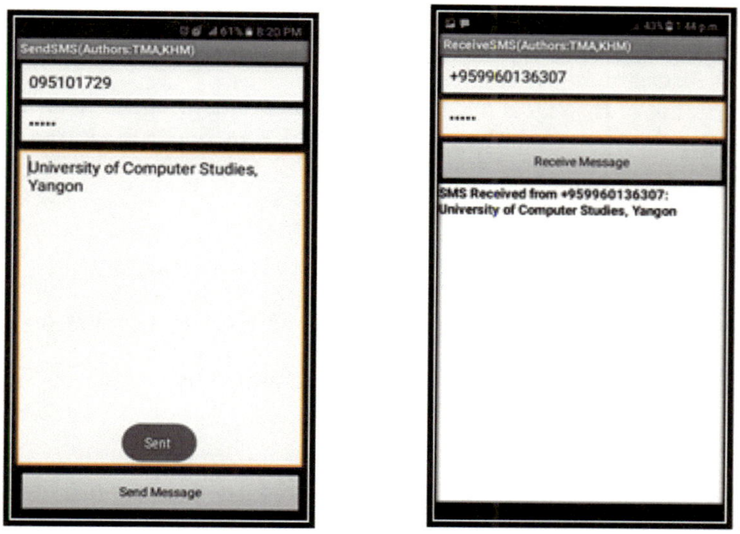

26

Figure 7.
System user interface.

the SMS message of the creator is shown in the window screen of the *ReceiveSMS* smart phone application (**Figure 7**).

7. Results and measurements

7.1 Implementation results of RC4

PlainText:01234567890123456789012345678890123456789012345678901234567890123456789012345678901234567890123456789012345678901234567890
Password:APPLE
Keystream:lvUùi+ESÑ;?é½þ0???Äã¥?)ß©Ñn??¸´O¹ÿa0_o?é6Ø?ao¬búG¢¦?)µÎEL?/
p6µÅ((*c:mÿÌ7íJÞO9«Ù[?¥¯Ï8U•?oJ½q??°slr@0)?VãG?W¿o»ª
CipherText:X]DfÍ\rkè¢Û?Ê«¢«ýÓ?¦ë?5æV¦2??|?ÊWg%_«Ûì¡WX?[Ê8u??£?ö||$´
47F?üPXÉûÔzï}
?ìm©??ÿg?¬Z|?I² ?A_Fu6>¥gÑ's•a?W??
Password:APPLE
Keystream: hlvUùi+ESÑ;?é½þ0???Äã¥?)ß©Ñn??¸´O¹ÿa0_o?é6Ø?ao¬búG¢¦?)µÎEL?/
p6µÅ((*c:mÿÌ7íJÞO9«Ù[?¥¯Ï8U•?oJ½q??°slr@0)?VãG?W¿o»ª
PlainText:01234567890123456789012345678890123456789012345678901234567890123456789012345678901234567890123456789012345678901234567890

The following statistical tests are applied to test the randomness of arbitrarily long binary sequences produced by the developed system based on RC4 pseudo-random number generators.

7.2 Frequency test

The objective of Frequency test is to decide whether the number of ones and zeros in an arrangement is just about the same as would be predictable for an actually random sequence [8].

Let the keystream 1011010101 be tested by the Frequency test. The result of Frequency test is SUCCESS because the numbers of occurrences of bits—zero and one—in the keystream are equal.

The description of the test is the tests change the sequence ε into a new sequence X, such that Xi = 2εi − 1 = ±1. The calculation of this sequence is assumed by

$$S_n = X_1 + X_2 + \ \dots \ + X_n. \tag{1}$$

If ε = 1011010101, then n = 10 and

$$S_n = 1 + (-1) + 1 + 1 + (-1) + 1 + (-1) + 1 + (-1) + 1 = 2.$$

The test statistic for the observed sum s_{obs} is assumed by

$$s_{obs} = \frac{|s_n|}{\sqrt{n}}. \tag{2}$$

$$s_{obs} = \left(\frac{|2|}{\sqrt{10}}\right) = 0.632455532$$

The P-value is assumed by

$$P - \text{value} = \text{erfc}\left(\frac{s_{obs}}{\sqrt{2}}\right) \tag{3}$$

erfc(z) is the complementary error function

$$\text{erfc}\left(\frac{.632455532}{\sqrt{2}}\right) = 0.527089$$

7.2.1 Decision rule

The verified sequence is recognized as random if the P-value ≥ 0.01, if not it is nonrandom. P-value = 0.527089 ≥ 0.01. Therefore, the sequence is random [8].

7.3 Runs test

The objective of Runs test is to define whether the number of runs of ones and zeros of various lengths is as predictable for a random sequence. In particular, this test defines whether the oscillation between such zeros and ones is too fast or too slow [8].

Let the keystream 1011010101 be tested by Frequency test. The description of the test is to calculate the pre-test proportion π of ones in the input sequence:

$$\pi = \frac{\sum_j \varepsilon_j}{n} \tag{4}$$

If ε = 1001101011, then n = 10 and $\pi = \frac{6}{10} = \frac{3}{5}$.

Define if the prerequisite Frequency test is approved: if it can be displayed that $|\pi - \frac{1}{2}| \geq r$, then the Runs test need not be executed. If the test is not appropriate, then the P-value is set to 0.0000. For this test, $r = \frac{2}{\sqrt{n}}$ has been predefined in the testcode. $|\pi - \frac{1}{2}| = 0.1 < r = \frac{2}{\sqrt{10}} = 0.63246$, and the test is not run. Since the observed value π is within the particular bound, the Runs test is appropriate.

Calculate the test statistic

$$V_n(obs) = \sum_{k=1}^{n-1} r(k) + 1 \tag{5}$$

$r(k) = 0$ if $\varepsilon_k = \varepsilon_k + 1$ and $r(k) = 1$ otherwise. Since $\varepsilon = 1\ 00\ 11\ 0\ 1\ 0\ 11$, then V10 (obs) = (1 + 0 + 1 + 0 + 1 + 1 + 1 + 1 + 0) + 1 = 7.

$$P - \text{value} = \text{erfc}\left(\frac{|v_n(obs) - 2n\pi(1-\pi)|}{2\sqrt{2n}\ \pi(1-\pi)}\right). \tag{6}$$

$$P - \text{value} = \text{erfc}\left(\frac{7 - 2 * 10 * \frac{3}{5}\left(1 - \frac{3}{5}\right)|}{2\sqrt{20}\ \frac{3}{5}\left(1 - \frac{3}{5}\right)}\right) = 0.147232$$

7.3.1 Decision rule

If the calculated P-value is <0.01, then define that the sequence is non-random. If not, define that the sequence is random. P-value = 0.147232 ≥ 0.01. The sequence is random [8].

Cryptographic algorithms of randomness testing are attacker and designer importance choice. Short sequences of at most 512-bit length are considered for block ciphers and hash functions. The National Institute of Standards and Technology guides to influence the properties of randomness of generators and sequences of statistical test suites. Some tests of this suite cannot be applied to short sequences and do not produce reliable test values. Most of the test suites are producing relatively short sequences that are not suitable for evaluation. Therefore, only the Frequency test and Runs Test approach to evaluate short sequences without tweaking the test. Apart from these tests in the test suite, other tests are not considered for short sequences that transmit SMS over mobile [9, 10].

7.4 Result of testing

Frequency test examines the numbers of occurrences of the bits in the keystream. Runs test examines the independence of the keystream bits.

Let the keystream 1011010101 be tested by Frequency test and Runs test. The result of Frequency test is SUCCESS because the numbers of occurrences of bits—zero and one—in the keystream are equal. The result of Runs test is FAILURE because the adjacent bits of the keystream are dependent.

Keystream : 1011010101
Frequency Test SUCCESS p_value = 0.527089
Runs Test FAILURE p_value = 0.005658

Let the keystream 1001101011 be tested by Frequency test and Runs test. The result of Frequency test is SUCCESS because the numbers of occurrences of bits—zero and one—in the keystream are equal. The result of runs test is SUCCESS because the adjacent bits of the keystream are independent.

Keystream : 1001101011
Frequency Test SUCCESS p_value = 0.527089
Runs Test SUCCESS p_value = 0.147232

In practical use, the following plain text is encrypted by using the following RC4 keystream. The confidentiality of the RC4 keystream is measured by using the Frequency test and Runs test. We found that their results are SUCCESS. Therefore, the confidentiality of RC4 stream cipher may be strong for SMS Security.

Plain Text:University of Computer Studies, Yangon (UCSY)
Password:APPLE
Keystream:hlvUùi+ESÑ;?é½þ0???Ääã¥?)ß©Ñn??¸´O¹ÿa0_o?é6
CipherText:=#?X,'¨ü??½_ðåæ°?×´z«Üg¸í'áÕ!Þ?wI,É°
Password:APPLE
Keystream:hlvUùi+ESÑ;?é½þ0???Ääã¥?)ß©Ñn??¸´O¹ÿa0_o?é6
Plain Text:University of Computer Studies, Yangon (UCSY)
RC4 Keystream(byte) : hlvUùi+ESÑ;?é½þ0???Ääã¥?)ß©Ñn??¸´O¹ÿa0_o?é6
RC4Keystream(bit):0110100001101100011101100101010111010010010101101
000101010100110011101100110000001010010000001101101110000001011
0011111101001111011000010011000000101111100011100011011110011010110
Frequency Test SUCCESS p_value = 0.763025
Runs Test SUCCESS p_value = 0.446643

8. Conclusion and future work

Nowadays, in this chapter, the pseudorandom number sequence made by RC4 stream cipher is measured by Frequency test and Runs test. The confidence of the pseudorandom number sequence is measured to be randomness with a confidence of 99% given to P-value of every single test. For that reason, it is suggested that the user should use the pseudorandom number sequence made by the RC4 steam cipher for data confidentiality of SMS message on Android.

Acknowledgements

I would like to express my gratitude to my invaluable university, University of Computer Studies, Yangon. Furthermore, I would like to thank all teachers at University of Computer Studies for giving me useful comments on my presentation.

I would like to express my gratefulness to IntechOpen Limited for suggestions, patience, and support while I wrote my paper in Myanmar.

Author details

Kaung Htet Myint
University of Computer Studies, Yangon, Myanmar

*Address all correspondence to: kolynn.2013@gmail.com

References

[1] Saxena N, Payal A. Enhancing security system of short message service for M-commerce in GSM. International Journal of Computer Science & Engineering Technology (IJCSET). 2011;**2**(4). ISSN: 2229-3345

[2] Verma SK, Ojha DB. An approach to enhance the mobile SMS security. Journal of Global Research in Computer Science. 2014;**5**(5). ISSN: 2229-371X

[3] Forouzan BA. Cryptography and Network Security. International ed. McGraw Hill; 2008. ISBN: 978-007-126361-0

[4] Singh V, Shrivastawa S. RC4 stream cipher design for data security. International Journal of Advance Research in Science and Engineering. 2017;**6**(5). ISSN: 2319-8346

[5] Medani A, Gani A, Zakaria O, Zaidan AA, Bahaa B. Review of mobile short message service security issues and techniques towards the solution. Scientific Research and Essays. 2011; **6**(6). ISSN: 1992-2248

[6] Ozeki NG. SMS Gateway. Available from: http://www.ozekisms.com/

[7] Katankar VK, Thakare VM. Short message service using SMS gateway. International Journal on Computer Science and Engineering. 2010;**02**(04)

[8] Rukhin AL, Bassham LE. NIST. A Statistical Test Suite for Random and Pseudorandom Number Generators for Cryptographic Applications. US: National Institute of Standards and Technology Special Publication 800-22; 2010

[9] Evaluation of randomness test results for short sequences. In: SETA 2010, 6th International Conference; Paris, France; September 13–17, 2010

[10] Agoyi M, Seral D. SMS security: An asymmetric encryption approach. In: IEEE 6th International Conference on Wireless and Mobile Communications (ICWMC 2010). Valencia, Spain; Sep 2010. pp. 448-452

Chapter 8

From Pillars to AI Technology-Based Forest Fire Protection Systems

Nikos Aspragathos, Eleftherios Dogkas, Pavlos Konstantinidis, Panagiotis Koutmos, Nefeli Lamprinou, Vassilis C. Moulianitis, Georgios Paterakis, Emmanouil Z. Psarakis, Evangelos Sartinas, Konstantinos Souflas, Georgios Thanellas, Georgios Tsiourlis, Nikitas Xanthopoulos and Panteleimon Xofis

Abstract

The importance of forest environment in the perspective of the biodiversity as well as from the economic resources which forests enclose, is more than evident. Any threat posed to this critical component of the environment should be identified and attacked through the use of the most efficient available technological means. Early warning and immediate response to a fire event are critical in avoiding great environmental damages. Fire risk assessment, reliable detection and localization of fire as well as motion planning, constitute the most vital ingredients of a fire protection system. In this chapter, we review the evolution of the forest fire protection systems and emphasize on open issues and the improvements that can be achieved using artificial intelligence technology. We start our tour from the pillars which were for a long time period, the only possible method to oversee the forest fires. Then, we will proceed to the exploration of early AI systems and will end-up with nowadays systems that might receive multimodal data from satellites, optical and thermal sensors, smart phones and UAVs and use techniques that cover the spectrum from early signal processing algorithms to latest deep learning-based ones to achieving the ultimate goal.

Keywords: artificial intelligence technology, fire risk assessment, detection and localization of the fire, motion planning

1. Introduction

Fire is an important ecological factor that has affected both the structure and distribution of numerous plant communities across the globe. Fire probably first appeared, as a natural disturbance factor, as soon as there was any existing terrestrial vegetation [1]. Prior to human Influence, the main ignition sources were lightning, volcanic and earthquake activity [2–4]. Fire was therefore a natural process occurring periodically in the natural vegetation succession cycle, contributing to the continued rejuvenation and promoting the productivity of many plant communities and ecosystems [5]. The periodicity of ecosystem burning was determined by the availability of a fuel load able to sustain a fire after a natural ignition event. Fuel load and consequently the flammability of an ecosystem increases with age ensuring that the ecosystem will be burned at a relatively mature stage. The effect of fire on landscape and ecosystems has been so determinative that the distribution, composition and structure of many biomes across the globe could not be explained by climate and soil alone and fire also needs to be taken into account [6, 7].

Later, fire became a very important human tool, widely used or misused for the improvement of living conditions. Archeological evidence from the Petralona caves of northern Greece and elsewhere indicate that fire has been used by man for at least half a million years [8]. The Paleolithic hunter and food gatherer used fire not only as a source of energy but also as a vegetation and landscape management tool. Both fire frequency and intensity increased dramatically with considerable impacts on the natural ecosystems and the Mediterranean flora. This change in fire characteristics has shifted the equilibrium between fire and ecosystem function, transforming fire from a natural ecological factor that initiates succession into a human-induced land degradation factor [9]. Despite the major human intervention in the relation between fire and ecosystem function, the Mediterranean ecosystems and vegetation types retain the ability to recover from fire quite rapidly [5], assuming fire frequency does not greatly exceed the natural return interval [10]. Wildfires, however, are not just a vegetation and ecosystem degradation factor but also a factor which can have significant social and economic consequences, especially when they occur in the rural urban interface.

The occurrence of wildfires and especially large ones is the result of the combined action of two driving forces, namely fuel availability and continuity and weather patterns. Over the last years, significant efforts have been made to study the historic and current trends in fire regime and identify the relevant significance of the above driving forces in determining the past and current fire regime [11–14]. The main conclusion of these studies is that there is an observed change in the fire regime of the Mediterranean Europe after the 1970s with a significant increase both in total number of fires (NF) and area burned (AB). There is a clear indication that wildfires have changed from fuel driven, before the 1970s, to weather driven after that date.

The decade of 1970s coincides with significant socioeconomic changes in the Mediterranean Europe, which may explain the sudden change in fire regime, better than the climatic changes and the observed increase in drought, which are more gradual and have also occurred in other climatically similar regions without associated changes in fire regime [12]. The abandonment of agricultural land in the mountainous and semi-mountainous areas, the decrease of livestock, the increased urbanization, a significant decrease to the use of wood as an energy source and the recovery of vegetation in the abandoned field, have all caused significant changes in the landscape structure and composition [15].

Another significant consequence of the changed fire regime in Mediterranean Europe is the expansion of fire to regions and ecosystems that are not considered fire prone and the vegetation components do not possess fire adaptive traits. An analysis of the fire characteristics in Greece for the year 2007 [16] revealed that 9.3% of the burned area falls in an altitudinal zone exceeding 1000 m, confirming previously reported similar trends of an increased occurrence of fires at higher altitudes. Furthermore, an analysis of landscape dynamics in the nature reserve of Dadia Forest National Park [17], revealed that during 2001–2011 large wildfires became a significant destruction factor, threatening the long-term sustainability of the reserve, although low intensity fires could be a mechanism for maintaining landscape heterogeneity.

The urbanization of population apart from land abandonment also led to an increased and urgent need for residential areas in the receptor cities, bringing the city borders close to semi-natural ecosystems and subsequently increasing the wildland-urban interface (WUI). Furthermore, the improvement of the economic status, led to the creation of settlements in forested areas, often maintaining extensive parts of semi-natural vegetation formations creating a rather deadly WUI. The most dreadful example of such a situation is the 2018 fire in eastern Attica which led to the death of 100 people and significant loss of properties and infrastructures. Finally, the increase of tourism in recent decades led to an increase of the residential zones in coastal areas increasing again the WUI and frequency of ignitions [18].

Despite the above changes in fire regime and the catastrophic incidences experienced in southern Europe, and Greece in particular, the wildfire management continues to rely almost exclusively on fire suppression and traditional means of fire detection. An analysis of the fire characteristics in Greece during the period 2007–2011 revealed some interesting findings regarding the efficiency of the currently applied wildfire management [16]. In 2007, the great majority of fire events burnt relatively small areas lower than 1000 ha. However, 11% of the events evolved to megafires resulting in the worst year ever recorded in terms of area burnt (more than 200,000 ha), although the number of fires was almost equal to the one in 2011, and according to the Forest Fires in Europe 2007 report [19], it was high but not the highest since 1980. The entire year of 2007, the summer period in particular, was characterized by extreme weather conditions with the highest temperatures recorded for almost a century, three consecutive heat waves from June to August and wind patterns that favored the spread and intensity of fires [20]. As stated in [20, 21], 2007 represents an example of how the weather conditions will be like by the end of the century as a result of climate change.

The almost 90% efficiency of the currently employed fire management in keeping most wildfires at relatively small sizes, even under the most favorable for fire spread conditions, reveals that this approach, which relies exclusively on fire suppression and traditional methods of early fire detection has reached its efficiency limits. This is because it is a small number of fire events which turn into megafires and destroy ecosystems, properties, infrastructures and most importantly have a high cost in human lives, and such events unfortunately are not avoided. Thus, a new approach is needed in wildfire management which will not rely exclusively on fire suppression, but it will utilize the technological advancements, the wide availability of remote sensing data and the large amount of research related to risk assessment and early fire detection.

In this chapter, after the brief fire's history, in Section 2, the basic forest fire monitoring systems technologies are presented. The vital requirements of an autonomous early forest fire detection system, its main modules and the necessary methods for wildfire management and risk assessment, smoke and fire detection based on images and video as well as navigational autonomy issues for UAVs are presented in Section 3. Finally, in Section 4 contains our conclusions.

2. Forest fire monitoring systems technologies

Forest fire monitoring system can be broadly classified into the following categories, each one strongly related with a corresponding technology [22]:

1. Human-based observation systems

2. Satellite-based systems

3. Wireless sensor networks based systems

4. Optical and thermal cameras based systems

2.1 Satellite-based systems

Regarding the satellite-based systems, there have been some initiatives for forest fire detection purposes. Specifically:

- the advanced very high-resolution radiometer (AVHRR) [23], launched in 1998

- the moderate resolution imaging spectroradiometer (MODIS) [24], launched in 1999, and the Visible Infrared Imaging Radiometer Suite (VIIRS) [25] that was launched in 2011 gave the capability for the use of a new generation of operational moderate resolution-imaging following the legacy of the AVHRR on NOAA and MODIS on Terra and Aqua satellites.

However, all these are not sufficient for early forest fire detection due to the fact that satellites follow orbits which deprive the demand for image acquisition of forests around the clock. In addition, beyond the ineffective forest fire scanning, the quality of images is heavily dependent on weather condition.

2.2 Wireless sensor networks based systems

There are numerous contributions in the literature, which are based on wireless sensor networks (WSN) technology. In particular:

- Sudha et al. [26] implemented a system based on IoT devices, which enables the continuous monitoring of a forest area, covering issues relative to fire detection and animal tracking.

- Toledo-Castro et al. [27] in order to minimize the false alarms rate of their own WSN system, introduced a fuzzy logic-based model were.

- Xu et al. [28] proposed a formula for sensor distribution in a forest area, solving problems of optimal coverage by taking into account the topology of the covered area.

- In terms of energy efficiency, Wang et al. [29] proposed an algorithm that minimizes the energy consumption of their WSN-based forest monitoring system.

- Finally, Kadri et al. [30] implemented a WSN-based system whose reliability, robustness and durability were examined, setting the foundations for future works.

Despite of their robustness and efficiency, WSN-based systems have their own limitations mainly the limited energy capacity, the relatively low communication speed and demanding installation and maintenance.

2.3 Optical and thermal sensors based systems

On the other hand, optical and thermal sensors based systems seem to be more popular than the satellite ones, because of the existence of a number of well-established practices introduced in systems EYEfi, FireWatch (Germany) and ForestWatch (Canada), FireHawk (South Africa), FireVu (England) and UraFire (France) [22, 31]. In all these systems, different kinds of fire detection sensors could be used:

- video cameras sensitive in visible spectra used for smoke recognition during the day and fire flame recognition during the night
- infrared (IR) thermal imaging cameras used for the detection of heat flux from the fire
- IR spectrometers which identify the spectral characteristics of smoke gases and
- light detection and ranging (LIDAR) systems which measure laser light backscattered by the smoke particles

In all those systems, automatic forest fire detection is based on smoke recognition during the day and flame recognition during the night. The main disadvantage of those optical-based systems is the high rate of false alarms, due to atmospheric conditions (clouds, shadows and dust particles), light reflections and human activities. Thus:

- their performance in terms of detection's speed was slower and less reliable than a trained human tower observer
- the detection performance depends on the fire's size and its distance from the optical/thermal sensor
- since the cameras are unable to take into account the topography of the land in localization calculations, their estimations might have large localization errors. Thus, even though there is a lot of image-based fire detection techniques, it is apparent that their performance is heavily dependent on the coverage of sensors as well as the topology and the forest's specific form

For all those reasons, it is common for these systems to be supervised by a human operator. After the fire alarm is generated and the suspicious region of the image is detected, the human operator confirms or discards the alarm. In cases where the human operator is not sure regarding a fire alarm, he can switch the system to manual operation and make additional inspections. Using the system in such a way, human operator efficiency is highly improved.

2.4 Unmanned aerial vehicles (UAV) based systems

By means of the modern sensory technologies, a forest fire can be quickly and accurately detected. For this, unmanned aerial vehicles (UAVs) can provide a full range of multi-sourced data for fire monitoring, which can provide an input to an algorithm which will determine the validity of a fire ignition event. UAVs can be classified, based on their configuration, as fixed wing, flapping wing and rotary wing models [23].

Fixed-wing UAVs are suitable for long-distance missions, but they need long runway to take-off and land. Flapping-wing UAVs are relatively new in research and do fly like birds and insects. On the plus side, they can perform vertical take-off and landing. Regarding the rotary wings UAVs, they are suitable for flight missions, which include hovering, maneuverability and easy control. However, rotary UAVs have a more limited flight time autonomy than the fixed-wing UAVs (about 30 min with no load and good weather conditions).

Generally, UAVs are used across the world for various civil applications in search and rescue, surveillance, journalism, and agriculture among others. According to Sherstjuk et al. [24], UAVs should fulfill some requirements for their autonomous operation in forest areas for early detection of wildfires:

- all-weather suitability
- self-localization
- navigational autonomy
- cooperation
- payload
- availability

Ideally, all UAVs should perform their missions around the clock even in the most difficult weather conditions. At this time, most UAVs are ranked according to max wind resistance, operating temperature, waterproof and magnetic interference resistance. This requirement is quite important and should be taken into account during the determination of an autonomous system based on UAVs.

3. Requirements of an early forest fire detection system

An early forest fire detection system should be able to notify firefighters as soon as possible in order to minimize the fire caused damage. Moulianitis et al. [32] defined the major requirements of an autonomous early forest fire detection system, as follows:

- robust continuous monitoring of the forest area (CMO)
- fast detection of fire (FDF)
- determination of the exact location of fire (ELF)
- early notification (ENO)
- minimization of faulty alarms (FA)

as well as all the following abilities: configurability, interaction ability, dependability, motion ability, perception ability and decisional autonomy, which should be incorporated in these kinds of forest monitoring systems. More specifically, configurability allows a forest system to be configured for every forest environment. Interaction ability enables the secure, non-faulty communication between components of system in order for fire fighters to monitor and specify the location of fire sources on time. Dependability specifies the level of trusting upon the system. Motion ability determines the capability of system to move towards fire, ensuring minimum flight time and consumption. Perception ability determines the level

Abilities \ Requirements	CMO	FDF	ELF	ENO	FA
Configurability	✓			✓	✓
Interaction Ability			✓	✓	✓
Dependability	✓				✓
Motion Ability	✓		✓	✓	✓
Perception Ability	✓	✓	✓	✓	✓
Decisional Autonomy	✓	✓	✓	✓	✓

Table 1.
Abilities of a potential early forest fire detection system [32].

of autonomy on fire detection and its exact location. Decisional autonomy corresponds to the capability of system to verify if an incident of a potential fire ignition is real and not a false estimation so as to notify local fire brigade. Finally, the above capabilities are classified to the requirements as follows, in **Table 1**.

3.1 Wildfire management and risk assessment

The increased amount of resources allocated in fire suppression, and yet the inability to prevent the catastrophic megafires, imposes the need to rethink the problem and pay serious attention to fire prevention and quick-fire detection. An important step towards a holistic approach in wildfire management is the identification of the most vulnerable areas and ecosystems, which can then be managed in a way that will prevent the evolvement of an ignition incident into a megafire. Advanced applications and methods which involve geographic information systems, remote sensing data and methods, geospatial statistics, existing knowledge on fire behavior in various fuel types, reliable records of the pyric history and weather data offer a great tool to land managers to plan for fire management under any scenario.

Wildland urban interfaces form particularly vulnerable areas today for the reasons described above. Although in these areas fires rarely turn into megafires, they inflict serious damage to human properties and often cost human lives. Various methods have been proposed to assess the relevant vulnerability of WUI to wildfires. Some of them focus on landscape structure, studying the vegetation and build environment spatial patterns, using typical methods of landscape ecology and landscape analysis [33]. Such methods, especially when they involve the existence of escape routes or fire barriers, can provide significant service in planning more fire resistant settlements in the WUI. Molina et al. in [34] proposed an ignition index for application in the WUI which integrates fuel components, such us fine fuel moisture content, physiographic parameters, weather data and flammability of vegetation components which results in a reliable estimation of the potential risk of fire and, therefore, it can help in prioritizing areas for strict protection and budget allocation.

The Fire Weather Index (FWI), a component of the Canadian Forest Fire Danger Rating System (CFFDRS) is another widely used method for estimating fire risk. It has been tested in several fire prone areas across the globe and, with some adjustments related to the peculiarities of each region, it provides relatively accurate estimations on the possibility of a wild fire [35, 36]. The FWI integrates some important aspects of wildfires regarding fuel moisture, weather conditions and fire behavior, resulting in an index which can then be classified into danger classes from low to extreme. This knowledge can be used in order to increase or decrease the degree of preparedness and alert.

Remote sensing data and methods are also often employed in an effort to manage wildfires in an effective manner, by providing mapping products with high spatial accuracy and predictive value in relation to potential fire behavior and fire risk [37–39]. Today, there is wide availability of remote sensing data and methods and many of them are offered free of charge. Sentinel 2 data at a spatial resolution of up to 10 m can be employed for land mapping at a high spatial scale, while its temporal resolution of 5 days ensures updatability of the resulting products. Landsat data on the other Hand, apart from landcover mapping, they can also be employed for the calculation of surface temperatures and moisture, both important determinants of fire behavior, during the vulnerable period.

Under the situation of increased fuel load and increased potential for intensive hot fires, the development and use of accurate tools for early assessment of fire risk and the potential behavior of fire is of particular importance. It could lead to the adoption of appropriate measures for managing the most vulnerable areas, towards decreasing the fuel load or developing appropriate strategies for the effective suppression of fire. Therefore, accurate fire propagation models can be used in the operational support of forest fires suppression, in the development of fire propagation scenarios, in the training of volunteer fire fighters, in the planning of actions to be taken by Civil Protection Agencies and in the decision support of local competent authorities. The fire behavior simulation models FARSITE [40] and Flammap [41, 42] are employed in many studies both for assessing potential fire behavior and fire danger with promising results. FARSITE is a two dimensional model which simulates fire behavior in both space and time under varying site and weather conditions. It is based on [43] fire spread model while it incorporates various other models from the international literature that deal with other aspects of fire behavior such as spotting, fire spread of ground and crown fires, etc. The great advantage of this model is that it allows the real time simulation of fire behavior, while at the same time it allows the simulation of the fire-fighting tactics and forces. The results of fire behavior simulation with FARSITE are spatial and non-spatial data regarding fire intensity and spread, flame height and others.

Unlike FARSITE, Flammap is a one-dimensional model which simulates fire behavior only in space independently of the prevailing weather and other conditions during the event. While FARSITE allows the simulation of fire behavior under real conditions, with the advantages mentioned above, Flammap allows the identification of potential hotspots with extreme fire behavior, in terms of intensity and rate of spread. Such hotspots might not be detected properly using FARSITE simply because at the particular time that fire passes from them, the weather conditions, among others, may not favor a fire with extreme behavior. Such spots, however, might be the ones that need careful management for fire prevention and future protection of an area.

It is now widely accepted that the effective management of wildfires needs the integration of several disciplines, including forest and landscape ecology, fire ecology, pyrology, environmental modeling, remote sensing and others in a supplementary manner. Furthermore, particular attention needs to be paid in the pyric history of a region, local knowledge, historical land uses and current trends in order to unravel the mysteries of wildfire and increase the effectiveness of fire prevention and fire suppression. The aim of an effective wildfire management strategy should aim not in the complete elimination of wild fires, which is practically impossible, but in the restriction of their ecological, economic and social cost.

3.2 Image/video-based fire/smoke detection algorithms

In most existing monitoring systems, the following problems should be addressed:

- detection of fire/smoke using images captured from a stationary camera
- identification of fire/smoke using images captured from the camera of a UAV

In both problems, images should be analyzed for deciding whether there is a fire or smoke in the scene in order appropriate actions be activated. The choice of the cameras type is of vital importance for the system's operation.

The most used types of cameras is optical or thermal. There are many well-known techniques in the literature that use one of these type of cameras [44–47] for solving the above-mentioned problems. However, in order to obtain a more accurate and robust system, that can detect and identify fire and smoke, under general photometric conditions, in both daylight and night, the joint use of the optical and thermal cameras is proposed. The aforementioned problems can be decomposed into the following strongly related sub-problems:

1. Pixel-based segmentation using temperature and color
2. Pixel-based segmentation by exploiting the temporal information (video)
3. Compactification of segmented regions using spatial information
4. Compactification of segmented regions by exploiting the temporal information and finally
5. The classification using image/video learning-based techniques

that we are going to analyze in the following subsections.

3.2.1 Pixel-based image segmentation using temperature and color

In this section, we are going to describe an image segmentation method, based on temperature and color images, respectively. This method can be applied independently:

- in each pixel of thermal image for the fire detection problem and
- in each pixel of the optical image for both smoke and fire detection

3.2.1.1 Thermal camera-based detection

Thermal camera is used mainly for fire detection and identification, by exploiting the high temperature of fire and its variability with respect to other objects existing in the forest scene. If the flame is "visible" from the thermal camera, it produces high intensity values in the infrared spectra. Thus, by applying a simple hard thresholding rule on each pixel of the image, an intensity-based segmentation into its "hot" and "cold" regions can be easily achieved.

3.2.1.2 Optical camera-based detection

In most real fire and smoke detection systems, optical cameras are used. Most well-known fire and smoke detection techniques are color-based. The basic idea behind these approaches is to adopt a color model for extracting pixels with high probability of being fire or/and smoke. Many color models are well known and have been proposed in the literature. However, the most used, that we are also going to adopt here, are RGB, HSI and YCbCr.

- For flame color identification the use of RGB and HIS color space is proposed in [45, 46]. However, since the most distinctive flame's feature is its chromaticity and not its luminance, the use of YCbCr color space is also proposed in [48], where the separation between luminance and chrominance is high.

- A simple well-known rule which is used for the solution of the smoke detection problem is the following one:

    - High luminance and low chrominance in an image indicate probable smoke's presence [47].

Based on that rule in [49] the use of the RGB color model rules along with a simple fuzzy classification scheme for smoke color segmentation, is proposed.

The above-mentioned techniques can, safely, be used for pixel-based classification in both UAV's and Stationary camera, are easily programmable and perform on a real time basis. The joint use of optical and thermal camera results in a system with high accuracy in the flame detection. However, the unsatisfactory results in smoke classification, which is of vital importance for the system because of smoke's early appearance in camera's view, makes the use of temporal information essential for the system.

The main limitation of optical camera is its behavior in illumination variation during different times of the day. Smoke for example, which is an essential factor for early fire detection, because of its faster appearance in camera's view, is not visible during night, due to the low luminance of the scene. However, this fact is compensated by the improvement of the sensitivity, due to low luminance and temperature, in both modalities in the fire case.

3.2.2 Pixel-based segmentation using temporal information

The shortcomings of IR imaging in long distances and the fact that color information cannot be used by itself to detect fire, because of the variability in color, density, lighting and background, make video fire detection techniques an important component of the system.

The first step in many object detection and tracking algorithms, which is also useful in fire and smoke detection case, is the background subtraction. This step is vital for the efficient operation of the next steps and the whole algorithm. Background subtraction is used both in fire and smoke detection, the output of the algorithm which is the moving parts of the image are further analyzed by color and temperature detection for the presence of fire or smoke.

There are many techniques used for background subtraction, the most known of them are the following:

- Running Gaussian average

- Temporal median filter

- Mixture of Gaussians

- Kernel density estimation

- Sequential Kernel density approximation

- Eigen-backgrounds

Each one of these techniques has different requirements in memory and complexity. Most of them are based on a model in which every pixel is considered as statistically independent from the others. The basic steps in all background subtraction methods are the following:

1. background estimation and

2. pixel's classification in background/foreground.

Because in a forest the background changes dynamically with time, many background objects such as trees and leaves are moving. Thus, the most common choice in such cases is the Mixture of Gaussian model, proposed in [50] which can adapt in multiple backgrounds with high performance. According to this technique, every pixel's intensity is modeled by a mixture of Gaussian distribution, composed by κ different gaussian distributions. Each one of the κ distributions describes a different object of the scene that can be classified as background or foreground. An important issue in background subtraction techniques is the selection of the parameters to be updated. In the smoke detection case, it is important for the system to operate robustly at day and night, independent of weather changes and at different seasons, so it is necessary to update the parameters based on the scene changes, in such a way that the smoke occurrence is not learned as background.

3.2.3 Compactification of segmented regions using spatial information

3.2.3.1 Pixel-based case

Ideally, the result of the image segmentation algorithms (independently if they come from color or temperature image processing) we would like to be compact binary images that classify the fire/non-fire and smoke/non-smoke like pixels. Although a certain number of non-fire or/and non-smoke pixels can be removed using simple nonlinear filtering, it is still difficult to solve all possible problems that can be inhered using just color and background information (e.g., such a difficult problem is to distinguish a moving object that has a similar color to a real fire). Therefore, these binary masks are usually analyzed by exploiting somehow the spatial information. To this end, for the identification of the connected components of these pixels, morphological operators and graph-based techniques are used [51]. The simplest way to reduce the false alarms in both fire and smoke detectors, is to eliminate isolated pixels that are falsely classified as fire and smoke, respectively. This idea sounds good in most cases, except special scenarios, regarding fires that can be captured in just a few pixels of image because of their very long distance from the optical or thermal sensor. In such a case, the elimination of isolated pixels could be achieved by using a median filter in the segmented mask obtained from the above-mentioned methods. However, because of the median filtering:

- the boundaries of the fire are smoothed, and

- denoising emerges some non-fire regions to be falsely detected as fire ones.

In order to overcome both problems, the inhered connectivity of the fire is imposed, that is only the pixels corresponding to a region with area larger than a specified threshold are kept. However, in many cases fire is not fully connected and in practice, the areas of different objects in the scene are ordered and further analyzed using different descriptors [52, 53]. Concerning the classification of flames, since they have varying colors even within a small area, spatial color difference analysis [51] that focuses on this characteristic, can be used for the separation of fire-colored objects from true fires. Specifically:

- range filters [54],
- variance/histogram analysis [55] and
- spatial wavelet analysis [53],

are some of the commonly used tools for analyzing the spatial color variations in fire's pixel values.

On the other hand, for the smoke classification problem these techniques are not always efficient for the following reasons:

- smoke regions often do not show high spatial color variation as flames do,
- textured smoke-colored moving objects are difficult to distinguish from smoke thus increasing the false alarms.

In general, smoke in a fire is gray and this reduces the color variation in the background. Therefore, in YUV color space the reduction of the dynamic range of chrominance color components U and V after the appearance of smoke in the field of view of camera, is expected.

3.2.3.2 Block-based case

The spatial information can be also exploited in a block-based way. To this end images are cut in blocks of size *MxM* and properly annotated, thus forming a dataset. This dataset can be used to further test the color-based algorithms, in a block-based classification scenario. To this end, the following M^2 hypotheses:

H_i:{*i* of M^2 fire/smoke pixels are enough for classifying the candidate block as a fire/smoke one, i = 1,2,..., M^2.

are formed and the classification problem can be easily solved as a classical hypothesis testing problem.

3.2.4 Compactification of segmented regions using temporal information

A dynamic texture or pattern in video, such as smoke, flames, water and leaves in the wind can be simply defined as a texture with motion [47]. Although dynamic textures are easily detected by human eyes, they are difficult to discern using computer vision methods as the spatial location and extent of dynamic textures can vary with time and they can be partially transparent. Dynamic texture techniques are also applied to the flame and smoke detection problem [47]. Ordinary moving objects in video, such as walking people, have a pretty stable and almost periodic motion over time, thus they can be easily separated from fire or smoke. On the other hand, flame and smoke regions exhibit chaotic boundary

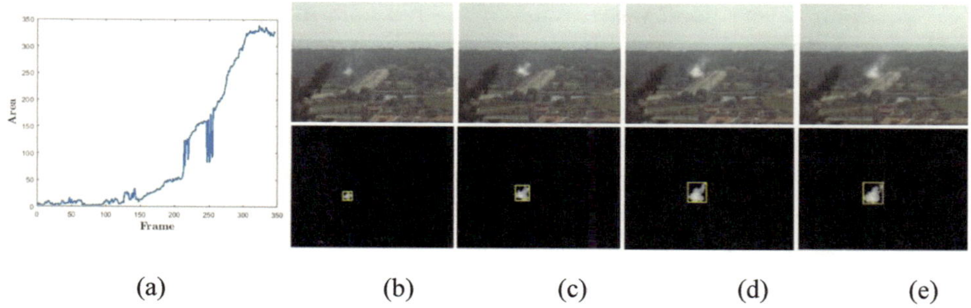

Figure 1.
Change in area of smoke's region as time evolves (a) in successive video frames (b–f).

contours. Therefore, disorder analysis of boundary contours of a moving object is useful for fire and smoke detection. Some well-known metrics that are used for this reason are:

- randomness of area size [54, 55]
- boundary roughness [52] and
- boundary area disorder [56].

The above-mentioned metrics, can be combined with any well-known tracking method to follow the motion of the supposed fire (or smoke), in order to classify it. More specifically, in each frame of the given sequence of images a clustering-based scenario can be applied, thus forming several possible classes and:

- a linear Kalman filter, or
- an extended Kalman, or even
- a particle filter

based tracking scenario can be used for solving the detection problem.

In **Figure 1(a)**, we can see the change in area of smoke as time evolves in a smoke region in a video. It is evident that the smoke area drastically increases in successive video frames (b–f) and this fact can be used for its separation from other moving objects. The linear Kalman filter can be used for that purpose and its performance is expected to be excellent. Other factors that can be exploited for the smoke's separation problem are the wind's velocity and its direction. These quantities can be safely used in a tracking system for predicting the smoke's motion.

3.2.5 Image/video learning-based techniques for classification

In this section, we are going to concentrate ourselves on linear and nonlinear learning-based techniques [57–63] in order to solve the above defined detection problems. In particular:

- A linear discriminator, tailored to the problem at hand, is a PCA based one [59, 62] with its performance in both fire and smoke detection problems to be very promising. In particular, the performance of the technique in the fire detection problem is very good (sensitivity: 95%—false alarms: 29%), while its performance in the smoke detection problem can be considered as promising (sensitivity: 85%—false alarms: 30%).

- Other powerful classification techniques such as KNN and SVM [63, 64] can be used for solving the detection problems with SVM-based classifier performing better. More precisely, the results we obtained for the fire detection problem were excellent (sensitivity: 99.5%—false alarms: 1%) while the corresponding for the smoke detection problem were quite satisfactory (sensitivity: 85%—false alarms: 30%).

- Finally, recently deep convolutional neural networks have been used [65, 66] for solving successfully the forest fire detection problem and the obtained results are promising.

Some results on images for the fire detection obtained from the application of the block-based PCA and SVM classifiers are shown in **Figure 2**.

3.2.6 A video learning-based technique

The above-mentioned methods can also be applied for the classification problem on a sequence of images (video). They can easily be generalized by using spatio-temporal 3-D blocks, though we could also extract the necessary information from the video using region covariance descriptors of the blocks [47] for its efficient solution. A popular approach for the classification of the multi-dimensional feature vectors obtained from each candidate flame or smoke blocks is again based on SVM classifiers, typically using Radial Basis Function (RBF) kernels. Many frames of fire and non-fire video sequences must be used for training these SVM classifiers, otherwise the number of false alarms (false positives or true negatives) might be significantly large [67].

3.3 Navigational autonomy

Autonomous navigation of UAV depends mainly on drone's ability to localize itself. In outdoor applications, UAVs can execute used-defined waypoint mission, using geo-localization of their spatial positions. This GPS approach is ideal for missions with limited presence of obstacles and tolerable small error in positioning accuracy. Nevertheless, this GPS localization method is not functional in indoor environments or outdoor environments in which navigation is heavily based

Figure 2.
Fire detection using the block-based PCA approach (top row) and block-based SVM approach (bottom row).

on local events. To deal with this problem, simultaneous localization and mapping (SLAM) techniques are employed. SLAM is the computational problem of constructing or updating a map of unknown environment while simultaneously keeping track of a robot's location within it. Several works have been published in that field [25], but it remains an open research area. Yang et al. [68] proposed an algorithm for the detection of landing sites, using monocular camera. An alternative SLAM approach was proposed by [69], in which SLAM algorithm estimation comes from fusion of data such as GPS, orientation sensor and monocular camera. Regarding the applications of SLAM in forest environments with UAVs, there is a scarcity of research footprint, which either highlights a new research area or confirms that UAV forest monitoring is quite functional with GPS localization accuracy.

Despite self-localization attribute of UAVs, autonomous navigation is based on motion planning. Given that UAVs are able to follow user defined waypoints, an outdoor autonomously navigated UAV should be able to generate routes (a list of waypoints) between current position and target position, ensuring a collision free route automatically. This can be achieved through motion planning, a process of breaking down a desired movement task into discrete motions that satisfy movement constraints and possibly optimize some aspect of the movement. In literature, there are many UAV 3D path planning algorithms, which can be divided in five categories: (a) Exact and approximate cell Decomposition, (b) control-based methods, (c) potential fields, (d) bioinspired algorithms and (e) randomized planning.

Exact and approximate cell decomposition technique splits the workspace into discrete cells corresponding to the obstacle free portion of the environment, resulting a graph roadmap, where the vertices represent the individual cells and edges indicate adjacency among the cells. Control-based methods are based on motion equations of drone, in order to navigate a drone along a specified trajectory. Potential fields represent obstacles with vector repulsive forces and goal position with a vector attractive force. So, the drone navigates in workspace using gradient descent to follow potentials to the goal. Bio-inspired algorithms originate from mimicking biological behavior to deal with problems. This path planning method leaves out the process of constructing complex environment models, and proposes a strong searching method to converge to the goal stably. Finally, randomized planning is achieved through random generation of nodes in a graph, such as rapidly exploring random tree [70] and probabilistic roadmap [71]. In literature, there are many variants of randomized planning algorithms which incorporate optimal search algorithms, such as Dijsktra's algorithm [72], A* [73] and D* [74] respectively.

Despite the great contribution of motion planning in navigational autonomy, there is a trade-off between effectiveness and energy consumption. There are two ways to execute a motion planner, offline and online. The first one requires a computer, in which motion planner is executed, and a telemetry link between computer and UAV. This method is ideal for small outdoor areas due to limited range of antenna radio waves. The latter one includes an installed mini-computer on drone, which will be able to generate routes without any transmission link. Online motion planning is quite effective for long-distance routes, but requires more energy for the operation of installed computer.

Another set of important parameters for an efficient outdoor path planning algorithm, is time and energy indexes. Sometimes, minimum time paths come from minimum Euclidean distance between start and goal position, while some others do not. This happens due to wind forces which impede the waypoint mission execution. To handle that problem, [75–78] proposed path planning approaches which extends path planning capabilities to a wind efficient navigation model, which is both time and energy optimal, while extend the capabilities of outdoor motion planning based on GPS positioning as Prasad et al. [79]. Thanellas et al. [77] show among others, that wind information can offer energy efficient paths, which are time optimal. This technique can be proved beneficial to outdoor environments in which drones execute waypoint missions without any knowledge of environmental factors which drain battery level.

These works prove that wind aware path planning methods are beneficial to outdoor environments. However, there is a problem of real-time wind prediction. More specifically, there are wind predictions which are not updated on time to predict wind variances for optimal wind path planning. Oettershagen et al. [15] support that onboard processing of a wind map can minimize the delay of wind data acquisition. There is still a lot of work to be done, but wind path planning is for sure an important parameter for long paths.

Sometimes the effective covering of a forest area depends from the multiple monitoring or sequential deployment of UAVs. In these cases, UAVs should be able to coordinate their behavior and to cooperate with each other in order to solver their tasks optimally. Additional information that aims the coordination of the UAVs can be provided by static cameras or other static sensors. The information of the sensors can be used to update the flight plans of the UAVs towards better covering and early detection of wildfire incidents.

The UAVs should be able to carry all required sensors and systems for fire perception purposes. Optical and thermal cameras are the most common sensors used to detect smoke and fire. On board computers are used for signal processing and feature recognition. Light weight weather stations are mounted on the drone in order to detect locally the direction and the velocity of the wind in order to adapt the flight plan of the UAV for energy and time efficiency.

All UAVs should be equipped with onboard communication devices that guarantee receiving commands from a ground command center, sending information back to it, as well as exchanging information with the other UAVs.

4. Conclusions

In this chapter, a brief history of fire and forest fire protection systems was presented. The basic forest fire monitoring systems technologies were reviewed and satellite, WSN and optical/thermal cameras based systems were emphasized. In addition, the vital requirements of an autonomous early forest fire detection system, its main modules and methods for wildfire management and risk assessment, smoke and fire detection based on images and video as well as navigational autonomy issues for UAVs were also presented.

Acknowledgements

This research was supported by the EU and Greek State on behalf of the program "Forest Monitoring System for Early Fire Detection and Assessment in the Balkan-Med Area (SFEDA)", MIS: 5013503.

Author details

Nikos Aspragathos[1], Eleftherios Dogkas[1], Pavlos Konstantinidis[2], Panagiotis Koutmos[1], Nefeli Lamprinou[3], Vassilis C. Moulianitis[4], Georgios Paterakis[1], Emmanouil Z. Psarakis[3]*, Evangelos Sartinas[3], Konstantinos Souflas[1], Georgios Thanellas[1], Georgios Tsiourlis[2], Nikitas Xanthopoulos[1] and Panteleimon Xofis[5]

1 Department of Mechanical Engineering and Aeronautics Department, University of Patras, Patras, Greece

2 NAGREF Forest Research Institute, Thessaloniki, Greece

3 Department of Computer Engineering and Informatics, University of Patras, Patras, Greece

4 Department of Product and Systems Design Engineering, University of the Aegean, Syros, Greece

5 Department of Forestry and Natural Environment Management, Eastern Macedonia and Thrace Institute of Technology, Drama, Greece

*Address all correspondence to: psarakis@ceid.upatras.gr

References

[1] Pausas JG, Keeley JE. A burning story: The role of fire in the history of life. Bioscience. 2009;**59**:593-601

[2] Edwards D. Fire regimes in the biomes of South Africa. In: Booysen Pd, Tainton NM, editors. Ecological Effects of Fire in South African Ecosystems. Berlin-Heidelberg: Springer-Verlag; 1984. pp. 19-37

[3] Kruger LM, Midgley JJ, Cowling RM. Resprouters vs reseeders in South African forest trees; a model based on forest canopy height. Functional Ecology. 1997;**11**:101-105

[4] Whelan RJ. The Ecology of Fire. Cambridge: Cambridge University Press; 1995

[5] Trabaud L. Postfire plant community dynamics in the Mediterranean Basin. In: Moreno JM, Oechel WC, editors. The Role of Fire in Mediterranean-Type Ecosystems. Vol. 107. New York: Springer-Verlag; 1994. pp. 1-15

[6] Bond WJ, Keeley JE. Fire as a global 'herbivore': The ecology and evolution of flammable ecosystems. Trends in Ecology and Evolution. 2005;**20**:387-394

[7] Bond WJ, Woodward FI, Midgley GF. The global distribution of ecosystems in a world without fire. New Phytologist. 2005;**165**:525-538

[8] Naveh Z. Fire in the Mediterranean—A landscape ecological perspespective. In: Fire in Ecosystems Dynamics. 1990. pp. 1-20

[9] Naveh Z, Dan J. The human degradation of Mediterranean landscapes in Israel. In: Di Castri F, Mooney HA, editors. Mediterranean Type Ecosystems. Origin and Structure. Vol. 7. Heidelberg, Berlin: Springer-Verlag, New York; 1973. pp. 373-390

[10] Blondel J, Aronson J. Biodiversity and ecosystem function in the Mediteranean Basin: Human and non-human determinants. In: Davis GW, Richardson DM, editors. Mediterranean-Type Ecosystems. The Function of Biodiversity. Vol. 109. Berlin, Heidelberg: Springer-Verlag; 1995. pp. 43-119

[11] Dimitrakopoulos AP, Vlahou M, Anagnostopoulou CG, Mitsopoulos ID. Impact of drought on wildland fires in Greece: Implications of climate change? Climatic Change. 2011;**109**:331-347

[12] Pausas JG, Fernandez-Munoz S. Fire regime changes in the Western Mediterranean Basin: From fuel-limited to draught-driven fire regime. Climatic Change. 2012;**110**:215-226

[13] Koutsias N, Xanthopoulos G, Founda D, Xystrakis F, Nioti F, Pleniou M, et al. On the relationships between forest fires and weather conditions in Greece from long-term national observations (1894-2010). International Journal of Wildland Fire. 2013;**22**:493-507

[14] Turco M, Bedia J, Di Liberto F, Fiorucci P, von Hardenberg J, Koutsias N, et al. Decreasing fires in Mediterranean Europe. PLoS One. 2016;**11**(3):e0150663

[15] Vacchiano G, Garbarino M, Lingua E, Motta R. Forest dynamics and disturbance regimes in the Italian Apennines. Forest Ecology Management. 2017;**388**:57-66

[16] Kontoes C, Keramitsoglou I, Papoutsis I, Sifakis NI, Xofis P. National scale operational mapping of burnt areas as a tool for the better understanding of contemporary wildfire patterns and regimes. Sensors. 2013;**13**:11146-11166

[17] Xofis P, Poirazidis K. Combining different spatio-temporal resolution images to depict landscape dynamics and guide wildlife management. Biological Conservation. 2018;**218**:10-17

[18] Keeley JE, Fotheringham CJ, Morais M. Reexamining fire suppression impacts on brushland fire regimes. Science. 1999;**284**:1829-1832

[19] Joint Research Center. Forest fires in Europe 2007. EU. 2008

[20] Founda D, Giannakopoulos C. The exceptionally hot summer of 2007 in Athens, Greece—A typical summer in the future climate? Global Planetary Change. 2009;**67**:227-236

[21] Tolika K, Maheras P, Tegoulias I. Extreme temperatures in Greece during 2007: Could this be a "return to the future"? Geophysical Research Letters. 2009;**36**:1-5

[22] Alkhatib A. A review on forest fire detection techniques. International Journal of Distributed Sensor Networks. 2013;**10**:1-12

[23] Gachoki NM, Muhia A, Kiio MN. A review of quad-rotor UAVs and their motion planning. In: Sustainable Research and Innovation Conference. 2017. pp. 117-121

[24] Sherstjuk V, Zharikova M, Sokol I. Forest fire monitoring system based on UAV team, remote sensing, and image processing. In: IEEE 2nd International Conference on Data Stream Mining & Processing (DSMP). 2018. pp. 590-594

[25] Li J, Bi Y, Lan M, Qin H, Shan M, Lin F, et al. Real-time simultaneous localization and mapping for uav: A survey. In: Proceedings of International Micro Air Vehicle Competition and Conference. 2016. pp. 237-242

[26] Sudha BS, Yogitha HR, Sushma KM, Bhat P. Forest monitoring system using wireless sensor network. International Journal of Advances in Scientific Research and Engineering. 2018;**4**:127-130

[27] Toledo-Castro J, Caballero-Gil P, Rodríguez-Pérez N, Santos-González I, Hernández-Goya C, Aguasca-Colomo R. Forest fire prevention, detection, and fighting based on fuzzy logic and wireless sensor networks. Complexity. 2018;**2018**:1-17

[28] Xu YH, Sun QY, Xiao YT. An environmentally aware scheme of wireless sensor networks for forest fire monitoring and detection. Future Internet. 2018;**10**(10):102

[29] Wang Y, Dang G. Forest fire monitoring system and energy saving algorithm. In: 14th Int. Conf on Computational Intelligence and Security. 2018. pp. 273-276

[30] Kadri B, Bouyeddou B, Moussaoui D. Early fire detection system using wireless sensor networks. In: 2018 International Conference on Applied Smart Systems (ICASS). IEEE;2018. pp. 1-4

[31] Stipaničev D, Vuko T, Krstinić D, Štula M, Bodrožic L. Forest Fire Protection by Advanced Video Detection System—Croatian Experiences. Citeseer; 2009

[32] Moulianitis VC, Thanellas G, Xanthopoulos N, Aspragathos NA. Evaluation of UAV based schemes for forest fire monitoring. In: International Conference on Robotics in Alpe-Adria Danube Region. 2018. pp. 143-150

[33] Marzano R, Camia A, Bovio G. Wildland-urban interface analyses for fire management planning. In: General Technical Report PSW-GTR-208. 2008

[34] Molina JR, Martin T, Silva FRY, Herrera MA. The ignition index based on flammability of vegetation improves planning in the wildland-urban interface: A case study in Southern Spain. Landscape and Urban Planning. 2017;**158**:129-138

[35] Dimitrakopoulos, Bemmerzouk AM, Mitsopoulos ID. Evaluation of the Canadian fire weather index system in an eastern Mediterranean environment. Meteorological Applications. 2011;**18**:83-93

[36] Ager AA, Preisler HK, Arca B, Spano D, Salis M. Wildfire risk estimation in the Mediterranean area. Environments. 2014;**25**:384-396

[37] Keramitsoglou I, Kontoes C, Sykioti O, Sifakis N, Xofis P. Reliable, accurate and timely forest mapping for wildfire management using ASTER and hyperion satellite imagery. Forest Ecology and Management. 2008;**255**:3556-3562

[38] Pan J, Wang W, Li J. Building probabilistic models of fire occurrence and fire risk zoning using logistic regression in Shanxi Province, China. Natural Hazards. 2016;**81**:1879-1899

[39] Sanchez SY, Martinez-Grana A, Frances FS, Picado MM. Mapping wildfire ignition probability using sentinel 2 and LiDAR (Jerte Valley, Cáceres, Spain). Sensors. 2018;**18**:826. DOI: 10.3390/s18030826

[40] Finney MA. FARSITE: Fire area simulator—Model development and evaluation. In: Res. Pap. RMRS-RP-4. Ogden, UT: U.S. Department of Agriculture, Forest Service, Rocky Mountain Research Station; 1998. p. 47

[41] Stratton R. Assessing the effectiveness of landscape fuel treatments on fire growth and behaviour. Journal of Forestry. 2004;**102**:32-40

[42] Finney MA. An overview of FlamMap fire modeling capabilities. In: USDA Forest Service Proceedings RMRS-P-41. 2006

[43] Rothermel RC. A mathematical model for predicting fire spread in wildland fuels. In: USDA Forest Service Research Paper INT-115. Ogden, Utah. 1972

[44] Xu Z, Xu J. Automatic fire smoke detection based on image visual features. In: Computational Intelligence and Security Workshops. IEEE; 2007. pp. 316-319

[45] Celik T, Demirel H, Ozkaramanli H, Uyguroglu M. Fire detection using statistical color model in video sequences. Journal of Visual Communication and Image Representation. 2007;**18**(2):176-185

[46] Chen TH, Wu PH, Chiou YC. An early fire-detection method based on image processing. In: International Conference on Image Processing (ICIP'04. 2004). Vol. 3. IEEE; 2004, October. pp. 1707-1710

[47] Çetin AE, Dimitropoulos K, Gouverneur B, Grammalidis N, Günay O, Habiboğlu YH, et al. Video fire detection—Review. Digital Signal Processing. 2013;**23**(6):1827-1843

[48] Celik T, Demirel H. Fire detection in video sequences using a generic color model. Fire Safety Journal. 2009;**44**(2):147-158

[49] Çelik T, Özkaramanlı H, Demirel H. Fire and smoke detection without sensors: Image processing based approach. In: 15th European Signal Processing Conference. 2007. pp. 1794-1798

[50] Stauffer C, Grimson WEL. Adaptive background mixture models for real-time tracking. In: CVPR. 1999, June. p. 2246

[51] Yuan F. A fast accumulative motion orientation model based on integral image for video smoke detection. Pattern Recognition Letters. 2008;**29**(7):925-932

[52] Toreyin BU, Dedeoglu Y, Cetin AE. Contour based smoke detection in video using wavelets. In: European Signal Processing Conference. 2006

[53] Yasmin R. Detection of smoke propagation direction using color video sequences. International Journal of Soft Computing. 2009;**4**(1):45-48

[54] Qi X, Ebert J. A computer vision based method for fire detection in color videos. International Journal of Imaging. 2009;**2**(S09):22-34

[55] Borges PVK, Izquierdo E. A probabilistic approach for vision-based fire detection in videos. IEEE Transactions on Circuits and Systems for Video Technology. 2010;**20**(5):721-731

[56] Xiong Z, Caballero R, Wang H, Finn AM, Lelic MA, Peng PY. Video-based smoke detection: Possibilities, techniques, and challenges. In: IFPA, Fire Suppression & Detection Research & Applications—A Technical Working Conference. 2007

[57] Ding C, He X. K-means clustering via principal component analysis. In: 21st International Conf on Machine Learning. ACM; 2004. p. 29

[58] Ding C, Zhou D, He X, Zha H. R1-PCA: Rotational invariant L1-norm principal component analysis for robust subspace factorization. In: 23rd International Conference on Machine Learning. ACM; 2006, June. pp. 281-288

[59] Aharon M, Elad M, Bruckstein A. K-SVD: An algorithm for designing overcomplete dictionaries for sparse representation. IEEE Transactions on Signal Processing. 2006;**54**(11):4311

[60] Tropp JA, Gilbert AC. Signal recovery from random measurements via orthogonal matching pursuit. IEEE Transactions on Information Theory. 2007;**53**(12):4655-4666

[61] Zhang T, Ghanem B, Liu S, Xu C, Ahuja N. Low-rank sparse coding for image classification. In: IEEE International Conference on Computer Vision. 2013. pp. 281-288

[62] Rosas-Romero R. Remote detection of forest fires from video signals with classifiers based on K-SVD learned dictionaries. EAAI. 2014;**33**:1-11

[63] Cheng G, Han J. A survey on object detection in optical remote sensing images. ISPRS Journal of Photogrammetry and Remote Sensing. 2016;**117**:11-28

[64] Russo AU, Deb K, Tista SC, Islam A. Smoke detection method based on LBP and SVM from surveillance camera. In: International Conf. on Computer, Communication, Chemical, Material and Electronic Engineering, IEEE. 2018. pp. 1-4

[65] Zhang QX, Lin GH, Zhang YM, Xu G, Wang JJ. Wildland forest fire smoke detection based on faster R-CNN using synthetic smoke images. Procedia Engineering. 2018;**211**:441-446

[66] Muhammad K, Ahmad J, Lv Z, Bellavista P, Yang P, Baik SW. Efficient deep CNN-based fire detection and localization in video surveillance applications. IEEE Transactions on Systems, Man, and Cybernetics: Systems. 2018;(99):1-16

[67] Amiaz T, Fazekas S, Chetverikov D, Kiryati N. Detecting regions of dynamic texture. In: International Conference on Scale Space and Variational Methods in Computer Vision. Berlin, Heidelberg: Springer; 2007, May. pp. 848-859

[68] Yang T, Li P, Zhang H, Li J, Li Z. Monocular vision SLAM-based UAV autonomous landing in emergencies and unknown environments. Electronics. 2018;**7**(5):73

[69] Munguía R, Urzua S, Bolea Y, Grau A. Vision-based SLAM system for unmanned aerial vehicles. Sensors. 2016;**16**(3):372

[70] Yang K, Sukkarieh S. Real-time continuous curvature path planning of UAVs in cluttered environments [C]/ mechatronics and its applications. In: 5th International Symposium on IEEE. 2008. pp. 1-6

[71] Yan F, Liu YS, Xiao JZ. Path planning in complex 3D environments using a probabilistic roadmap method. International Journal of Automation and Computing. 2013;**10**(6):525-533

[72] Musliman I A, Rahman A A, Coors V. Implementing 3D network analysis in 3D-GIS. International Archives of ISPRS. 2008;**37**(part B)

[73] De Filippis L, Guglieri G, Quagliotti F. Path planning strategies for UAVs in 3D environments. Journal of Intelligent & Robotic Systems. 2012;**65**(1-4):247-264

[74] Carsten J, Ferguson D, Stentz A. 3d field d: Improved path planning and replanning in three dimensions. intelligent robots and systems. In: 2006 IEEE/RSJ International Conference on IEEE. 2006. pp. 3381-3386

[75] Techy L, Woolsey CA. Minimum-time path planning for unmanned aerial vehicles in steady uniform winds. Journal of Guidance, Control, and Dynamics. 2009;**32**(6):1736-1746

[76] Al-Sabban WH, Gonzalez LF, Smith RN, Wyeth GF. Wind-energy based path planning for electric unmanned aerial vehicles using Markov decision processes. In: IEEE/RSJ International Conference on Intelligent Robots and Systems. 2012

[77] Thanellas GA, Moulianitis VC, Aspragathos NA. A spatially wind aware quadcopter (UAV) path planning approach. In: 10th IFAC Symposium on Intelligent Autonomous Vehicles, IAV. 2019

[78] Oettershagen P, Achermann F, Müller B, Schneider D, Siegwart R. Towards fully environment-aware UAVs: Real-time path planning with online 3D wind field prediction in complex terrain. arXiv preprint arXiv:1712.03608; 2017

[79] Prasad G, Vijayaganth V, Sivaraj G, Rajasekar K, Ramesh M, Raj RG, et al. Positioning of UAV using algorithm for monitoring the forest region. In: 2nd International Conference on Inventive Systems & Control. 2018. pp. 1361-1363

Chapter 9

Vehicle Tracking Using Video Surveillance

Sandesh Shrestha

Abstract

In numerous applications including the security of individual vehicles as well as public transportation frameworks, the ability to follow or track vehicles is very helpful. Using computer vision and deep learning algorithms, the project deals with the concept of vehicle tracking in real-time based on continuous video stream from a CCTV camera to track the vehicles. The tracking system is tracking by detection paradigm. YOLOv3 object detection is applied to achieve faster object detection for real-time tracking. By implementing and improving the ideas of Deep SORT tracking for better occlusion handling, a better tracking system suitable for real-time vehicle tracking is presented. So as to demonstrate the achievability and adequacy of the framework, this chapter presents exploratory consequences of the vehicle following framework and a few encounters on handy executions.

Keywords: object detection, YOLOv3, multiple object tracking, Deep SORT, vehicle re-identification

1. Introduction

Innovation has changed the manner in which individuals convey and work with one another. Vehicle tracking frameworks, as the name recommend, permit the following of vehicles in its most fundamental capacity. They utilize a mix of innovations to keep ongoing tabs on the situation of a vehicle or to develop a background marked by where a vehicle has been. These frameworks are utilized in an assortment of enterprises, and they likewise a key piece of most stolen vehicle recuperation methodologies. Vehicle tracking is a significant innovation for the safety of the fleet of vehicles as well as for the conventional driver. This is to turn out to be much progressively significant for road safety as the innovation turns out to be progressively available and cheap. There are two kinds of vehicle tracking, every one of which is valuable in explicit circumstances.

Passive vehicle tracking: these trackers normally utilize a GPS gadget to record the situation of a vehicle after some time. At the point when the tracker is expelled, the information can be moved to a PC and broke down. These following frameworks are valuable for fleet management, yet they likewise have different applications.

Active vehicle tracking: increasingly intricate tracking frameworks transmit the area of a vehicle progressively. For fleet management and dispatch purposes, this information is normally observed from a focal area. This kind of framework can likewise be utilized for stolen vehicle recuperation.

The proposed system tracks vehicles esp. cars on the road or highway by constant surveillance by the CCTV cameras installed at a certain elevation (on the bridge). The object detection algorithm is run on each frame of the video stream obtained in real-time from the CCTV camera. If the object detection system detects a vehicle (car), the system starts tracking each detected vehicle on each frame. The proposed system can also be utilized in the autonomous vehicle industry to escort so particular vehicle. In light of the tracked vehicle, autonomous vehicles can make decisions.

In Computer Vision, tracking is one of the most significant fields. Tracking is the issue of estimating the position of an object over continuous image sequences. This is likewise additionally isolated into two subcategories-single object tracking and multiple object tracking. Both of them two require marginally various methodologies.

1.1 Challenges in tracking systems

It is very crucial to know the challenges we need to take care of during tracking. Some of the common and major challenges are as follows.

- *Object occlusion*: If the target object is occluded or blocked by other objects in a sequence of images, then it not only becomes difficult to detect the object but to update future images too if it becomes visible again.

- *Fast movement*: Cameras, such as on smart-phones, often suffer from jittery movement. This produces a blurred effect and, sometimes, the object becomes completely absent from the frame. Therefore, sudden changes in the motion of cameras also lead to problems in tracking applications.

- *Change of shape*: If we are targeting non-rigid objects, changes in the shape or the complete deformation of an object will often lead to being unable to detect the object and tracking failure.

- *False positives*: In a scene with multiple similar objects, it is hard to match which object is targeted in subsequent images. The tracker may lose the current object in terms of detection and start tracking a similar object.

These challenges can make our tracking application fail and give the wrong estimate of an object's location.

The bigger challenge is to deal with these problems in a real-time scenario. One of the ongoing accessible tracking frameworks is Simple Online and Real-time Tracking (SORT) [1] that attempted to beat these difficulties. It is a straightforward system to track people progressively. SORT uses the Kalman filter on each frame of the video. To solve the association problem in visual tracks, Hungarian calculation is utilized. But their proposed framework is appropriate only for tracking humans in various appearance scenes. This system still involves identity switch problem and does not handle occlusion.

Simple Online and Real-time Tracking with Deep Association Metric (Deep SORT) [2] is an improvement over SORT. It used appearance features from deep convolutional neural network (CNN) for handling occlusion during tracking the people. Deep SORT uses a cost matrix dependent on both motion information and appearance features from the CNN model to abstain from missing tracking as a result of tracking or missed detection of people. Their framework incorporates a convolutional neural system for an individual's evident highlights trained on a person re-identification dataset. This framework also is applicable only for tracking humans.

This paper presents a unique technique for detecting vehicles and tracking them. Using motion features and appearance features, a distinctive approach is provided to track the vehicles in real-time. This framework utilizes a convolutional neural system to obtain the appearance features to track the vehicle (car). A separate vehicle re-identification database has been created to train with an improved CNN model.

1.2 Major contribution

The proposed system ameliorated the tracking performance in the MOT problem. It presents the visual tracking system for vehicles on the road using the similar principles used for human tracking. Given better equipment assets, the proposed framework beats the current best in class frameworks as far as accuracy and performance. The major contributions are as follows.

1. Improved real-time tracking performance by replacing Faster R-CNN by YOLOv3 which is a faster object detector with similar accuracy.

2. Created a vehicle re-identification dataset from scratch necessary to train the CNN model for obtaining appearance features.

3. Modified the CNN model of Deep SORT for better tracking performance.

The remainder of the paper is classified into various areas. Section 2 gives the detailed foundation of past methodologies for MOT frameworks. Section 3 describes the methodology alongside complete design. Segment 4 illuminates experimentation and evaluation. The last segment concludes the proposed framework for vehicle tracking.

2. Background

The research community has begun to focus on tracking every single object in distinct settings with the enhancement in multi-object detection. The entire MOT problem can be considered as an association problem in which the basic goal is to associate the objects detected. Following object detection using some object detection algorithm, tracking is implemented. In this segment, the following systems will be reviewed.

- Object detection algorithms since the first step is to detect the vehicles before tracking.

- Already available tracking systems.

2.1 Review of object detection systems

At the beginning of the 1990s, object detection was completed utilizing template matching algorithms [3], where a format or template of the particular item is slid over the information picture to locate the most ideal match in the info picture. In the late 1990s, geometric appearance-based object detection got the center of attention [4, 5]. The essential spotlight was on the tallness, width, angles, and other geometric properties in these techniques.

During the 2000s, the object identification model was moved to low-level features dependent on some statistical classifiers, for example, local binary design (LBP) [6], oriented gradient histogram [7], scale-invariant feature transform [8], and covariance [9]. Detection and classification of objects based on extraction characteristics engaged machine learning based on extracted features.

Handcrafted traditional characteristics have been used for object detection for many years in the field of computer vision. But, with the advancement in deep learning following the notable results of the challenge of image classification in 2012 [10], CNNs are being used for this purpose. Researchers moved their attention to object detection and classification after the achievement of object classification in [10]. Deep convolutional neural networks operate extremely well in terms of edges, texture, and appearance to extract local and global features.

The research community has shifted in the latest years towards region-based object detection networks. In various apps such as video description [11], this sort of object detection is used. Convolutional characteristics are obtained over suggested areas in region-based algorithms for object detection, followed by region categorization into a particular class.

With the appealing results of AlexNet [10], Girshick et al. [12] suggested the concept of object detection using a convolutional neural network. They used a selective search to propose regions where you can find prospective objects [13]. They called their network of object detection as a neural network (R-CNN) area convolution. The fundamental flow of the neural network area convolution (R-CNN) can be defined as follows.

- Regions are suggested using the selective search [13] for each item in the input image.
- Proposed areas are resized to the same coherent size to classify the proposition into predefined classes based on regional CNN features obtained.
- The softmax layer was substituted by the linear SVM classifier to train the system on fixed-length CNN features.
- Ultimately, a bounding box regressor is used to locate the object perfectly.

Despite being a major breakthrough in the field of object detection, the suggested R-CNN has some important shortcomings.

- Training procedures are quite slow because R-CNN has to train separately in different phases.
- Selective searching, which is itself a slow method, proposes regions.
- It is expensive to train the separate SVM classifier as CNN features are extracted for each region, making SVM training even more challenging.
- Detection of objects is slow because CNN characteristics are extracted for each test image for individual suggestions.

Kaiming He et al. [14] suggested spatial pyramid pooling (SPP) to solve the problem of feature extraction for each proposition. The fundamental concept was to recognize the input of any size by the convolutional layers; fully connected layers force input to be set size to make the multiplication of matrix feasible. Following the last convolutional layer, they used SPP layer to obtain the fixed size characteristics to feed into a fully connected layer. Comprehensively enhanced performance was obtained with SPPNet R-CNN. For suggestions of distinct dimensions, SPPNet extracts convolutional functions on the input picture only once. This network enhances testing efficiency, but it does not enhance R-CNN training performance. In addition, the weights of convolutional layers in front of the SPP layer cannot be altered which limits the method of fine-tuning.

R-CNN's primary contributor, Girshik [15], suggested that Fast-ECNN address some R-CNN and SPPNet issues. Fast R-CNN uses the concept of sharing convolutional computation for various suggested areas. It adds a region of interest (ROI)-pooling layer to create fix-sized features of individual proposals after the last convolutional layer. ROI-pooling layers' fix-size features are supplied to the stack of fully connected layers that further divided into two branch networks: one serves as the network for object classification and the other for bounding box regression. They asserted that R-CNN's general training step efficiency was improved by three times and that for testing was improved by ten times.

While Fast R-CNN has significantly enhanced R-CNN's efficiency, it still utilizes selective search as a regional proposal network (RPN). The Regional Proposal phase consumes the time that operates as the Fast R-CNN bottleneck. Modern advances in object location using deep neural networks [16] have driven Ren et al. [17] to use CNN to replace slow regional proposal processes using selective search. They suggested effective RPN proposals for objects. RPN and Fast R-CNN share, respectively, the convolutional layers for region proposal and region classification in Faster R-CNN. Faster R-CNN is a purely convolutional neural network with no handcrafted features that use a fully convolutional neural network (FCN) for region proposal. They asserted that for the test stage, Faster R-CNN could function at 5 fps.

Redmon et al. [18] suggested a new method for detecting objects called You Only Look Once (YOLO). The region proposal phase was totally dropped; YOLO divides the entire image into grids and predicts detection on candidate regions bases. YOLO splits the entire picture into the grids of $S \times S$. Each grid has a probability of class C, B as the locations of the bounding box, and a likelihood for each box. Removing the RPN step enhances detection efficiency; YOLO can detect the objects while operating around 45 fps in real-time. YOLOv2 [19] and YOLOv3 [20] are the later improvements on YOLO. YOLOv3 processes images at 30 fps; as accurate as other state-of-art object detectors but $3\times$ faster.

2.2 Review of multiple object tracking

Multiple scientists concentrated on motion and spatial characteristics to track multiple objects [21, 22]. To capture the associations between various detections [23, 24], some of the scientists concentrated on appearance features.

There are some traditional techniques that create a frame-by-frame prediction. These traditional methods require multiple hypothesis tracking (MHT) [25] and the JPDAF filter [26]. To track the identified items, both of these ancient methodologies involve a lot of computation. The complexity of these methodologies improves exponentially as the number of trackable objects rises, which makes them very slow to be used in complicated environments for online applications. A single state hypothesis is produced in JPDAF based on the relationship between individual measurement and probability of association. In MHT, consideration is given to a full set of hypotheses for monitoring followed by post pruning for tractability.

By offering JPDA approximation, Rezatofighi et al. [27] made an attempt to enhance the efficiency of JPDAF. They have utilized the latest advances in solving an integer program's m-best solution. This system's primary benefit is to make JPDA less complicated and more tractable. They redefined the technique for calculating an individual JPDAF assignment as a solution to a linear program. Another team of researchers, Kim et al. [23], used features based on appearance to track the target. By pruning MHT's graph to achieve state-of-the-art performance, they improved the MHT. They used the least regularized squares to increase the MHT methodology's effectiveness.

Compared to the legacy implementations, these two improvements perform quite well, but these two techniques still have a great deal of delay in the decision-making phase making these techniques unsuitable for real-time applications. By raising the individual density, these techniques involve big computational resources.

Some scientists have been working on the theory of graphs to track people. Kayumbi et al. [28] suggested a multi-camera viewing algorithm to discover the trajectories of football players based on distributed sensing algorithms. Their algorithm begins with mapping the camera view plane to the floor plane's virtual top perspective. Finally, to track each person on the ground plane, they utilized graph theory.

Some online tracking techniques use individual appearance features to track [29, 30]. These models extract apparent individual look features. Both schemes provide precise descriptors of the appearance to guide the association of data. The first system incorporates individuals' temporal appearance along with the features of the spatial appearance. After tuning the parameters in each iteration, their appearance model is learned by implementing the incremental assessment. Markov decision process (MDP) is used in the second system to map the detected object's age in Markov chain terms. MDP determines the tracks based on the target's present status and history.

Some of the scientists have recently been working on simple online tracking and trying to create live stream tracking in real-time [1, 2]. These schemes are referred to as simple online and real-time tracking, and simple online and real-time tracking with a deep metric association. Two consecutive variants of the same methodology are both of these schemes. Kalman filter is used in both schemes to discover the target's motion features. These systems used as the key tracking parameters like intersection over union, central position, height, width, and velocity. Convolutional features for target appearance are also used in Deep SORT along with movement features to decrease missing tracks in various frames after occlusions and missed detections. This framework has been used only for human tracking.

The proposed approach improves the performance of tracking by redefining the CNN model to decrease the number of identity switches and applies it to track vehicles in real-time.

3. Methodology and framework

Deep SORT is tracking by detection paradigm. In tracking by detection, each frame uses an object detector to find possible instances of objects and then matches those detections with corresponding objects in the preceding frame. In the MOT problem setting, there are many objects to track in each frame. There are two steps to a generic technique of solving this-Detection and Association. In real-world applications, prior bounding box detections are necessary, so tracker could be combined with a detector. First, all the objects in the frame are detected. Single or multiple detections may be available. Once we have frame detections, comparable detections are matched with the prior frame, to get the tracking for an object, the matched frames are carried through the sequence.

This generic method can be further divided into three steps.

1. A technique of identification of objects based on CNN is used to calculate detections. Faster-RCNN is used in the Deep SORT paper [2] to conduct the initial detection on every frame. This scheme uses YOLOv3.

2. An estimation model is an intermediate phase before the data association. This utilizes the status of each track as an eight-quantity vector, i.e., bounding box center (x, y), box height (h), box aspect ratio (γ), and its derivatives with time as velocities. The Kalman filter is used to model these states as a dynamic system. If for a limit of successive frames there is no detection of a tracking object, it is regarded to be out of frame or lost. The new track is initiated for a newly detected box. In addition, the bounding box descriptors are computed using a pre-trained CNN.

3. In the final step, given the predicted states from Kalman filtering using the previous information and the newly detected box in the current frame, an association is made for the new detection with old object tracks in the previous frame. This is computed using Mahalanobis distance between the detection and the position of the track predicted by the Kalman filter and Cosine distance obtained from the appearance feature set.

3.1 System flowchart

The tracking procedure consists of many complex concepts and calculations. All the processes involved are explained below in detail.

3.1.1 State estimation

The tracking scenario is defined on eight-dimensional state space (u, v, γ, h, $\dot{x}$, $\dot{y}$, $\dot{\gamma}$, $\dot{h}$) where (u, v) is the bounding box position, h is the height and γ is the aspect ratio and their respective velocities in image coordinates. A standard Kalman Filter is used with constant velocity motion and a linear observation model is used to solve the velocity components.

3.1.2 Trajectory processing

This mainly says when the trajectory terminates and when a new trajectory is generated. First, for each track, there is a threshold of 30 frames for recording the time from the last successful match to the current time. When the value is greater than the threshold set in advance, the track is deleted. A new trajectory may be generated for the detectors that have no matching success. However, since these detections may be some false alarms, the newly generated trajectory is marked with the state 'tentative'. Then it is observed whether the consecutive matching is successful in the next 3 consecutive frames. If so, it is considered as a new trajectory, marked as 'confirmed', otherwise, it is considered a false trajectory, and the status is marked as 'deleted'.

3.1.3 Motion matching

Matching naturally refers to the match between the currently active track and the current detection. The so-called effective trajectory refers to the trajectories that are still alive, i.e., the trajectories whose states are tentative and confirmed. The degree of motion matching is plotted using the Mahalanobis distance between the detection and the position of the track predicted by the Kalman filter. The Mahalanobis distance takes into account state estimation uncertainty by assessing how many standard deviations the detection is away from the mean track location.

$$d^{(1)}(i,j) = \left(d_j - y_i\right)^T S_i^{-1}\left(d_j - y_i\right) \tag{1}$$

indicates the motion matching degree between the jth detection and the ith trajectory, where S_i is the covariance matrix of the observation space at the current time predicted by the Kalman filter. (y_i, S_i) is the projection of the ith track distribution into measurement space and d_j is the jth bounding box detection.

The 0.95 quantile of the inverse Chi-square distribution is used as the threshold for the Mahalanobis distance to exclude unlikely associations which is denoted by

$$b^{(1)}(i,j) = (1)\left[d^{(1)}(i,j) \le t^{(1)}\right] \tag{2}$$

that evaluates to 1 if the association between the ith track and jth detection is admissible. The corresponding Mahalanobis threshold for four-dimensional measurement space is $t(1) = 9.4877$.

3.1.4 Appearance matching

The use of the Mahalanobis distance matching metric alone can lead to serious conditions such as ID Switch, especially when the camera motion may cause the Mahalanobis distance metric to be invalid, so this time should be remedied by apparent matching. For each detection, including the detections in the track, the deep network is used to extract the feature vector. A gallery is built for each tracking target, storing the feature vectors of the last 100 frames that each tracking target successfully associates with. The second metric is calculated as the minimum cosine distance between the feature set of the last 100 successful associations of the ith track and the feature vector of the jth detection result of the current frame.

$$d^{(2)}(i,j) = min\left[1 - r^T r_k^{(i)} | r_k^{(i)} \varepsilon R_i\right] \tag{3}$$

where r is the appearance descriptor and R is the gallery.

If the above distance in Eq. (3) is less than the specified threshold, then the association is successful. The threshold is obtained from a separate training set.

$$b^{(2)}(i,j) = (1)\left[d^{(2)}(i,j) \le t^{(2)}\right] \tag{4}$$

To build the association problem, both metrics are combined using a weighted sum

$$c_{i,j} = \lambda d^{(1)}(i,j) + (1-\lambda) d^{(2)}(i,j) \tag{5}$$

where an association is admissible if it is within the gating region of both metrics

$$b_{i,j} = \prod_{m=1}^{2} b_{i,j}^{m} \tag{6}$$

The influence of each metric on the combined association cost can be controlled through the hyperparameter λ. For the situations where camera motion is present, set $\lambda = 0$.

3.1.5 Deep appearance descriptor

Person re-identification in video surveillance is a familiar issue in which a given query image is used to look at a huge gallery of images that have been gathered at distinct moments, lighting environments from different cameras that may contain the same individual. Direct classification is discouraged in this situation because the training set does not include people in the gallery gathered at the moment of the test. Thus, the re-identification model is used in which the goal is to learn a feature representation-from a separate training identities-suitable for performing nearest neighbor queries on images and identities provided at test time. The primary concept of a simple re-identification model is that the output of the CNN model without the last fully linked layer can give us the feature of a person image. Then we can calculate the cosine similarity of two features of the individual to get whether or not these two are the same individual.

The overview of the network architecture of the re-identification model used in Deep SORT framework is as follows (**Figure 1**).

The architecture of the network is shallow to promote quick training and to apply online tracking that involves quick computation of appearance features. Input images in RGB color space are rescaled to 128×64 and fed to the network. A series of convolutional layers then reduce the size of the feature map to 16×8 before fully connected layer Dense 10 extracts a global feature vector of 128. There are six residual blocks in the network. All the convolutions are 3×3 and the layer of the Max Pool is 2. Dropout and normalization used as regularization means. The Dense layer is introduced at a stage where the feature map still offers sufficient spatial resolution for re-identification. Before the application of the softmax classifier, *l2* normalization projects features on the unit hypersphere after the final Dense layer. Exponential Linear Units (ELUs) used as activation function and Cross-Entropy used as the loss function.

In this approach, the network architecture was modified to get more features and to increase the accuracy of the trained model which in turn would improve the tracking efficiency. The modified CNN architecture of the re-identification model is as follows (**Figure 2**).

Various architectures were tested, but this architecture resulted in maximum accuracy. In this model, one of the convolutional layers before Max Pooling is removed and eight residual blocks are used before Average Pooling to get 512 features. Then the Dense layer converts it into 256 feature vectors before applying *l2* normalization. The addition of many layers also degraded the performance because by going deep, the feature maps couldn't provide enough spatial resolution. Instead of ELU, the Rectified Linear Unit (ReLU) activation function was used in this model.

3.1.6 Cascade matching

When a target is occluded for a long time, the uncertainty of the Kalman filter prediction is greatly increased and the observability in the state space is greatly reduced. In a case where two tracks compete for the matching with the same detection, the track that is occluded for a longer period often has more uncertainty in tracking the predicted position because the position information is not updated for a long time, i.e., the covariance is larger. The Mahalanobis distance calculation

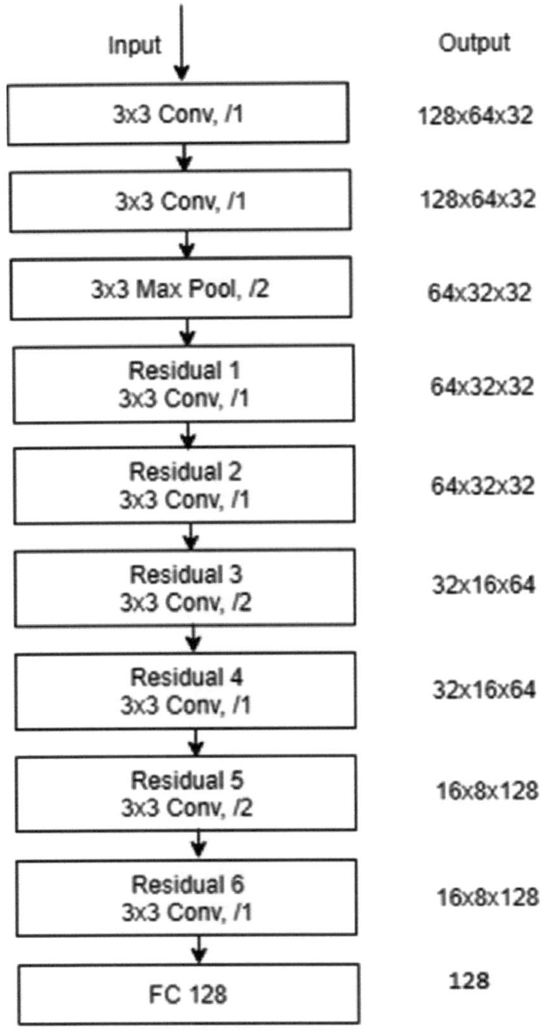

Figure 1.
The re-identification CNN architecture of Deep SORT framework.

uses the reciprocal of the covariance, so for large covariance, the Mahalanobis distance is smaller, so the detection result is more likely to be associated with the trajectory with a longer occlusion time. This undesired effect often destroys the continuation of the tracking. Thus, cascading matching is used to give priority to more frequently occurring targets.

In cascade matching, the set of indices of track T and detection D as well as the maximum age of A_{max} are provided as input. We calculate the price matrix of the association and the matrix of admissible associations are calculated according to Eqs. (5) and (6). Then iteration is done over track age n to solve a linear assignment problem for increasing age tracks. Next, a subset of T_n tracks not connected with the last n frames detection are picked. After that, the linear assignment is solved between T_n tracks and unmatched U detections. Finally, the set of matches and unmatched detections are updated. This matching cascade provides priority to lower age tracks, i.e., tracks that have recently been seen.

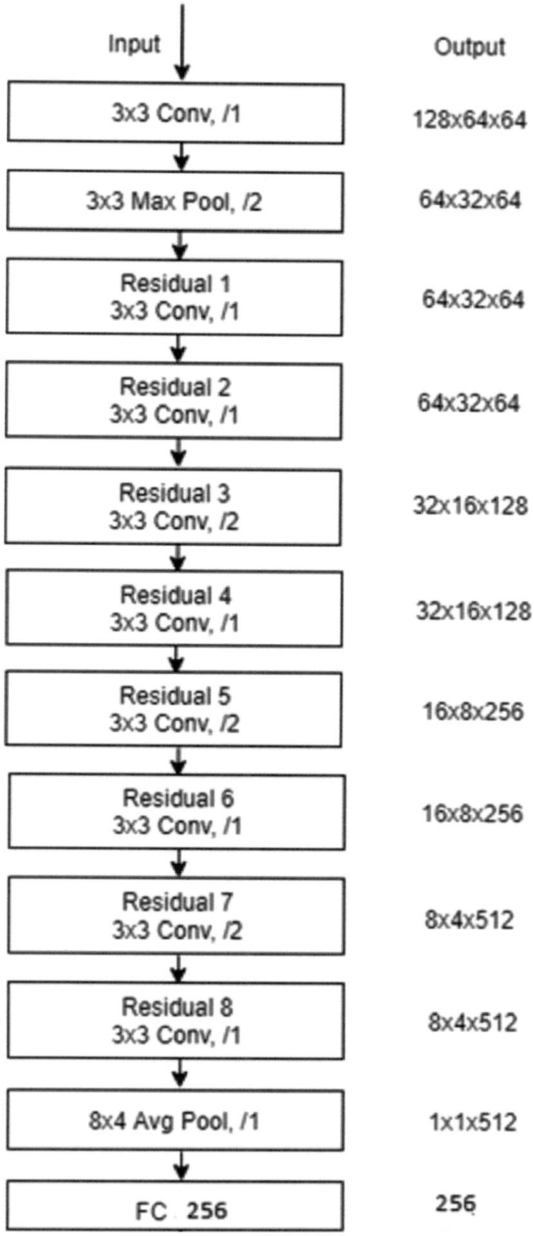

Figure 2.
The modified CNN architecture of re-identification model.

3.1.7 Intersection over union (IoU) matching

IoU-based matching of unconfirmed and age = 1 unmatched trajectories is also performed at the final stage of the match. This can alleviate large changes due to apparent mutations or partial occlusions. Of course, there are advantages and disadvantages. This may also cause some newly generated tracks to be connected to some old tracks. But this is less.

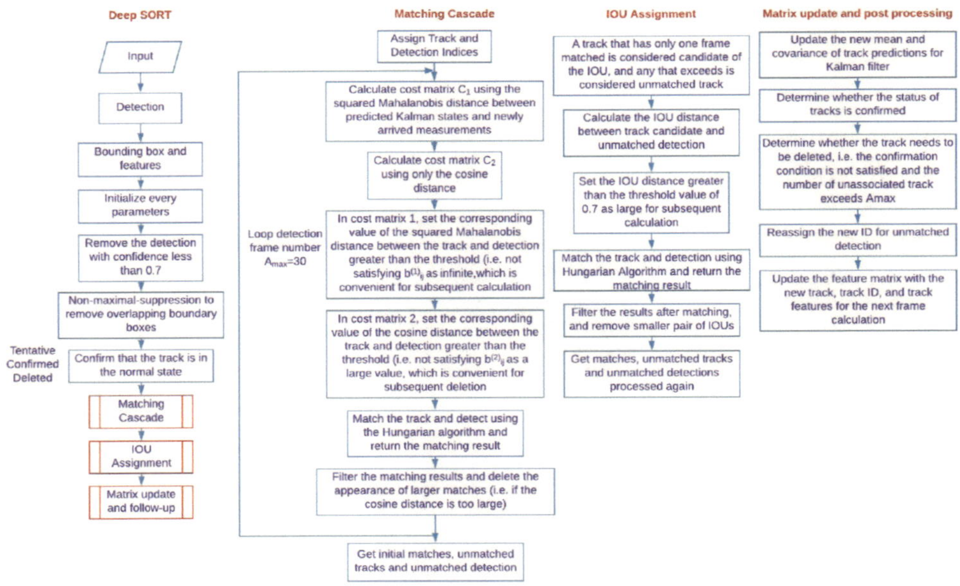

Figure 3.
Flowcharts involved in the tracking process.

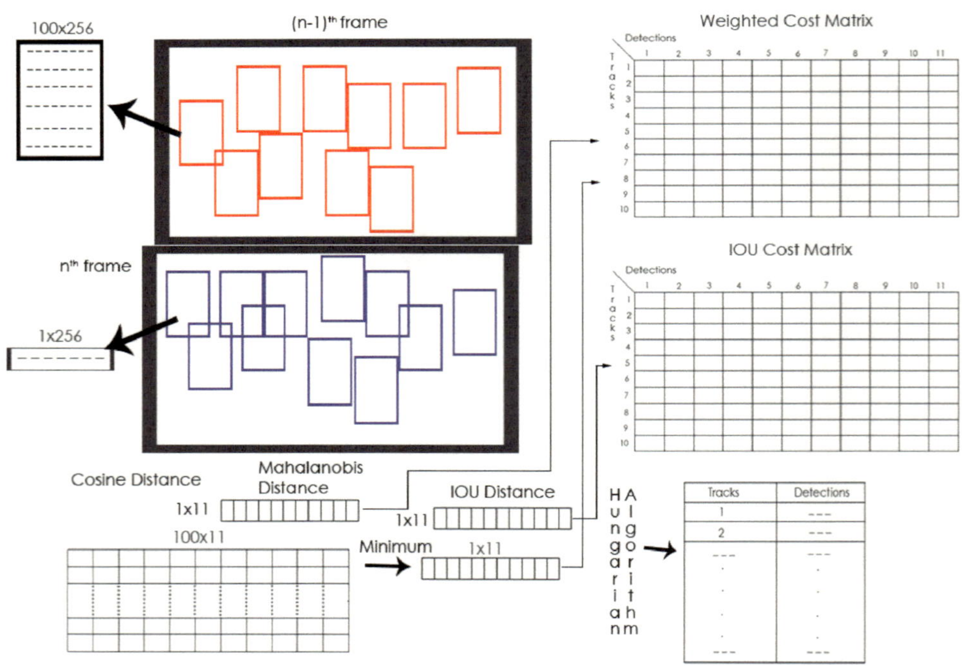

Figure 4.
Pictorial representation of matching during the tracking process.

The complete system flowchart is explained as follows (**Figures 3** and **4**).

1. Read the position of each detected object in each frame and the feature of each detection.

2. Filter the detection frame according to the confidence level, i.e., delete the detection frame and features with insufficient confidence.

3. Perform non-maximum suppression on the detection frame to eliminate multiple frames on one target.

4. *Prediction:* Use Kalman filtering to predict the position of the target in the current frame.

Perform Kalman filter equations:

$$x(k) = Ax(k-1) \tag{7}$$

and

$$p(k) = Ap(k-1)A^T + Q \tag{8}$$

where $x(k)$ is status information for the target (mean), $x(k\ 1)$ is the information of the target in the previous frame [center x, center y, aspect ratio, height, 0, 0, 0, 0]; $p(k\ 1)$ is the estimated error of the target (covariance); A is the state transition matrix; Q is the system error or noise matrix.

5. *Update:* Update the Kalman tracker parameters and feature set, and additionally assess the target disappearance and the new target appearance.

   a. Match results with tracking prediction results.

      - A tracker must distinguish between confirmed states and unconfirmed the status.
      - Tracker with confirmed status is set for cascading matching.
      - For a plurality of trackers of the same disappearance time, calculate a cosine distance matrix between the depth feature of each target newly detected by the current frame and the feature set saved by each tracker. If there are 11 detection targets in the current frame, there are 10 trackers, and each tracker has retained the depth features (256 features) of the previous 100 frames. The calculated cost matrix size is 10×11, and the calculation process is first for each tracker. For 100 features, calculate ($1 -$ cosine distance) between the 11 new detection target features of the current frame, get 100×11 matrix, then find the minimum cosine distance for each detection block, get 1×11 matrix, and store it in cost matrix corresponding row, indicating the minimum cosine distance between the current tracker and the current detection block.
      - Calculate the Mahalanobis distance between the predicted position of the Kalman filter and the detection frame. The specific process is to convert each detection frame from [x, y, w, h] to [center x, center y, aspect ratio, height]. For each tracker, i.e., each row in cost matrix, calculate the Mahalanobis distance between the prediction result and the detection result. Assuming the frame has 11 detection results, there is a matrix with the distance of 1×11, for the current row in the cost matrix. The position where the Mahalanobis distance is greater than the specified threshold is assigned as 1e+5.
      - Set the element larger than the max distance in cost matrix to cost matrix > max distance.

- Assign the track to the correct detection by minimum cost matching of the cost matrix using the Hungarian algorithm.

- After the assignment is completed, the unassociated detections and tracks are regarded as unmatched detections and unmatched tracks. The matching whose cost matrix value is greater than max distance (threshold) is also sent to the unmatched detections and the unmatched tracks.

- Further matching of unconfirmed and unmatched tracks is based on the IoU. The specific implementation is to calculate the cost matrix. The elements larger than the max distance in cost matrix are set to the max distance, and then the Hungarian algorithm is used to assign the cost matrix as input. After the assignment is completed, the outputs are still matched, unmatched detections, unmatched tracks.

b. *To do parameter updates:*

- The process of parameter updating is to calculate the Kalman filter formulas

$$K(k) = p(k)H^T\left(Hp(k)H^T - R\right) - 1 \tag{9}$$

$$x(k) = x(k-1) + K(k)(z(k) - Hx(k-1)) \tag{10}$$

$$p(k) = (1 - K(k)H)p(k-1) \tag{11}$$

where $K(k)$ is the Kalman gain, z is the measurement vector, H is the measurement matrix and R is the measurement noise.

- After the parameter update is completed, the feature is inserted to the tracker feature set and corresponding parameters are re-initialized.

a. *Delete unmatched tracker:* The unmatched tracker indicates that although the new location is predicted, the detection box does not match.

b. *Initialize unmatched detection to a new tracker:* If there is no matched detection, indicating that a new target to be tracked appears, at this point, a new Kalman filter is initialized, and a new tracker is initialized.

c. Delete the trackers to be deleted.

d. Update the feature set of the remaining trackers.

3.2 Dataset preparation and training

To create the vehicle re-identification dataset, thousands of pictures of vehicles on the road were collected. A CCTV camera was placed at the bridge which captured the images of the vehicles on the road below the bridge. Over 33,000 images of 600 different vehicle models were collected. The collected vehicle images were cropped into 128 $\times$ 64 pixels. The training and test sets were split into 80 and 20% respectively and both were structured as follows by placing all the images of one vehicle into a specific numbered folder for both training and test sets. The sample of the images used in the dataset is shown below (**Figure 5**).

The system was run on the hardware with the following configurations.

CPU: Intel Core i7-7770k, 4.3 GHz
Motherboard: Asus Prime Z270-A and Asus ROG-SYRIX-GTX1080TI
RAM:16 GB

The experimentation steps were performed using Python programming language with PyTorch framework.

The following configurations were used for training both the original re-identification model and the modified re-identification model.

batch size = 64
no. of epochs = 40
learning rate = 0.1
momentum = 0.9
weight decay = 5e−4
learning decay = 0.1 after every 20 epochs

The training loss and validation loss obtained for the original model at the end of 40th epoch was 0.51079 and 0.59293 respectively and the training accuracy and validation accuracy was 85.843 and 83.234 respectively. Similarly, the training loss and validation loss obtained for the modified model at the end of the 40th epoch was 0.10816 and 0.16873 respectively and the training accuracy and validation accuracy was 95.747 and 92.705 respectively.

The training charts obtained are as follows (**Figures 6** and 7).

Figure 5.
Sample images in the dataset to train the vehicle re-identification model.

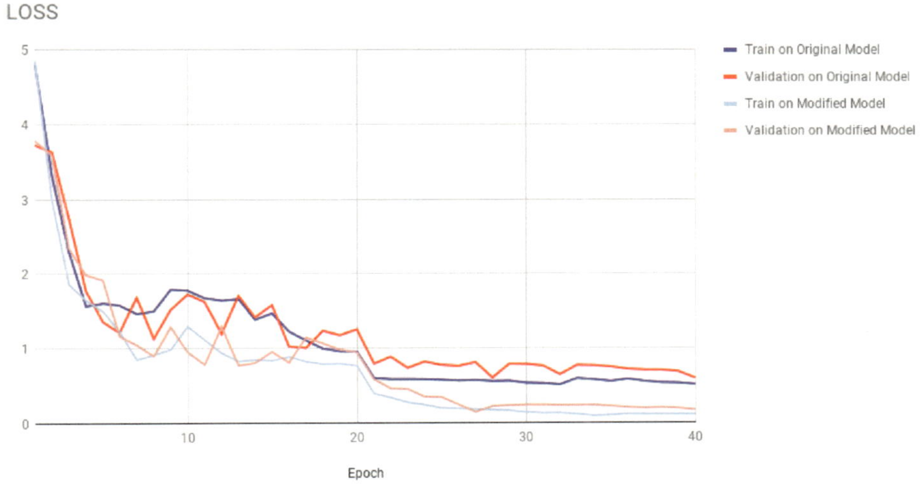

Figure 6.
Training loss and validation loss for both the original and modified re-identification networks.

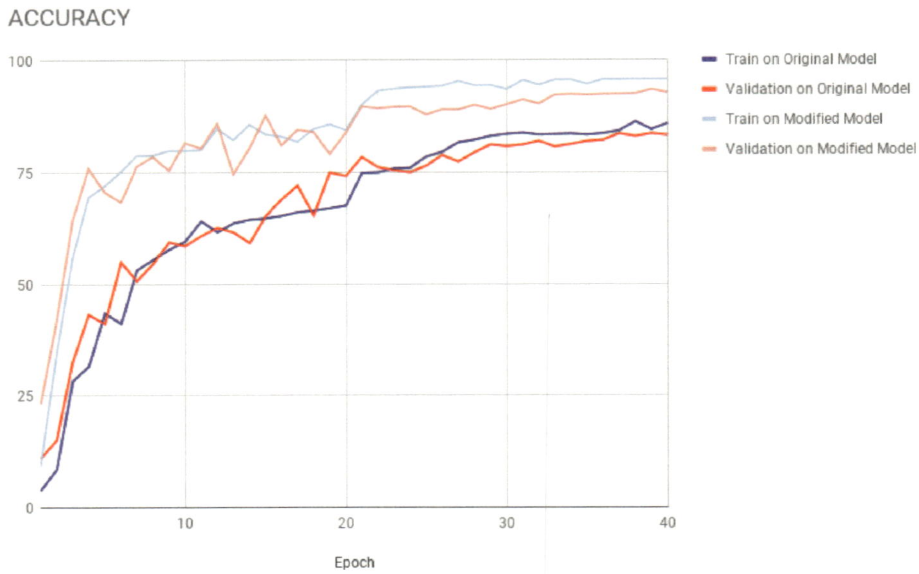

Figure 7.
Training accuracy and validation accuracy for both the original and modified re-identification networks.

4. Evaluation and results

The system was evaluated by feeding the live stream video from the CCTV camera as well as the pre-recorded videos. With the help of pre-recorded videos, both the original model and the modified model were tested. There was far less number of identity switches obtained for the modified model than the original model while running on the same video. The confidence score of YOLOv3 detector was set to 0.7. During occlusions as well as missed detections from the detector, tracking performance of the system with the modified re-identification model was far superior. The test videos can be viewed at the following hyperlinks. The first four videos are the outputs when tested with the original model and the last four are those when tested with the modified model.

The system runs at 15–20 fps on the hardware configuration mention in Section 3.2.

The system was evaluated based on detection accuracy and performance as well. MOT-16 [31], based on the CLEAR metrics [32] intuitively expresses the evaluation in two numbers- MOTP and MOTA.

Tracking system	MOTA	MOTP (%)	MT (%)	ML (%)	ID	FM	FP	FN
Deep SORT	61.4	79.1	32.8	18.2	781	2008	12852	56668
Proposed system	64.2	81.7	37.3	16.2	686	1985	12266	54089

MOTA: summary of general accuracy of tracking in relation to false positives, false negatives and identity switches; MOTP: summary of overall tracking precision in terms of bounding box overlap between ground-truth and reported location; MT: mostly tracked-percentage of ground-truth tracks with the same label for at least 80% of their lifetime; ML: mostly lost- percentage of ground-truth tracks tracked for a maximum of 20% of their life span; ID: identity switch-number of times the reported identity of a ground-truth track changes; FM: fragmentation-number of times a track is interrupted by a missing detection.

Table 1.
Evaluation table.

- *Multiple object tracking precision* (*MOTP*): expresses how well exact positions of persons are estimated.
- *Multiple object tracking accuracy* (*MOTA*): shows how many mistakes the tracker made in terms of misses, false positives, mismatches, failures to recover tracks.

The comparison of tracking with the original model with the proposed model in terms of MOTA and MOTP is as shown in **Table 1**. The accuracy and precision of tracking are increased in the proposed system and the number identity switches are significantly decreased.

5. Conclusion

The major objective of this work is to track the moving vehicles (cars) on the road. The various concepts of deep learning and computer vision have been utilized for this purpose. Track by detection framework was applied for real-time vehicle tracking. YOLOv3 object detection system was used to detect the vehicles and the concepts of Deep SORT algorithm was applied for tracking. By modifying the re-identification model of the original Deep SORT system and training the network on the vehicle dataset developed from scratch, the proposed system enhances the tracking performance by reducing the number of identity switches. With the use of more powerful hardware, the performance of the system can be enhanced.

The pros and cons of the proposed system are as follows.

5.1 Pros

- Fast-tracking due to the use of a very fast object detector.
- It can be used for real-time applications.
- High accuracy and reduced identity switches.

5.2 Cons

- The limitations include the necessity of both CPU and GPU. CPU is slow in calculations and is not suitable for real-time processing. Similarly, GPU cannot drive itself. It needs CPU to be controlled. Hence this system becomes a bit expensive in terms of cost and power consumption.
- The effectiveness of the algorithm is reduced if the bounding boxes from the object detector are too big because too much background information is captured in the features.
- The performance is slightly reduced during dark conditions.

Acknowledgements

This work is carried out at Internet of Things (IoT) Lab, Industrial Systems Engineering Department, Asian Institute of Technology, Thailand.

Videos

Videos available at: https://mts.intechopen.com/download/index/process/451/authkey/330918e03cab6552f87b111634173303

Author details

Sandesh Shrestha
Asian Institute of Technology, Pathumthani, Thailand

*Address all correspondence to: sandeshshrestha45@gmail.com

References

[1] Bewley A, Ge Z, Ott L, Ramos F, Upcroft B. Simple online and realtime tracking. In: 2016 IEEE International Conference on Image Processing (ICIP). IEEE.; 2016. pp. 3464-3468

[2] Wojke N, Bewley A, Paulus D. Simple online and realtime tracking with a deep association metric. In: 2017 IEEE International Conference on Image Processing (ICIP). IEEE.; 2017. pp. 3645-3649

[3] Jain AK, Zhong Y, Lakshmanan S. Object matching using deformable templates. IEEE Transactions on pattern analysis and machine intelligence. 1996; **18**(3):267-278

[4] Mundy JL. Object recognition in the geometric era: A retrospective. In: Toward Category-level Object Recognition. Berlin, Heidelberg: Springer; 2006. pp. 3-28

[5] Ponce J, Hebert M, Schmid C, Zisserman A, editors. Toward Category-Level Object Recognition. Vol. 4170. Berlin, Heidelberg: Springer; 2007

[6] Ojala T, Pietikinen M, Menp T. Multiresolution gray-scale and rotation invariant texture classification with local binary patterns. IEEE Transactions on Pattern Analysis & Machine Intelligence. 2002;**1**(7):971-987

[7] Dalal N, Triggs B. Histograms of oriented gradients for human detection. In: IEEE Computer Society Conference on Computer Vision and Pattern Recognition, 2005. CVPR 2005; Vol. 1; IEEE; 2005. pp. 886-893

[8] Lowe DG. Distinctive image features from scale-invariant keypoints. International Journal of Computer Vision. 2004;**60**(2):91-110

[9] Tuzel O, Porikli F, Meer P. Region covariance: A fast descriptor for detection and Classification. In: European Conference on Computer Vision. Berlin, Heidelberg: Springer; 2006. pp. 589-600

[10] Krizhevsky A, Sutskever I, Hinton GE. Imagenet classification with deep convolutional neural networks. In: Advances in Neural Information Processing Systems. 2012. pp. 1097-1105

[11] Khan G, Ghani MU, Siddiqi A, Seo S, Baik SW, Mehmood I, et al. Egocentric visual scene description based on human-object interaction and deep spatial relations among objects. Multimedia Tools and Applications. Vol. 77. Springer; 2018. pp. 1-22

[12] Girshick R, Donahue J, Darrell T, Malik J. Rich feature hierarchies for accurate object detection and semantic segmentation. In: Proceedings of the IEEE Conference on Computer Vision and Pattern Recognition. 2014. pp. 580-587

[13] Uijlings JR, Van De Sande KE, Gevers T, Smeulders AW. Selective search for object recognition. International Journal of Computer Vision. 2013;**104**(2):154-171

[14] He K, Zhang X, Ren S, Sun J. Spatial pyramid pooling in deep convolutional networks for visual recognition. In: European Conference on Computer Vision. Springer; 2014. pp. 346-361

[15] Girshick R. Fast r-cnn. In: Proceedings of the IEEE International Conference on Computer Vision. 2015. pp. 1440-1448

[16] Zhou B, Khosla A, Lapedriza A, Oliva A, Torralba A. Learning deep features for discriminative localization. In: Proceedings of the IEEE Conference on Computer Vision and Pattern Recognition. 2016. pp. 2921-2929

[17] Ren S, He K, Girshick R, Sun J. Faster r-cnn: Towards real-time object detection with region proposal networks. In: Advances in Neural Information Processing Systems. 2015. pp. 91-99

[18] Redmon J, Divvala S, Girshick R, Farhadi A. You only look once: Unified, real-time object detection. In: Proceedings of the IEEE Conference on Computer Vision and Pattern Recognition. 2016. pp. 779-788

[19] Redmon J, Farhadi A. YOLO9000: Better, faster, stronger. In: Proceedings of the IEEE Conference on Computer Vision and Pattern Recognition. 2017. pp. 7263-7271

[20] Redmon J, Farhadi A. Yolov3: An incremental improvement. arXiv preprint arXiv:1804.02767. 2018 Apr 8

[21] Dicle C, Camps OI, Sznaier M. The way they move: Tracking multiple targets with similar appearance. In: Proceedings of the IEEE International Conference on Computer Vision. 2013. pp. 2304-2311

[22] Yoon JH, Yang MH, Lim J, Yoon KJ. Bayesian multi-object tracking using motion context from multiple objects. In: 2015 IEEE Winter Conference on Applications of Computer Vision. IEEE.; 2015. pp. 33-40

[23] Kim C, Li F, Ciptadi A, Rehg JM. Multiple hypothesis tracking revisited. In: Proceedings of the IEEE International Conference on Computer Vision. 2015. pp. 4696-4704

[24] Bewley A, Ott L, Ramos F, Upcroft B. Alextrac: Affinity learning by exploring temporal reinforcement within association chains. In: 2016 IEEE International Conference on Robotics and Automation (ICRA). IEEE; 2016. pp. 2212-2218

[25] Reid D. An algorithm for tracking multiple targets. IEEE Transactions on Automatic Control. 1979;**24**(6):843-854

[26] Fortmann T, Bar-Shalom Y, Scheffe M. Sonar tracking of multiple targets using joint probabilistic data association. IEEE Journal of Oceanic Engineering. 1983;**8**(3):173-184

[27] Rezatofighi AM, Zhang Z, Shi Q, Dick A, Reid I. Joint probabilistic data association revisited. In: Proceedings of the IEEE International Conference on Computer Vision. 2015. pp. 3047-3055

[28] Kayumbi G, Mazzeo PL, Spagnolo P, Taj M, Cavallaro A. Distributed visual sensing for virtual top-view trajectory generation in football videos. In: Proceedings of the 2008 International Conference on Content-based Image and Video Retrieval. ACM; 2008. pp. 535-542

[29] Yang M, Jia Y. Temporal dynamic appearance modeling for online multi-person tracking. Computer Vision and Image Understanding. 2016;**153**:16-28

[30] Xiang Y, Alahi A, Savarese S. Learning to track: Online multi-object tracking by decision making. In: Proceedings of the IEEE International Conference on Computer Vision. 2015. pp. 4705-4713

[31] Milan A, Leal-Taix L, Reid I, Roth S, Schindler K. MOT16: A benchmark for multi-object tracking. arXiv preprint arXiv:1603.00831. 2016

[32] Bernardin K, Stiefelhagen R. Evaluating multiple object tracking performance: The CLEAR MOT metrics. Journal on Image and Video Processing. 2008;**2008**:1

Chapter 10

Intelligent Routing Mechanisms in IoT

Gowrishankar Subhrahmanyam and Ishwari Ginimav

Abstract

Wireless sensor networks (WSN) works on battery in order to communicate with each other. Energy consumption is the challenging issue in WSNs. In the recent few years, green communications have become a major concern in communication research and industries. Its major goal is to minimize the energy consumed by the nodes of the WSN. In order to save energy we need to switch off the extra components that are not in use during low traffic period. The technique where the unused extra components are switched off is called as sleep-scheduling, and the routing algorithm used to implement this is called as sleep-scheduling routing algorithm. In WSN the network is divided into multiple clusters. In each cluster one of the sensor nodes is elected as cluster head (CH) and other sensor nodes act as cluster members (CM). The cluster head collects the data form all the other nodes, removes the redundant data and transmits it to the destination. As the amount of workload is much more on the cluster head, the energy consumed by the cluster head is also more. Therefore to equalize the energy consumption among all the nodes, cluster head rotation is done. This chapter deals with different energy consumption techniques.

Keywords: wireless sensor networks, green IOT, sleep scheduling algorithms, LEACH algorithm, FCM algorithm, GA algorithm, PSO algorithm, GSO algorithm, MINEN algorithm

1. Introduction

Wireless sensor networks (WSNs) [1] is one of the widely used technologies in twenty-first century. The main objective of WSN is that it provides wireless communication by low-cost sensor networks which consumes very limited power and provides multiple functionalities. A typical sensor node [1] is made of four basic components namely: a sensing unit, a processing unit, a communication unit, and a power unit as shown in **Figure 1**. Peer-to-peer communication exists between two nodes. Multi-hop communication exists when the nodes are not in the radio range and the communication between the two nodes is carried out via intermediate nodes. Therefore the WSN provides a very good feature of adding and removing nodes. In WSN the network is divided into number of clusters. In each cluster one of the sensor nodes is elected as cluster head (CH) and other sensor nodes act as cluster members (CM). The cluster head collects the data form all the other nodes, removes the redundant data and transmits it to the destination. As the amount of

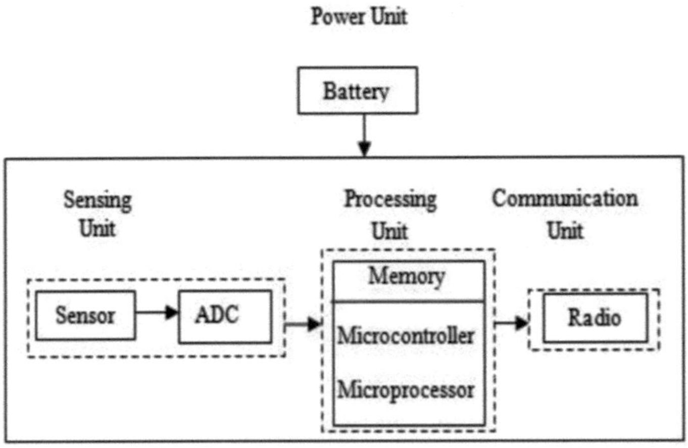

Figure 1.
Sensor node structure.

workload is much more on the cluster head, the energy consumed by the cluster head is also more. Therefore to equalize the energy consumption among all the nodes, cluster head rotation is done. Energy consumption is the challenging issue in WSN. This chapter deals with different energy consumption techniques.

1.1 Characteristics of WSNs

Following are some of the important characteristics of WSN:

1.1.1 Dynamic network topology

Whenever there is an addition or deletion of nodes the network topology keeps updating. This feature adds to the flexibility to the network.

1.1.2 Application specific

The design of the network keeps changing based on requirement of the application.

1.1.3 Energy constrained

As the nodes are movable, there is always a constraint on energy consumption.

1.1.4 Self-configurable

Nodes are set haphazardly in the network. Once the nodes are deployed, they configure themselves based on the requirement.

1.2 WSN routing protocol

In order to reduce the energy consumption in WSN, the different routing protocols are implemented which define how the packets have to be transmitted from source to destination **Figure 2** shows the classification of routing protocols in WSNs [1].

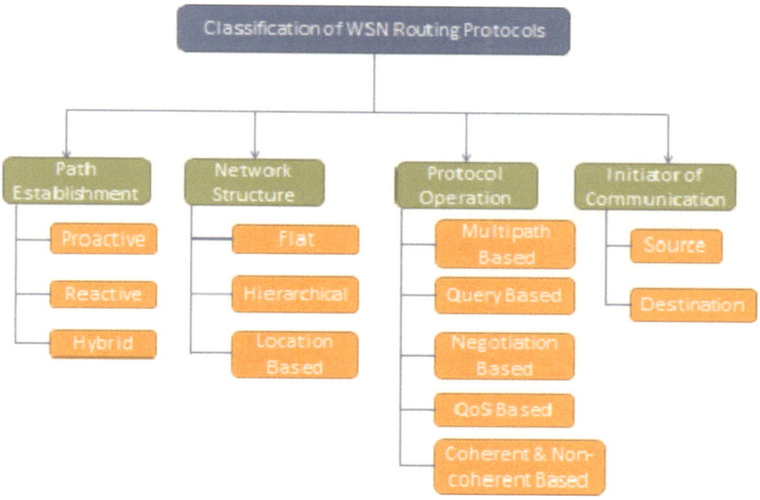

Figure 2.
Classification of WSN routing protocols.

1.3 Green IoT

In the recent few years, green communications have become a major concern in communication research and industries. Its major goal is to reduce the energy consumption of the WSNs. This has come into picture in order to reduce the pollution in the environment and cost. Recent studies have identified that ICT contributes to 10% of the global CO_2 emissions and its contribution is increasing day by day. Research also says that 30–37% of greenhouse gases (GHG) is been produced by ICT sector. Nowadays technology has improved so much that the bandwidths are made larger so as to handle large amount of traffic loads. In such cases in order to save energy we need to switch off the extra components that are not in use during low traffic period. To achieve this, a routing algorithm should be implemented such that the unused network components will be switched off. The technique where the unused extra components are switched off is called as sleep-scheduling, and the routing algorithm used to implement this is called as sleep-scheduling routing algorithm. This chapter mainly deals with sleep-scheduling base green routing algorithm [2].

2. Different routing algorithms

2.1 LEACH routing algorithm

LEACH protocol [3] is an implementation of hierarchical routing protocol. It is a self-organized and self-adaptive routing protocol. The LEACH protocol works on around technique in which each round is considered as one unit. Every round is made out of cluster set-up phase and steady-state for lowering excess energy cost. The steady-state phase is sometimes more time consuming than the set-up phase. One advantage is that the nodes are allowed to elect themselves as the cluster head if required. When a node becomes a cluster head, it can generate a TDMA time schedule for the sensing nodes during which the non-cluster head nodes are turned off when it is not in its time interval. **Figure 3** shows the timeline operation of LEACH.

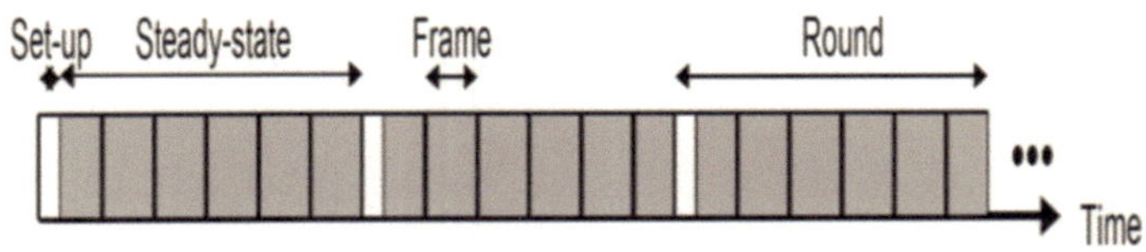

Figure 3.
Timeline showing operation of LEACH.

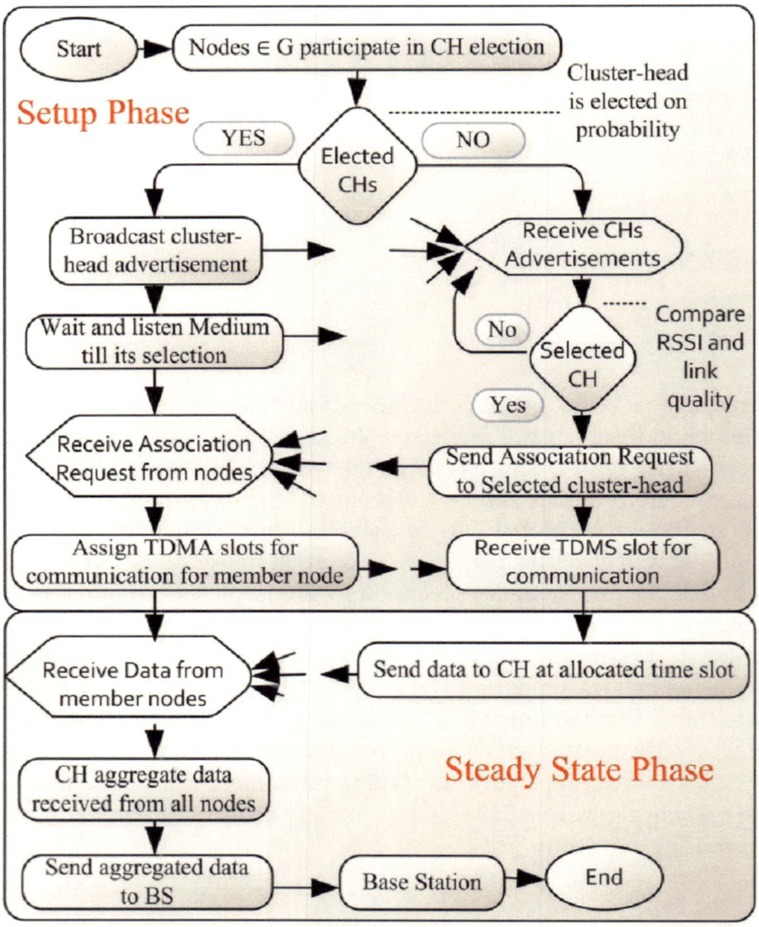

Figure 4.
Flow of LEACH routing protocol.

Figure 4 shows the flow of the LEACH routing protocol [4]:

i. **The set-up phase**: At first the LEACH protocol will randomly generate a number between 0 and 1 for each node and selects a cluster head based on that random number generated. A threshold value is identified which is given by the threshold function T(n). If the randomly generated number of a node is less than the threshold value, then that node will be selected as cluster head node.

$$T(n) = \begin{cases} \dfrac{p}{\left(1 - p * \left(rmod\dfrac{1}{p}\right)\right)} & , n \in G \\ 0 & , n \in G \end{cases} \tag{1}$$

where P is the probability of the cluster-head and G is the set of nodes that have never been chosen as cluster-head nodes before 1/p round.

Once the cluster head is selected, it will send a CDMA message to all other normal nodes to inform them to join the cluster head node. Later the cluster head will use TDMA so that every node which is connected to the cluster head node will be allocated time duration for data transmission.

i. **The steady-state**: In this phase the normal node will sense the data and transmit the data to the cluster head node. The data is processed at the cluster head node and then transmitted to the base station.

2.2 Fuzzy C-means

The distribution of nodes in a cluster is one of the major concerns in WSNs. The distribution of nodes in a cluster is said to be good if the distance between a particular node and its cluster head is reasonably small. The communication cost between a node and its cluster head and the unbalanced load distribution can be optimized if the distribution of the nodes in the cluster is maintained and managed properly. Let us consider a network with N nodes. Our main job is to group the nodes into different c clusters in such a way that the distance between every node and its cluster head is reasonably small. It is the responsibility of the base station to compute the cluster head based on the geographical location of the nodes in the WSN. In order to achieve this fuzzy technique [5] is implemented to find the cluster center.

The equation for the distribution of nodes into a cluster is given as:

$$f_{obj} = \sum_{j=1}^{c} \left[\sum_{i=1}^{N} u_{ij}^{m} d_{ij}^{2} \right] \tag{2}$$

where u_{ij} is the degree of belongingness of node i to cluster j, d_{ij} is the Euclidean distance of node I from the centroid of cluster j, and m is the fuzzy control parameter.

There is an update in the value of membership degrees in the course of iteration according to the equation given below:

$$u_{ij} = \frac{1}{\sum_{l=1}^{c} \left[\frac{d}{d_{il}}\right]^{\frac{2}{m-1}}} \tag{3}$$

The equation mentioned below is used to compute the centroid of each cluster:

$$c_j = \frac{\sum_{i=1}^{N} pos(node_i)}{\sum_{i=1}^{N} u_{ij}^{m}} \tag{4}$$

The FCM clustering algorithm is given below:
Algorithm: Fuzzy C-Means (FCM) Algorithm [5]
1. Input: Position of nodes
2. Output: Center of the cluster
3. begin
4. initialize Uf
5. repeat
6. for cluster j=1 to c do
7. $c_j \leftarrow$ compute cluster centroid
8. end for
9. update Uf
10. until the algorithm converges
11. return c
12. end

2.3 Genetic algorithm

Genetic algorithm is based on the genetic principle and concept of natural selection and evolution. Speaking in terms of biology, every individual is obtained by the combination of the parent chromosomes. GA technique follows the same principle of chromosomes. Here in order to evaluate the chromosome we identify a function called as fitness function for the concerned problem. Now the new population is obtained by the operations such as selection, crossover, repair and mutation. The main aim of selection mechanism is to identify the best individuals (parents) for crossover and mutation. The role of crossover mechanism is to exchange the genetic materials of the parents and provide the best genes to the offspring. The role of mutation mechanism is to provide new genetic materials to the offspring.

We explain the routing of data with the help of chromosomes technique, which provides genetic information for genetic algorithm. Consider a given network, on which every chromosome contains a set of genes and each chromosome presents a particular arrangement of nodes in the routing chain [6]. A chromosome is made out of three different things as shown in **Figure 5**, a gene index, gene value and the gene. The gene index tells the location of the node in the network, the gene value tells the node's identification number (ID) and the gene tells what gene is present in the node. Let us consider the size of chromosome to be some integer N, where every integer is unique and lies between 1 and N. These integers represent the individual gene values that make up the chromosome.

i. **Initial population**: A population is nothing but a collection of chromosomes represented as P = (C_1,C_2, …., C_r). In this algorithm the first ever population

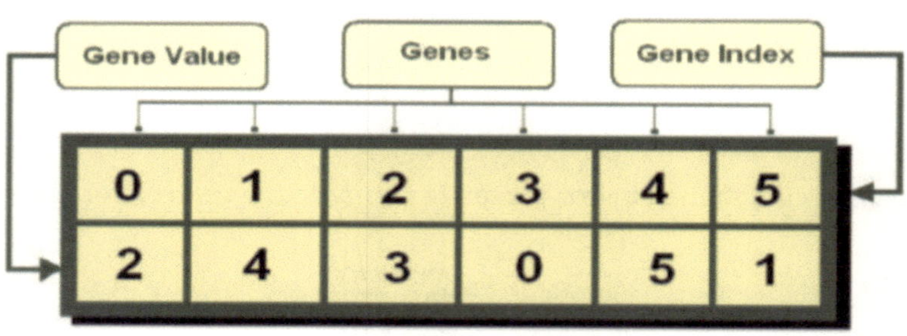

Figure 5.
A chromosome containing 6 genes.

or the initial population is randomly generated by using some random function in order to initiate the algorithm. The initial population consists of r chromosomes. Greater the value of r, greater the probability of getting a better solution. When the chromosomes are generated, care should be taken to see that the distance between two consecutive gene values do not exceed the threshold distance for communication d_{TH}.

ii. **Parent selection**: In this process, each time a selection of two chromosomes is done out of r chromosomes, which will decide which two chromosomes will be mating in order to obtain the offspring.

iii. **Generation**: With the help of crossover and mutation mechanism the new generation will be created.

iv. **Crossover**: Crossover mechanism indicates what combination of the two parent chromosomes can achieve the best offspring.

v. **Evaluation and fitness**: While creating a new generation care should be taken to see that the new generation meets the survival of the fittest condition.

vi. **Repair and mutation**: *Repair:* If the newly generated offspring violates the constraints imposed on it then the algorithm should be repaired in order to obtain the optimized solution. *Mutation:* Mutation adds variation to the next generation. *Cooling schedule:* Anneal temperature is one of the important parameter. In this case, every time the θ of particles approaches a better solution there is a decrement in the value. Let θ_i be the initial temperature and θ_f be the final temperature and t be the cooling time, then the equation of are cooling schedule will be as given below:

$$\theta(t) = \theta_f + (\theta_i - \theta_f) * \alpha^t \tag{5}$$

Algorithm: Genetic Algorithm [6]
Inputs: A set of N sensor nodes along with their position co-ordinates
Output: An ordered sequence of the nodes.
Step 1: Generate the initial population.
Step 2: for N times do:
    Step 2.1: Parents should be selected
    Step 2.2: Crossover has to be performed
    Step 2.3: Evaluate the offspring in order to select or reject it
    Step 2.4: Repair selected offspring and perform mutation
    Step 2.5: Store the produced offspring for the next generation
Step 3: Mark the best offspring in the generation as C_{best}
    Increment generation number by 1
Step 4: Find the anneal temperature
    if $\theta(t) \geq \theta_f$ go to step 2
    else required sequence is represented by C_best

2.4 Particle storm optimization

The particle swarm optimization (PSO) [7] technique was based on the real time activity such as a group of birds flying together, a group of fishes swimming together and the search techniques used by them in order to search their food. It is seen from nature that all winged creatures like birds, fishes etc., all travel together in groups in search of food without colliding with each other. This is possible because they all keep a track of the group (cluster) they belong to and they also have the knowledge of the speed and distance of all the members of that group. So based on the information they have, they change their speed and location.

PSO comprises of a swarm of a predefined size (say N_P) of particles. Every element has a solution to every multi-dimensional issue. For a particular set of particles the dimension D is equal. A particle P_i, $1 \leq i \leq N_P$ has position X_{id}, $1 \leq d \leq D$ and velocity V_{id} in the dth dimension of the hyperspace. The notation for demonstrating the ith particle P_i of the inhabitants as follows:

$$P_i = [X_{i,1}, X_{i,2}, X_{i,3}, \ldots., X_{i,d}] \tag{6a}$$

Each particle is associated with a fitness function which will tell about the superiority of the solution. In order to update the individual position and velocity every particle P_i monitors its individual best called P$best_i$ and global best called G$best$ to achieve the global best position. In every repetition, its position X_{id} and velocity V_{id} and in the dth dimension is modified utilizing the accompanying conditions.

$$\begin{aligned} V_{i.d}(t) &= w * V_{i.d}(t-1) + c_1 * r_1 * \left(Xpbest_{i.d} - X_{i.d}(t-1)\right) \\ &\quad + c_2 * r_2 * \left(Xgbest_d - X_{i.d}(t-1)\right) X_{i.d}(t) \\ &= X_{i.d}(t-1) + V_{i.d}(t) \end{aligned} \tag{6b}$$

where w is the inertial weight, c_1 and c_2 are the non-negative constants called acceleration factor, and r_1 and r_2 are the two different consistently disseminated arbitrary numbers in the range [0, 1].

The modified procedure is iteratively repeated until either an acceptable G$best$ is attained or a fixed number of iterations t_max are achieved.

Description of the algorithm

- Every particle is associated with position and velocity.
- Every particle is aware of its position and the value associated with it.
- Every particle is aware of its best position (P$best$) it has ever achieved and the value associated with it.
- Every particle has the knowledge about its neighbors and their best position (G$best$) and the value associated with it.

Every move of a particle is made by considering one of the three possible choices:

- First one is to follow its own way as it has decided.
- Second is to go back to its previous best position.
- Third is to go to the previous best position or current best position of its neighbor.

Algorithm: PSO Algorithm [8]
1. Initialize a population of particles
2. Do
3. for each particle P_i with position $X_{i.d}$ do
4. if ($X_{i.d}$ is better than $Pbest_i$) then
5. $Pbest_i \leftarrow X_{i.d}$
6. end_if
7. end_for
8. Define G*best* as the best position found so far by any of P's neighbors
9. for each particle P do
10. $V_{i.d} \leftarrow$ Compute_velocity($X_{i.d}$, $Pbest_i$, $Gbest_i$)
11. $X_{i.d} \leftarrow$ update_position($X_{i.d}$, $V_{i.d}$)
12. end_for
while (a stop criterion is not satisfied):

$$V_{i.d}(\mathrm{t}) = V_{i.d}(\mathrm{t}-1) + c_1 * r_1 * (Pbest_i - X_{i.d}(\mathrm{t}-1)) + c_2 * r_2 * (Gbest_i(\mathrm{t}-1) - X_{i.d}(t-1)) \quad (6c)$$

$$X_{i.d}(t) = X_{i.d}(t-1) + V_{i.d}(t) \quad (7)$$

2.5 Genetic swarm optimization

The combination of GA and PSO algorithms is the GSO algorithm [7], which formulated to obtain better performance. The main elements of GA are crossover and mutation. The strength of the algorithm is that it is more robust and adaptable. It does not bother about whether the problem is continuous or discrete. What it is bothered is only about whether the problem can be solved. Therefore GA overcomes all the problems and puts light on optimizing the process. One drawback of GA is that its convergence speed very low, whereas, the convergence speed of PSO is relatively fast as compared to GA. Therefore we combine GA and PSO in order to overcome the drawbacks. GSO is different from PSO. The GSO algorithm has combined the core elements of GA i.e., crossover and mutation with PSO so as to improve the performance and efficiency of the PSO algorithm. **Figure 6** shows the process of GSO [7].

Step 1: The numbers of particles are initialized (circuit codes are initialized).

Step 2: The speed and position of each particle are randomly initialized.

Step 3: In this step we calculate the fitness value of each particle. We then store the fitness value and current position of every particle in its P*best*. Later the comparison of all the P*best* is done and we store the fitness value of the best individual in G*best*.

Step 4: The particle swarm is randomly selected and crossover and mutation operations are executed on it.

Step 5: The next step is to update the speed and position with all the particles with expressions (1) and (2).

Step 6: Finally we judge whether the algorithm is achieving convergence criteria. If the convergence touchstone is achieved, then output the best person and the search algorithmic program, otherwise repeat step 3.

When compared with the GA and PSO, GSO mainly has the following advantages:

i. There is much of the improvement in the search capability.

ii. The convergence rate of the partial area is also enhanced.

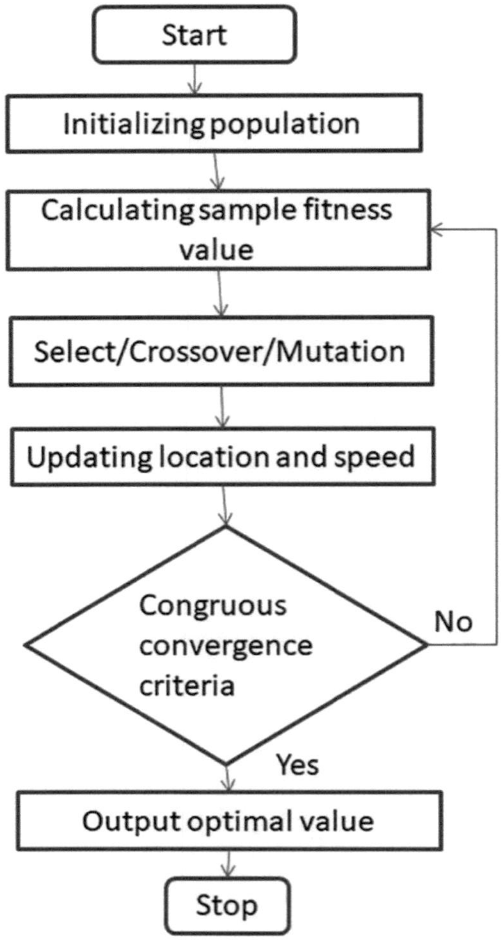

Figure 6.
Process of GSO algorithm.

iii. The partial area convergence stagnation phenomenon which existed in the search appendage of GA is avoided in this algorithm.

iv. The searching truth can be improved and the number of successful evolution looping can be reduced.

2.5.1 Objective function design of GSO

In the optimization process, the fitness functions are given from objective functions. The outcome of the fitness function called the fitness value, shows the ability of individual and adapts to the evolution. By optimizing the process one can achieve greater value, stronger adaptability and preserving the evolution. It should be taken into consideration that the difference among the individual fitness value cannot be neither too big nor too small so as to overcome the fall in local optimum and avoid slow convergence. The fitness value is directly related to the algorithm accuracy and convergence speed. We can get the target function y based on the truth table. The fitvalue is given by:

$$fitvalue = \sum_{1}^{2^n} fitnumber_i \tag{8}$$

where $fitnumber_i$ is the ith input combination of truth table suffice for the evolution circuit and n is the number of inputs.

2.6 Minimum energy

Figure 7 shows the flow of minimum energy (MINEN) [9] algorithm. The steps involved in the execution of the protocol are as follows:

- Firstly we identify the nodes that are not participating in the current scenario. This step is optional.
- Secondly, we form a cluster and elect a cluster head so that all the communications can be carried out through the cluster head.
- Third is, designing a DAG which shows the connections between all the cluster heads and the weights associated with each link.
- Finally, we run the Dijikstra's algorithm to find the shortest path between the cluster head and the base station.

In order that the protocols performs well, we need to make some assumptions such as:

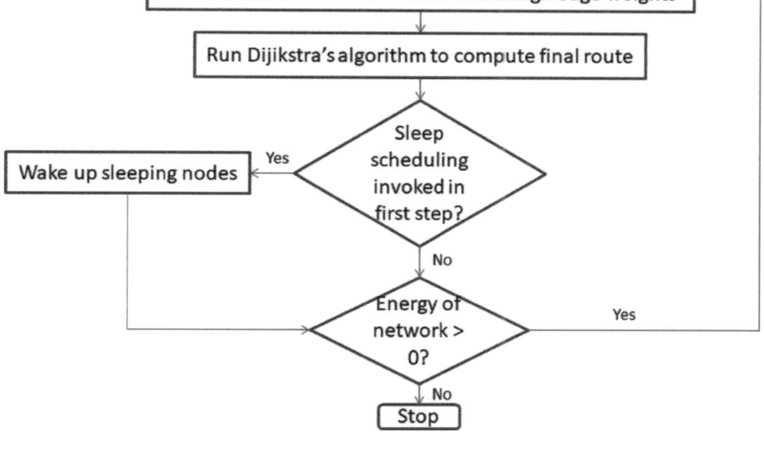

Figure 7.
Flow of MINEN algorithm.

- All devices have the same level of energy at the beginning.
- In an IoT network there is only one base station located that is static in nature.
- It is assumed that the base station is provided with infinite energy, i.e., there is no risk of the base station to be shut down due to lack of energy.
- It is assumed that one round of communication is calculated as the time period between the election of new cluster head and successful transmission of messages from all the cluster heads to the base station.

We now explicitly detail each step of the algorithm shown in **Figure 7** in the following subsections [9]:

```
Algorithm: MINEN Algorithm [9]
1. Run the GSO sleep scheduling algorithm (optional)
2. Run clustering algorithm to create cluster set C
3. for c ∈ C do
4.        for i ∈ c.devices do
5.              if d.current energy ¡ i.current energy then
6.                    d = i
7.        Set c.cluster head = d
8. Create graph G connecting all cluster heads
9. Run Dijkstra on G
10. Return routing path
11. if sleep scheduling is invoked in step 1 then
12.        Wake up all sleeping devices in the network
13. if Number of active devices > 0 then
14.        goto step 1
15. else exit
```

3. Conclusions

In this chapter we have discussed about MINEN routing protocol [9] for IoT-WSN. MINEN protocol is primarily based on cluster rule where the amount of energy utilized is distributed equally among all the devices of the network. This is achieved by using the clustering method, giving every node a chance to become a cluster head and reducing the energy utility of transmitting and receiving messages as well as devices with low residual energies across the link. This is done following the various steps such as, designing a DAG wherein the nodes of the graph act as cluster head. Then we assign appropriate weights to the links of the graph taking into consideration the energy required for transmitting and receiving messages over a link called the energy spent (E_{sf}). The E_{sf} factor enforces energy load balancing across several links. The Dijikstra's Algorithm is used to find the minimum cost (energy) between the cluster head of the sender to the base station. Further, the GSO sleep scheduling algorithm along with the MINEN algorithm would help in enhancing the energy conservation effort. MINEN performs better when compared to GSO. MINEN overcomes the drawback of the two most widely used algorithms namely LEACH and FCM, in terms of network coverage, number of alive nodes and energy dynamics.

For future work we can extend the protocol such that it can be used in IoT where all the nodes are mobile nodes and there no end-to-end path between the source and the destination. Further, work can be progressed to achieve better performance by improving the sleep scheduling algorithm. Work can also be done to improve the clustering techniques, where clusters result in more energy conservation.

Abbreviations

ADV	advertisement message
CDMA	code division multiple access
CH	cluster head
CM	cluster member
DAG	directed acyclic graph
FCM	fuzzy C-mean
GA	genetic algorithm
GHG	greenhouse gases
GMM	Gaussian mixture model
GPS	global positioning system
GSO	genetic swarm algorithm
ICT	information and communication technology
IoT	Internet of Things
LEACH	low energy adaptive clustering hierarchy
MAC	media access control
MEMS	microelectronic mechanical systems
MINEN	minimum energy
PSO	particle swarm algorithm
TDMA	time division multiple access
WSN	wireless sensor networks

Author details

Gowrishankar Subhrahmanyam*† and Ishwari Ginimav*†
BMS College of Engineering, Bengaluru, India

*Address all correspondence to: gowrishankar@bmsce.ac.in and ishwari29@gmail.com

† These authors contributed equally.

References

[1] Dhawan H, Waraich S. A comparative study on LEACH routing protocol and its variants in wireless sensor networks: A survey. International Journal of Computers and Applications. 2014;**95**(8):0975-8887

[2] Dabaghi F, Movahedi Z, Langar R. A survey on green routing protocols using sleep-scheduling in wired networks. Journal of Network and Computer Applications. 2017;**77**:106-122

[3] Yadav L, Sunitha C. Low energy adaptive clustering hierarchy in wireless sensor network. International Journal of Computer Science and Information Technologies. 2014;**5**(3):4661-4664

[4] Available from: https://www.researchgate.net/publication/324943617_Analysis_of_Energy_Efficient_Hierarchical_Routing_Protocols_in_Wireless_Sensor_Networks/figures?lo=1

[5] Tamene M, Rao KN. Fuzzy C-means clustering algorithm for optimization of routing protocol in wireless sensor networks. International Journal of Computer Science and Network. 2016;**5**(2):2277-5420

[6] Chakraborty A, Mitra SK, Naskar MK. A Genetic Algorithm Inspired Routing Protocol for Wireless Sensor Networks

[7] Junbin Z, Jinyan C, Yafeng M, Tianzhen M. Genetic algorithm particle swarm optimization based hardware evolution strategy. WSEAS Transactions on Circuits and Systems. 2014;**13**:274-283

[8] Sudarmani R, Shankar Kumar KR. Particle swarm optimization based routing protocol for clustered heterogeneous sensor networks with mobile sink. American Journal of Applied Sciences. 2013;**10**(3):1546-9239

[9] Vashishth V, Chhabra A, Khanna A, Sharma DK, Singh J. Energy Efficient Routing Protocol for Wireless Internet-of-Things Sensor Networks. Vol. 22019: 1-12

Section 3

Emerging Devices for Intelligent Systems

Chapter 11

Graphene Nanowire Based TFETs

Jayabrata Goswami, Anuva Ganguly, Anirudhha Ghosal and Jyoti Prakash Banerjee

Abstract

The present work is aimed at improving the performance potential of tunnel field effect transistors (TFETs), where the carriers are transported by the process of band to band tunneling. The nanoscale TFETs serves the purpose of ULSI integration with high speed and memory. The requirements of new device technology are challenging: for logical switching. In this paper, a p-channel graphene nanoribbon (GNR) TFETs has been analyzed and designed for low power and high performance digital switching application. The energy band diagram of the device is obtained from self-consistent iterative method for numerical solution of one-dimensional Poisson's equation subject to appropriate boundary conditions. It is observed that the optimized p^+ channel GNR TFET provides high ON–OFF current ratio, low sub-threshold slope for a channel length of 85 nm and channel width of 4 nm.

Keywords: TFET, GNR, sub-threshold slope

1. Introduction

The fundamental limitation of silicon MOSFETs in gigascale integration has led to the proposal of several non-classical transistors as the future replacement. In recent years tunnel field effect transistors (TFETs) are attracting the attention of researchers due to their low sub-threshold slope much below the thermionic limit of 60 mV/decade for silicon MOSFET at room temperature along with their low-voltage application and low power consumption. A low voltage tunnel transistor beyond CMOS logic was proposed by Seabaugh and Zhang [1] in 2010. It is reported that TFETs with Si as channel material exhibit low ON state current density (100 μA/μm) [2] due to large bandgap of Si. If a lower band gap material, Ge is used as channel material in TFETs, the ON state current increases to 850 μA/μm [3]. A heterojunction TFET with Si as channel material and lower bandgap semiconductors such as $In_xGa_{1-x}As$ as source material leads to improved performance of the device. Graphene is an emerged electronic material due to its highest carrier mobility and carrier saturation velocity at room temperature among all semiconductor materials [4]. However, the bulk graphene sheet is a semimetal with a zero bandgap and cannot be used for room temperature transistors with sufficient on/off ratio. The main challenge is to apply graphene for digital electronic or photonic applications. Graphene nanoribbons (GNRs) of sub-10 nm width are found to be semiconducting due to lateral confinement of the electron wave function in the transverse direction with a band gap inversely proportional to the conducting channel width. Further the low-energy electronic states of graphene have two non-equivalent mass less Direc spectrum. The confinement gap (ΔE) in GNRs is

IntechOpen

inversely related with the ribbon width (w_{GNR}) [5]. Thus GNR with narrow widths (15 nm) has been reported as a channel material for room temperature operation of Tunnel Field Effect Transistor (TFET) providing high ON–OFF current ratio [6]. In this paper authors used one dimensional Poisson equation to evaluate energy band diagram, surface potential subject to appropriate boundary conditions. The basic performance parameters of the device such as On-state current, On–Off current ratio, sub-threshold slope are calculated for high performance digital applications.

2. Structure of device

Figure 1 shows a p-channel tunnel field effect transistor using graphene nanoribbon(GNR) with highly doped source, channel and drain region, respectively. Here t_{OX} and t_{GNR} are represented as gate oxide and nanoribbon thicknesses, respectively. A high-k gate dielectric Y_2O_3 is chosen in between of gate and GNR. The channel of the GNR TFET is fully depleted for both in Off and On state of the device. A thin layer of graphene is deposited in the Si substrate to form a graphene nanoribbon as channel material of the device.

Now from numerical solution of following one-dimensional Poisson equation for the purpose of surface potential and energy band diagram of the device is obtained by [7]

$$\frac{d^2\varphi_{surf}(x)}{dx^2} - \frac{\varphi_{surf}(x) - V_{GS} - V_{BI}}{\lambda^2} = -\frac{q\rho(x)}{\varepsilon_{GNR}} \tag{1}$$

where V_{GS} is the gate to source potential, V_{BI} is the built-in potential, $\phi_{surf}(x)$ is the surface potential at position x, λ is the screening length for the particular device structure, ρ(x) is the total charge density and ε_{GNR} is the permittivity of GNR.

The screening length λ is

$$\lambda = \sqrt{\frac{\varepsilon_{GNR}}{\varepsilon_{OX}} t_{GNR} t_{ox}} = \sqrt{t_{GNR} t_{ox}} \tag{2}$$

The boundary conditions are for the calculation of energy band diagram as follows:

i. The electric field is zero at both side of the device i.e. source and drain ends.

ii. At the source-channel and drain-channel junction a continuous electric field potential exist.

iii. $\xi_F - \xi_C = qV_T$ at the source region and $\xi_V - \xi_F = qV_T$ at the drain region.

iv. At zero gate potential the Fermi level of the device is aligned with the valence band of the channel.

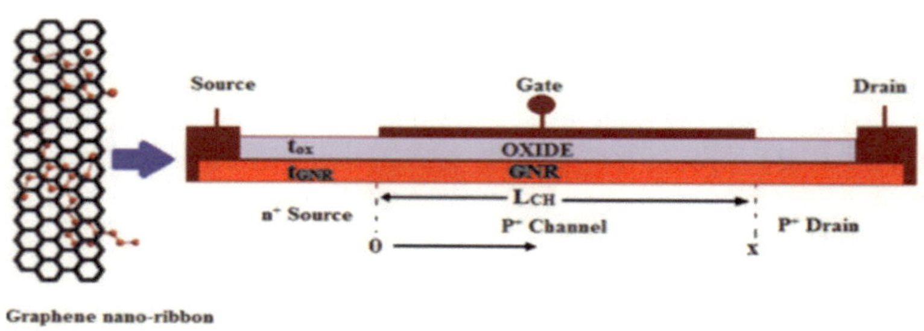

Figure 1.
P-channel tunnel field-effect transistor using graphene nanoribbon.

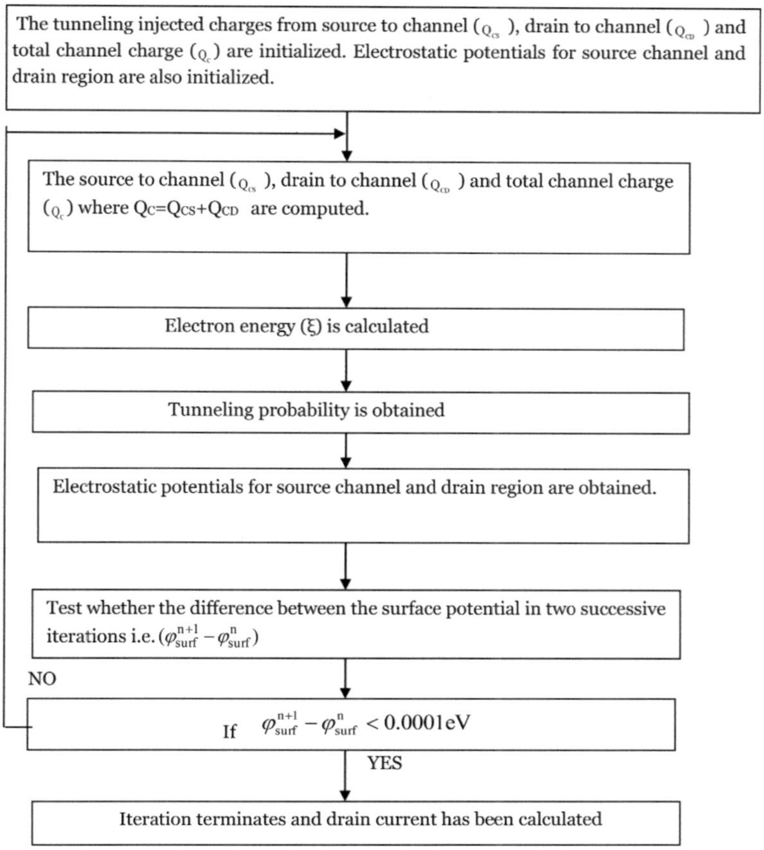

Figure 2.
Flow chart to obtain the energy band diagram and drain current.

The flow chart for the self-consistent iterative method to obtain the drain current is given in **Figure 2**.

The of tunneling probability as a function of energy is written as

$$T_S(\xi) = \exp\left(-2\int |k_x| dx\right) \tag{3}$$

where k_x is the wave vector.

The drain current is calculated from following Landauer's equation [8].

$$I_D = \frac{q}{\pi\hbar}\int (f_D(\xi) - f_S(\xi))T_S(\xi)d\xi \tag{4}$$

where f_s and f_d is the Fermi distribution function regarding source and drain regions and $\hbar$ is the reduce Plank's constant, respectively.

3. Results

The basic energy band diagram of the GNR PTFET is shown in **Figure 3a, b** in Off state (V_{GS} = 0 V, V_{DS} = −0.1 V) and On state (V_{GS} = −0.1 V, V_{DS} = −0.1 V) at channel length 85 nm and width 4 nm, respectively.

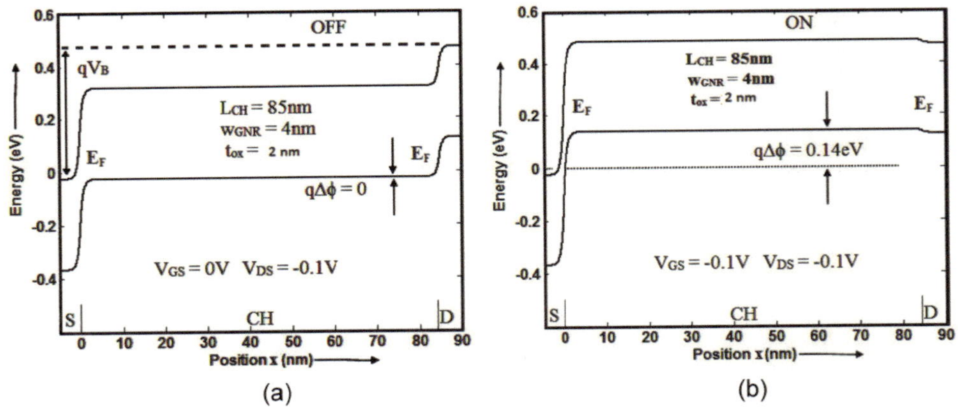

Figure 3.
(a) Off state energy band diagram of GNR PTFET. (b) On state energy band diagram of GNR PTFET.

Figure 3a shows that no band to band tunneling occurs in the OFF state. But in **Figure 3b** shows that the significant tunneling of carrier can occur properly in ON state of the device.

The OFF current for long channel GNR (L_{CH} = 85 nm) arises from thermionic emission over the barrier only and direct source to drain tunneling is negligibly small. Therefore The OFF current in GNR is written as [9]

$$I_{OFF} = \left[q^2 V_T/\hbar\pi\right] \exp\left(-V_B/qV_T\right) \quad (5)$$

where qV_B is the barrier height and qV_T is the thermal energy.

Figure 4a shows the ON/OFF current ratio versus gate to source bias (V_{GS}) for five different channel lengths from 45 to 85 nm in steps of 10 nm for fixed ribbon width of 4 nm and oxide thickness of 2 nm. It is observed that the ON–OFF current ratio increases with the increase of channel length. The ON–OFF current ratio increases from 2.34×10^3 to 4.96×10^4 at $V_{GS} = -0.1$ V when the channel length

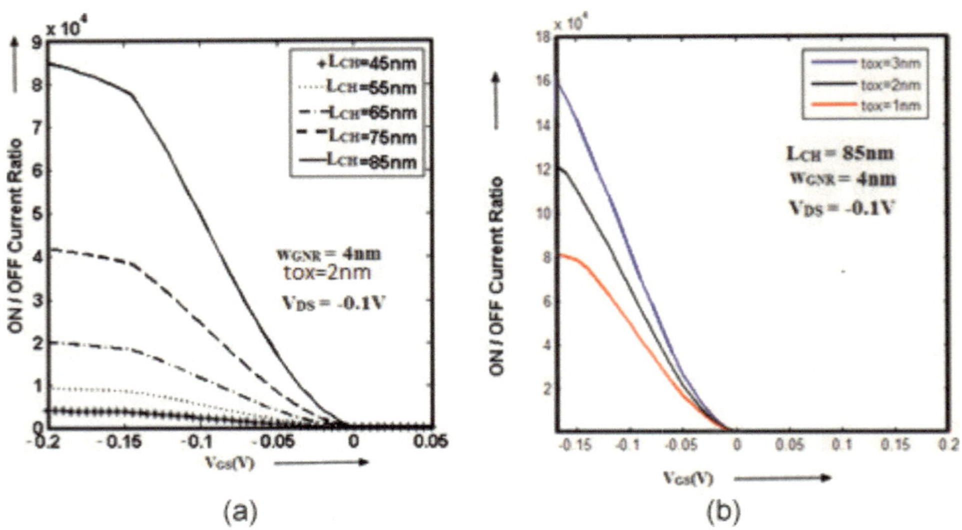

Figure 4.
(a) ON/OFF current ratio versus V_{GS} for different channel lengths of GNR PTFET. (b) ON/OFF current ratio versus V_{GS} for different oxide thicknesses of GNR PTFET.

increases from 45 to 85 nm. The ON–OFF current ratio reaches a maximum of 4.96×10^4 at $V_{GS} = -0.1$ V for L_{CH} = 85 nm. The higher ON–OFF current ratio for longer channel length at a particular gate-to-source bias and fixed ribbon width and oxide thickness can be explained as follows: In case of longer channel length, the total tunneling path length increases since tunneling takes place through all paths from source to drain. Thus tunneling probability will increase as seen from Eq. (3) so that drain current increases. **Figure 4b** shows the ON–OFF current ratio versus V_{GS} for three different gate oxide thicknesses (t_{ox}).

Figure 5 shows On state current versus gate to source voltage of GNR PTFET with different widths. The simulated results show that higher value of ribbon width on current is increase significantly.

Figure 6 shows the on state current density versus V_{GS} for GNR PTFET corresponding to L_{CH} = 85 nm, W_{GNR} = 4 nm and t_{ox} = 2 nm. The maximum on-state current density is found to be 590 μA/μm at V_{GS} = -0.1 V. The sub-threshold slope is given by

$$SS = \frac{\partial V_{GS}}{\partial(\log I_D)} \tag{6}$$

The sub-threshold slope is found 2.76 mV/decade from equation (6) at channel length 85 nm and ribbon width 4 nm, respectively.

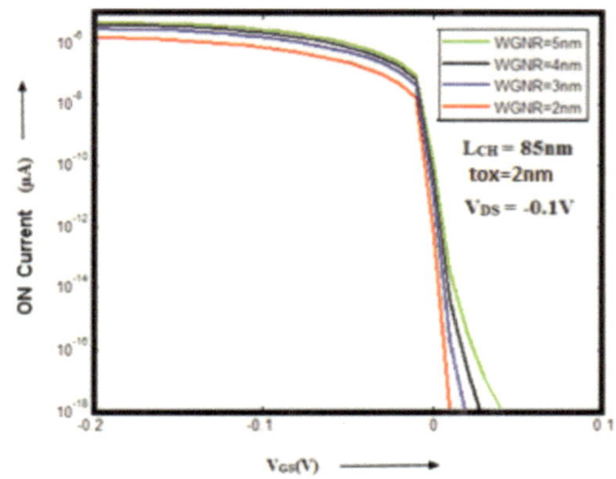

Figure 5.
ON current versus gate to source voltage of GNR PTFET for different GNR widths.

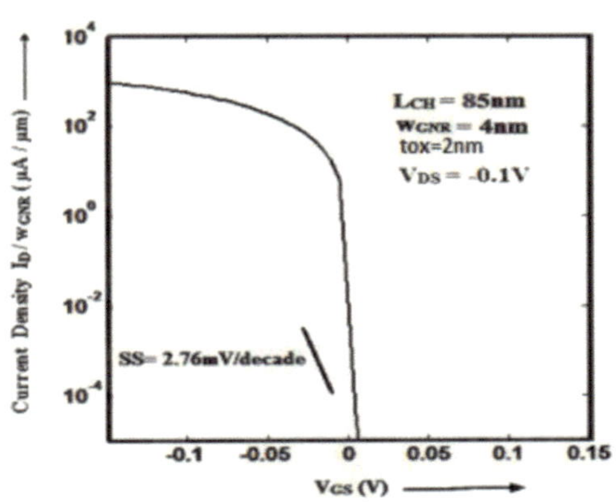

Figure 6.
ON-state current density versus gate to source voltage (V_{GS}).

Figures of merit	Parameter values with units
Current density I_D/w_{GNR}	590 (μA/μm)
Off state current	0.0092 (pA)
I_{ON}/I_{OFF} ratio	4.96×10^4
SS	2.76 (mV/decade)

Table 1.
Structural parameters of GNR PTFET.

Table 1 structural parameters of GNR PTFET:

$$L_{CH} = 85\text{ nm}, w_{GNR} = 4\text{ nm}, t_{ox} = 2\text{ nm}$$
$$V_{GS} = -0.1\text{ v}, V_{DS} = -0.1\text{ v}.$$

4. Conclusion

The results show that the graphene nanoribbon based Tunnel Field Effect Transistor (GNR-PTFET) provides higher on–off current ratio, lower sub-threshold slope for better switching in digital circuits using low voltage power supply. The values of I_{ON}/I_{OFF}, SS of the optimized device are 4.96×10^4, 2.76 mV/decade at channel width and length is 4 nm and 85 nm, respectively. Therefore this device is useful for high performance digital switching applications.

Acknowledgements

The author, Professor (Dr.) J. P. Banerjee (same as J. P. Bandyopadhyay) is grateful to the University Grants Commission, India for supporting the research through the award of an Emeritus Fellowship in the Institute of Radio Physics and Electronics, University of Calcutta.

Author details

Jayabrata Goswami*, Anuva Ganguly, Anirudhha Ghosal and Jyoti Prakash Banerjee
Institute of Radio Physics and Electronics, University of Calcutta, Kolkata, India

*Address all correspondence to: goswamijayabrata@gmail.com

References

[1] Seabaugh A, Zhang Q. Low-voltage tunnel transistors for beyond CMOS logic. Proceedings of the IEEE. 2010; **98**(12):2095-2110

[2] Nirschl T et al. The tunneling field-effect transistor (TFET) as an add-on for ultra-low voltage analog and digital processes. IEDM Technical Digest. 2004:195-198. DOI: 10.1109/IEDM.2004.1419106

[3] Bhuwalka KK, Schulze J, Eisele I. Performance enhancement of vertical tunnel field-effect transistor with SiGe in the δp+ layer. Japanese Journal of Applied Physics. 2004;**43**(7A):4073-4078

[4] Zhu Y, Murali S, Cai W, Li X, Suk JW, Poos JR, et al. Graphene and graphene oxide: Synthesis, properties, and applications. Advanced Materials. 2010;**22**:3906-3924. DOI: 10.1002/adma.201001068

[5] Zhang YB, Tang TT, Girit C, Hao Z, Martin MC, Zettl A, et al. Direct observation of a widely tunable bandgap in bilayer graphene. Nature. 2009;**459**:820

[6] Chen Z, Lin YM, Rooks MJ, Avouris P. Graphene nano-ribbon electronics. Physica E. 2007;**40**:228-232

[7] Appenzeller J, Knoch J, Bjork M, Riel H, Schmid H, Riess W. Toward nanowire electronics. IEEE Transactions on Electron Devices. 2008;**55**(11):2827-2845

[8] Fahad MS, Srivastava A, Sharma AK, Mayberry C. Analytical current transport modeling of graphene nanoribbon tunnel field-effect transistors for digital circuit design. IEEE Nanotechnology. 2016;**15**(1):39-50

[9] Sze SM. Physics of Semiconductor Devices. 2nd ed. New York: Wiley; 1981. p. 255.98

Chapter 12

Wax Deposition in Crude Oil Transport Lines and Wax Estimation Methods

Fadi Alnaimat, Mohammed Ziauddin and Bobby Mathew

Abstract

Petroleum industry is one of the major industries serving the energy demands. Flow assurance is essential for providing continuous fuel supply. Wax deposition is the main issue that affects flow assurance or reduces the efficiency of transporting crude oil. As the maintenance cost of repairing and troubleshooting is very high, addressing issues related to flow assurance becomes critical in the petroleum industry. This chapter will explore methods used for reducing, cleaning, and monitoring deposition of wax. Wax dissolved in the crude oil gets crystallized causing accumulation across the pipe walls once the bulk temperature of the crude oil gets lower than wax appearance temperature (WAT). Mechanical, thermal, chemical, and microbial methods highlighting general practice in the industry are discussed in this chapter. Next, the direct techniques providing information about the numerical wax deposition models used along with scientific measurement techniques are emphasized. Later, the indirect measurement techniques are discussed providing information about the external probing and nondestructive techniques to obtain information about wax layer deposition inside the pipe. The role of artificial intelligence and use of fuzzy logic for effective wax prediction or in developing the existing wax numerical models are emphasized in the last section.

Keywords: wax deposition, numerical model, wax appearance temperature, artificial intelligence, fuzzy logic

1. Introduction

Energy demands are continuously fulfilled by the petroleum industry. Transport lines of crude oil play an essential role in ensuring continuous supply of fuel, that is, providing flow assurance. As the maintenance cost of repairing and troubleshooting transport lines is very high, addressing issues related to flow assurance becomes critical in the petroleum industry. Crude oil consists of wax particles that are initially in the dissolved state and those get crystallized once the temperature of the pipe wall goes below certain temperature. The wax content in the crude oil is firstly in the dissolved form, and then it gets to precipitation and then gets crystallized causing accumulation across the pipe walls. This process is explained by molecular diffusion of wax particles toward the pipe wall when the temperature of the crude oil in bulk gets lower than wax appearance temperature (WAT) [1–8].

Wax deposition is a serious problem that causes reduction in the flow cross section, hence affecting flow assurance. In the subsea transport lines, the surrounding temperature drops very low which increases the crystallization, and wax deposition becomes more acute (as shown in **Figure 1**). With time, the crystallized wax particles get accumulated layer by layer and even can clog the pipe completely, which dramatically affect the maintenance work. Therefore, there are many methods used and studied by the industry as well research and development institutions in this direction to effectively find out the location of clog, to minimize the issue deposited wax, to remove the deposited layer of wax, and to predict the wax deposition inside the transport lines with time. All these efforts are taken to reduce wax deposition and mitigate in such a way that wax layer thickness can be predicted and addressed for maintenance once it reaches to a caution limit. Hence, predicting wax deposition can help in preventive maintenance and cost-effectiveness [9–13, 16].

This book chapter will discuss methods which are used for wax clearance, prediction, and estimation. This book chapter will cover mainly four sections: introduction, wax deposition issues and solutions, wax estimation methods, and role of artificial intelligence in wax prediction. The first section introduces the basic theory behind the wax deposition as a process and explains the main factors that are affecting wax deposition. The second section discusses four different methods adopted to tackle the problems related to wax deposition. The four different methods, mechanical, thermal, chemical, and microbial methods, will be discussed highlighting general practice in the industry. Further, their advantages and limitations are added in the same section. The third section is consisting of broad discussion which includes comparison of direct and indirect measurement techniques. The direct techniques are highlighting information about the numerical wax deposition models used along with scientific measurement techniques. On the other hand, the indirect measurement techniques are discussed knowing the external probing and examining techniques that can provide information about wax layer deposition inside the pipe. Finally, the role of artificial intelligence is discussed with benefits associated with the use of mapping information and using fuzzy logic for effective wax prediction or in developing the existing wax numerical models. Lastly, a brief conclusion is provided to reflect recent literature and hot topics in this direction.

Figure 1.
Sectional view of pipeline affected with acute wax deposition [14].

2. Wax deposition issues and solutions

The issue of wax accumulation is complex because many factors affect the wax deposition such as the wax concentration in crude oil, temperature of the surrounding, wax appearance temperature, pressure drop, viscosity of the oil, and bulk temperature of the oil. The main associated issues with wax deposition in crude oil transport lines are impact on flow assurance and a sudden clog that can lead to immediate actions of maintenance and repair. Deposited layer of wax can be observed as three sub-layers: the topmost layer is more granular and soft, the bottom layer is observed to have a strong bond with pipe wall and considered as close fitted layer, and the sandwiched layer in between the top and bottom has the mechanical impurities and high wax content. With time, the sedimented layers get hard and move from top layer stage to bottom layer stage, consequently reducing the effective flow cross section. The process of hardening of the bottom sedimented layer of wax is referred to as "aging." Hence, it is crucial to understand that with time, the wax deposition can cause difficulty in the cleaning process [11–13, 16–22].

To deal with wax deposition issues, conventionally in the industry, pigging process is used for cleaning the wax after inspecting externally or cleaning as a part of regular maintenance. In the pigging process (as shown in **Figure 2**), the deposited wax is scrapped by passing the pig device through the pipe such that its movement along the pipe causes its head to collect the deposited layer of wax. However, the pig device inserted has possibility to get stuck due to hard layer of wax or due to higher friction from the accumulated wax when cleaning longer distances. If the pig device gets stuck in between the pipe not near to inlet or outlet connections, then it becomes a hectic and complex issue to deal. In the following sections, common methods (mechanical, thermal, chemical, and microbial) are discussed in dealing with issues related wax deposition [16–22].

2.1 Mechanical methods

Mechanical removal of wax is considered as the oldest method used in the industry. This method includes the use of scrappers directly, use of scrappers in the tube, and use of "pig" device inside the pipe. Scrappers are used to scrap the tube wall and remove wax even when the well is under operation. Pipeline inspection gauge (pig) device is one of the broadly used old methods that have been used since a century in the industry [16–22].

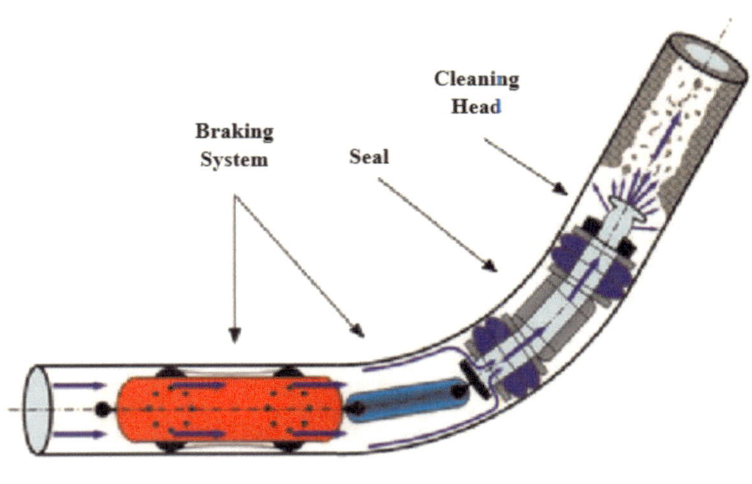

Figure 2.
Pigging process [15].

Both wax removal techniques are used for maintenance; however, they have disadvantages of plugging of perforation within the well when scrapping and when the pigging device gets stuck inside due to wax. All the mechanical methods are economical in comparison to other methods [16–22].

2.2 Thermal methods

Thermal methods are basically used to adhere to temperature medium or maintain temperature of crude oil for reducing wax buildup. Some of the common ways used are hot oiling or hot watering, cold flow, and surface coating. Using hot oil or hot water (temperature in range 65–105°C) pumping in the transport line, the deposited wax is melted. Using solid resin particles that are having melting point more than WAT in the oil facilitating slurry flow, the wax is prevented to deposit toward the wall. Using surface coating of thermal insulation material (like plastic), the wax inhibition is achieved shielding the drop of crude oil temperature. All these methods can be used; however, there is limitation of using hot oil or hot water depending on the heat capacity of the oil [16–22].

2.3 Chemical methods

Chemical methods are basically using chemical inhibitors, and these are added in the oil to reduce the wax deposition. These chemical wax inhibitors can be classified into three: detergents, dispersants, and wax crystal modifiers. Surface activators are the detergents and dispersants that sustain wax particles as suspended and dispersed such that reducing the wax particles to adhere to each other or the pipe walls or any solid surface. The surface activation agents also modify the solid surface of the pipe reducing shear and interaction of wax particles on the wall of the transport line. Other types of surfactants also modify the solubility by solubilizing nucleus and avert agglomeration of wax particles. Wax crystal modifiers are also referred to as pour point depressors because they allow the flow of oil at a minimum temperature at its own density and given conditions. Wax crystal modifiers have same structure as that of the wax particle, and they coprecipitate occupying on the crystal lattice of wax particle forming hydrocarbon chains. In this manner, they also act as encumber in the growth of wax crystals as they reduce the possibility of wax crystals to form 3D structures. All these wax inhibitors are effective but must be used before crude oil bulk temperature is above its WAT [16–22].

2.4 Microbial methods

Microbial method of treating wax is not common; however, it was found effective in few field testings. The action of bacterial culture is producing the biosurfactant which is reported to facilitate as wax inhibitor. The bacterial strains such as *Actinomyces* species have shown breakdown of heavy chain hydrocarbon fractions (from C_{15} to C_{20}) when treating the crude oil samples. Bacterial treatment was also noticed to induce crude oil lowering the WAT of the crude oil. This makes the crude oil with less susceptible condition for wax deposition. Using microbial method is an innovative approach, but it could be used in wells to have static culture [16–22].

3. Need for mitigation and wax estimation methods

Methods discussed earlier including mechanical, thermal, chemical, and microbial methods were to reduce and clean the wax deposited inside the transport lines.

Those methods served as the final solution for cleaning. However, as a part of mitigation and carrying out preventive maintenance, there is a need to have methods that can serve for estimating the wax buildup. This section is focused toward discussion of methods that can help in wax estimation. The direct measurement techniques are discussed to show the benefits associated with information process and about the numerical wax deposition models used along with scientific measurement techniques. The indirect measurement techniques are discussed to show how nondestructive testing can provide information through external probing and examining externally to know about wax layer deposition inside the pipe. Hence, this section is critically for knowing how the estimation is carried out regarding the wax layer thickness inside the crude oil transport line. Further, these are significant in avoiding sudden shutdowns due to blockage or complete closure with wax inside the pipe which can result in immediate maintenance cost [23].

3.1 Direct measurement techniques

Direct measurement techniques help in estimating the deposited wax layer thickness based on numerical assessment of deposition models as mentioned in the literature. The information about the parameters that are added to the model dependent on wax properties are measured using scientific measurement techniques. In other words, the estimation of the deposited wax layer is through wax deposition model but coupled with output obtained from scientific measurement. This section explains various wax deposition models highlighted in the literature and pointing out the most suitable model based on the assessment. In addition, different scientific measurement techniques are explained with respect to capability of each technique based on properties measured [23].

3.1.1 Wax deposition models

Mathematical modeling approach is widely adopted in order to predict and monitor wax deposition either through numerical estimation directly or using software that has back-end mathematical model. In this section, some of the main highlighted wax deposition models from the literature are discussed along with respective equations. Wax deposition models are developed based on consideration of assumptions and selection of parameters. Four main models are discussed in this

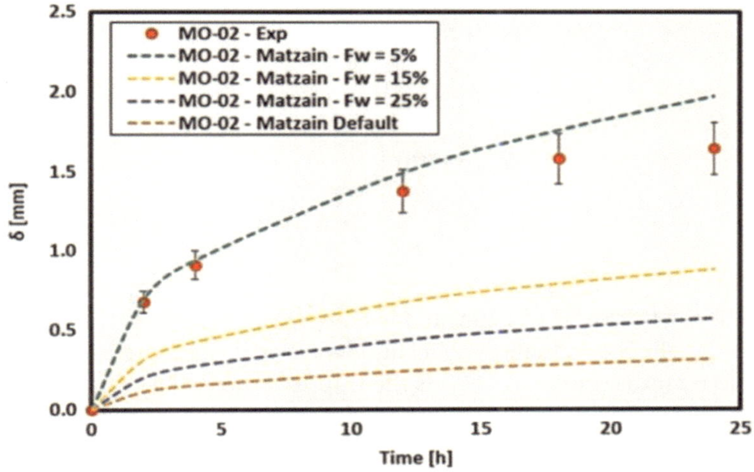

Figure 3.
Matzain's model comparison to experimental results [28].

book chapter: film mass transfer model (FMTM), equilibrium model (EM), Matzain's model (MM), and Venkatesan's model (VM) [24–28].

FMTM is developed based on the mass and heat transfer assumptions considering both transfers occur independently. EM is developed based on thermodynamic equilibrium along with consideration of concentration gradient in the model unlike FMTM. MM is a modified model of EM making it more effective by including the diffusion equation as empirical correlation as well as including the factors related to shear stripping and trapped oil factors. VM is developed mainly considering shear effect with two coefficients along with quantification of mass flux in the model [23, 29–36].

Among these four models, MM was found to be self-sufficient due to its fitting with experimental data. Also, it was due to consideration of oil entrapment and correlation of hear stripping effect in the model. The results obtained by testing MM model are shown in **Figure 3** [23, 29–36].

The mathematical equations governing with respect to all these four models are mentioned below [23, 29–36]:

$$k_L.2\pi r_i L\,[C_b - C_i(T_i)] = D_e(2\pi r_i L)\frac{dC}{dT}\bigg|_{T_i}\frac{dT}{dr}\bigg|_{r_i}^{+} + -\frac{d(\pi r_i^2)}{dt}\rho_w.L.\overline{F_w} \tag{1}$$

$$D_e = \frac{D_{wo}}{1+\alpha^2\frac{\overline{F_w}}{(1-\overline{F_w})}} \tag{2}$$

$$\alpha = 1 + K_\alpha.\overline{F_w} \tag{3}$$

where k_L is the convective mass transfer coefficient, D_{wo} is the diffusivity coefficient, F_w is the wax fraction percentage, C_b is the bulk soluble wax concentration, C_i is the interface soluble wax concentration, ρ_w is the wax density, α is the wax crystal aspect ratio, and K_α is the aspect ratio proportionality constant. K_α is used as fitting parameter [23, 29–36]:

$$\begin{gathered} D_{wo}.(2\pi r_i L)\frac{dC}{dT}\bigg|_{T_i}\frac{dT}{dr}\bigg|_{r_i}^{-} = D_e(2\pi r_i L)\frac{dC}{dT}\bigg|_{T_i}\frac{dT}{dr}\bigg|_{r_i}^{+} + -\frac{d(\pi r_i^2)}{dt}\rho_w.L.\overline{F_w} \\ \frac{d\delta}{dt} = -\frac{\pi_1}{1+\pi_2}\frac{D_{wo}}{\rho_o}\left[\frac{dC}{dT}\bigg|_{T_i}\frac{dT}{dr}\bigg|_{r_i}^{+}\right] \end{gathered} \tag{4}$$

$$\pi_1 = \frac{15.0}{1 - C_{Oil}/100} \tag{5}$$

$$\pi_2 = 1 + 0.055\,N_{SR}^{1.4} \tag{6}$$

$$C_{Oil} = 100\left(1 - \frac{R_e^{0.15}}{8}\right) \tag{7}$$

$$N_{SR} = \frac{\rho_o v_o \delta}{\mu_o} \tag{8}$$

where δ is the thickness of deposited wax, ρ_o is the density of the oil, μ_o is the viscosity of the oil, N_{RE} is the Reynolds number, N_{SR} is the dimensionless shear number in Matzain's correlation, C_{Oil} is the trapped oil coefficient, and π_1 and π_2 are the Matzain's empirical coefficients 1 and 2, respectively [23, 29–36]:

$$\dot{m}_w = k_L.2\pi r_i L\,[C_b - C_i(T_i)] - m\,\tau_W^n \tag{9}$$

where $\dot{m}_w$ is the rate of molecular weight of deposited wax; coefficients m and n are 0.8 x 10^{-12} and 1.9, respectively; and τ_W is the wall shear stress [23, 29–36].

3.1.2 Scientific measurement techniques

Scientific measurement techniques are coupled to wax deposition models practically because these techniques assist in providing information that is necessarily required as inputs for providing the output which is predicting the wax deposition thickness. Most dominating measurement techniques used in the industry for obtaining properties of crude oil samples are discussed here, which include near-infrared scattering (NIR), small-angle X-ray scattering (SAXS), X-ray diffraction (XRD), controlled stress rheometer (CSR), and cross-polarized microscope (CPM) [23].

3.1.2.1 Near-infrared scattering (NIR)

NIR is using the property of light scattering considered in a colloidal solution and obtains the physical properties. The near-IR range wavelength (low IR wavelength) attenuation spectra provide accurate results for obtaining WAT. The measurement deviation of ±2.5°C is observed when comparing the results obtained by CPM. This technique is effective with high-resolution results analyzed in 55 nm size window. It is also applicable if the oil sample is almost opaque to find out the WAT through delineation of radiation attenuation [37].

3.1.2.2 Small-angle X-ray scattering (SAXS)

Investigations carried out studying SAXS help in obtaining the radii of gyration. This technique can be used to study different fractions of crude oil at different operation temperatures. X-ray scattering at small angle can have issues related to low intensity. SAXS experimental results can be compared to calculations of scattering length density using chemical composition. This technique is applicable to obtain the size from radius of gyration and power law exponents providing details about physical properties of the crude oil sample [38–45].

3.1.2.3 X-ray diffraction (XRD)

This diffraction technique provides information about the crystal size of the wax by scattering in the time domain. XRD can help in understanding the wax structure capturing the wax deposition and aging. When using XRD it is important to understand that crystal size can also affect the diffraction. When the size of the crystal is below 0.1 μm, broadening of the diffraction peaks can be observed, and this broadening is as twice of the given angle. However, when the size of the crystal is above 0.1 μm, the diffraction characterizes Darwin width the same as the given angle of diffraction. XRD is suitable for characterizing crude oil samples studying the solid-solid transitions; hence, this method is effective in determining the crystal structure. But XRD has limitations in understanding the liquid-solid equilibrium, that is, identifying the crystallization from liquid to solid [46–50].

3.1.2.4 Controlled stress rheometer (CSR)

This technique utilizes the application of controlled stress on the sample with arrangement of parallel plate to obtain the strain exerted. In this manner, steady stress and steady deformation are obtained, and measurement of viscoelastic properties of the wax sample is achieved. For measurement, it is important to make sure that two parallel plates are set properly. The difficult part in measurement is that during measurement and when applying stress, it is crucial to make sure that the top layer does not slip. Slippage can affect the results, and when the wax weight percent is above 5%, slippage can be more prevailing when taking measurement [51–56].

3.1.2.5 Cross-polarized microscopy (CPM)

When analyzing the impact of cooling on the crude oil microstructure, CPM can be used. It can help in measuring WAT because cooling rate provided to the crude oil sample can be controlled and morphology can be observed with time. CPM provides information about the wax precipitation as wax appearance can be noticed with high resolution in small-size dimensions up to 0.5 μm. The volume of the sample stored for testing is very small, and CPM is sensitive to film thickness of sample which is dependent on the concentration of the sample [57–64].

3.2 Indirect measurement techniques

Indirect measurements here are referred to techniques which are evaluating the wax deposition experimentally by assessing physical quantities such as volume, temperature, pressure, electric capacitance, and ultrasonic signals. Change in volume is evaluated such that the resulted difference is the volume fraction of deposited wax. Similarly, the difference in pressure is also considered accounting for deposited wax. Both methods are intrusive, hence limiting its application in the industry. Therefore, nondestructive techniques are to interest which includes techniques that use temperature sensing, electrical capacitance measurement, and ultrasonic assessment [23]. Firstly, applying the temperature-based techniques, thermal sensing utilizes the heating pulse applied externally, and its transient response can assist in real-time assessment and monitoring of wax deposition [65, 66]. The investigation by [67] collected information about wax thickness inside the transport pipeline by observing the acoustic signals after providing the heat pulse externally. Signals obtained were Fourier transformed to observe frequency domain and extract information correlating to deposited wax layer thickness. Secondly, electrical capacitance measurement widely known in the literature as electric capacitance tomography (ECT) is effective in providing high-quality images by applying complex algorithms. ECT examined on the nonmetallic transport pipe experimentally showed that online monitoring of wax deposition can be achieved. Thirdly, ultrasonic measurement technique is also applied externally, and the information provided by the decaying time of ultrasonic signals can be correlated to deposited thickness of wax. Overall, many investigations are in the direction of exploring capabilities of ECT; however, few studies focused on nondestructive testing related to temperature-based prediction and related to ultrasonic decay time measurements [68–70].

4. Role of artificial intelligence

The trend of research and development in the oil and gas industry is shifting toward utilization of artificial intelligence (AI) algorithms and machine learning concepts. Based on the respective operating conditions, making the systems equipped with AI can enhance the decision-making capabilities. Some of the commonly used AI algorithms are evolutionary algorithm (EA), artificial neural network (ANN), swarm intelligence (SI), and fuzzy logic (FL). More than one AI algorithm can be applied if needed. Adaptive Neural Fuzzy Inference System (ANFIS) provided information about best has condensation ratios using ANN and FL assessment helped in continuous optimization of wax deposition model [23]. Among the recent studies, the work of [71] used statistical model considering the dependent and independent variables for wax deposition prediction. The dependent variable considered is viscosity of the crude oil, whereas the independent variable considered is pressure. By plotting pressure/viscosity versus pressure plot (as shown in **Figure 4**), the linear boundary limits were kept, and if the actual plot goes above the upper limit, it implicates high potential of wax deposition. The ANN model was developed based on backpropagation neural network (BNN). BNN uses two loops, a forward and a backward loop. The forward loop helps in processing the information inputs to outputs, whereas the backward loop does opposite from output to input. The backward loop processes information along with the weight error correction to take as input to forward loop. In the manner, the continuous operation of forward to backward and backward to forward loops, backpropagation algorithm gets trained. Hence, BNN is also referred as learning algorithm due its adjusting weights confined in the neural network.

More concise modeling is observed from work of [72] which consisted of ANN mathematical model for predicting rate of wax deposition. After observing that deposition rate of wax experimentally to be nonlinear, Kolmogorov theorem was applied; it virtually approximates nonlinear function to linear using two-layer ANN with certain error limit. The mapping structure for predicting wax deposition rate is shown below in **Figure 5**. The input variables (viscosity, shear stress, temperature gradient, and concentration gradient) and output variable is wax deposition rate. Comparison of the results with determined set showed that linear regression model was having correlation of 0.78, whereas ANN model had 0.97.

The work of [73] used ANFIS model to predict thickness of deposited layer of wax considering single-phase turbulent flow. Five-layered ANFIS model was considered consisting of input variable as Reynolds number, wax concentration (%), time, temperatures (outside, inside, and pipe wall), and temperature-driven force

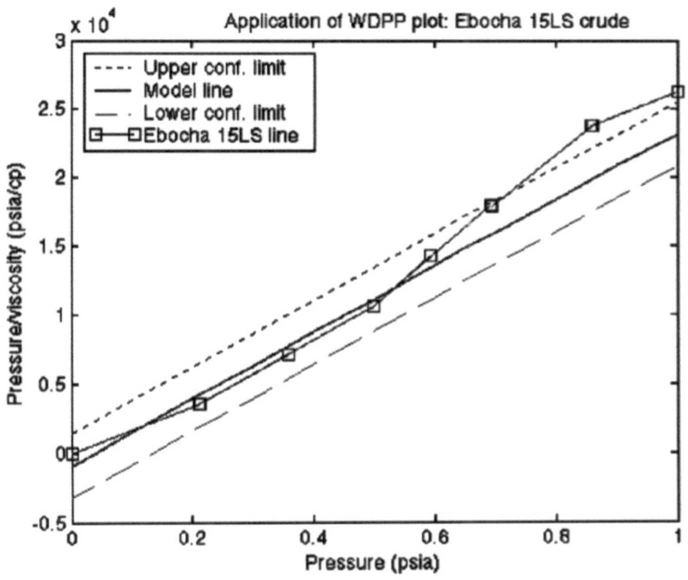

Figure 4.
Prediction of potential to wax deposition [71].

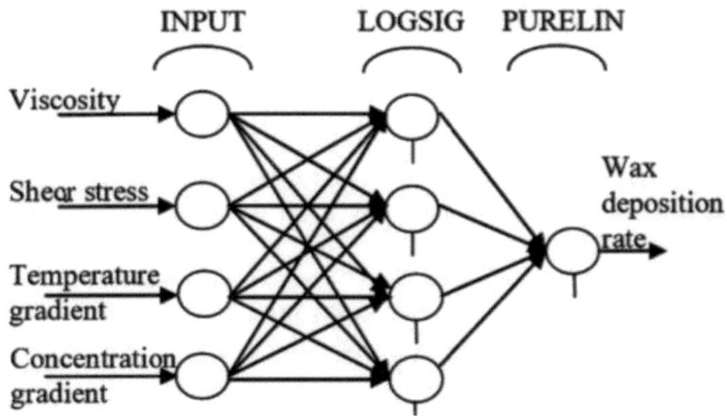

Figure 5.
Structure of ANN model for wax deposition rate prediction [72].

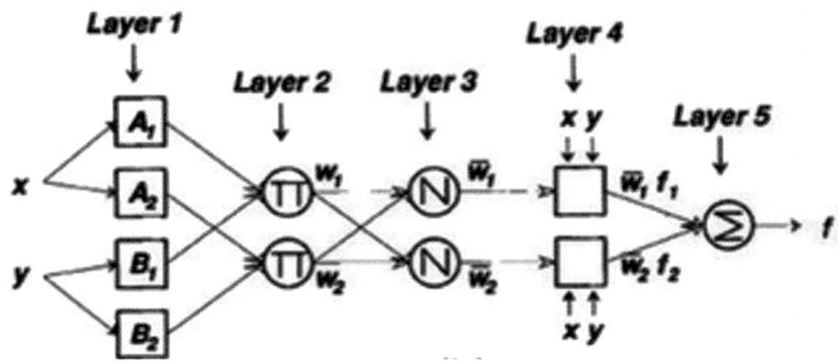

Figure 6.
Structure of ANFIS model for wax thickness prediction [73].

(ratio of gradient temperature wall and outside to bulk temperature of the oil). As shown in the **Figure 6**, the ANIFS model has five layers, and respective equations governing output of the model are mentioned below. The first-order fuzzy logic is applied using if/then rule. Considering if "A_i" belongs to "x" and "B_i" belongs to "y," then "f_i" the output function can be represented with combination of the parameters ("p_i," "q_i," and "r_i"):

$$f_i = p_i x + q_i y + r_i (i = 1, 2, ..., n) \tag{10}$$

In the first-layer equation representation, the combination can be calculated as its membership degree (μ) for labels set "A_i" and "B_i":

$$O_i^1 = \mu_{A_i}(x) \quad \textit{and} \quad O_i^1 = \mu_{B_i}(y) \;\; (i = 1, 2, ..., n) \tag{11}$$

In the second-layer equation representation, it can be shown with product of the membership degrees:

$$O_i^2 = W_i = \mu_{A_i}(x) * \mu_{B_i}(y) \qquad (i = 1, 2, ..., n) \tag{12}$$

In the third-layer equation representation, the calculation of the weighted ratio from each variable with respect to total weight is

$$O_i^3 = \overline{w_i} = \frac{w_i}{\sum_{i=1}^{n} w_i} \qquad (i = 1, 2, \ldots, n) \tag{13}$$

In the fourth layer equation representation, the adaption is achieved at this layer identifying this layer as defuzzification layer, where the learning rule is applied on this layer (i.e., minimizing the error). The summation of the weight applied with function is a resultant referring to the output layer, which is the fifth layer:

$$O_i^4 = \overline{w_i} f_i = \overline{w_i}\left(p_i x + q_i y + r_i\right) \qquad (i = 1, 2, \ldots, n) \tag{14}$$

$$O_i^5 = \sum_{i=1}^{n} \overline{w_i} f_i \qquad (i = 1, 2, \ldots, n) \tag{15}$$

The prediction of the deposited thickness of wax using this model resulted in close agreement with experimental values. The mean square error values comparing to experimental results was to three digit accuracy (0.00077034) and high value of correlation (0.9858).

5. Conclusions

In brief, this chapter explores different methods used in the industry and research for predicting and monitoring wax deposition. The information discussed introduces the process of wax deposition and wax deposition models as a theoretical background. Observing the recent literature, the role of artificial intelligence is discussed which is to serve in effective and precise prediction of wax deposition. Hence, artificial intelligence for application of nondestructive data collection assessment helps in developing the wax deposition models to incorporate the updated oil sample information periodically to ensure that the wax predictions are reliable.

Acknowledgements

The authors would like to acknowledge the funding support provided by United Arab Emirates under the grant numbers 31N265 and 31R168.

Conflict of interest

The authors have no conflict of interest.

Notes/thanks/other declarations

The authors would like to thank the United Arab Emirates University for providing necessary research facilities.

Author details

Fadi Alnaimat*, Mohammed Ziauddin and Bobby Mathew
United Arab Emirates University, Al Ain, United Arab Emirates

*Address all correspondence to: falnaimat@uaeu.ac.ae

References

[1] Wang Z, Yu X, Li J, Wang J, Zhang L. The use of biobased surfactant obtained by enzymatic syntheses for wax deposition inhibition and drag reduction in crude oil pipelines. Catalysts. 2016; **6**(61):1-16

[2] Ajienka J, Ikoku C. The effect of temperature on the rheology of waxy crude oils. In: Society of Petroleum Engineers, Vols. SPE-23605-MS. 1991. pp. 1-68

[3] Sun G, Zhang J, Ma C, Wang X. Start-up flow behavior of pipelines transporting waxy crude oil emulsion. Journal of Petroleum Science and Engineering. 2016;**147**: 746-755

[4] Aiyejina A, Chakrabarti DP, Pilgrim A, Sastry M. Wax formation in oil pipelines: A critical review. International Journal of Multiphase Flow. 2011;**37**:671-694

[5] Jaworski AJ, Meng G. On-line measurement of separation dynamics in primary gas/oil/water separators: Challenges and technical solutions—A review. Journal of Petroleum Science and Engineering. 2009;**68**:47-59

[6] Oh K, Jemmett M, Deo M. Yield behavior of gelled waxy oil: Effect of stress application in creep ranges. Industrial and Engineering Chemistry Research. 2009;**48**(19): 8950-8953

[7] Pedersen KS, Rønningsen HP. Influence of wax inhibitors on wax appearance temperature, pour point, and viscosity of waxy crude oils. Energy & Fuels. 2003;**17**:321-328

[8] Theyab MA, Diaz P. Experimental study of wax deposition in pipeline – Effect of inhibitor and spiral flow. International Journal of Smart Grid and Clean Energy. 2013;**2**:1-10

[9] Chala GT, Sulaiman SA, Japper-Jaafar A. Flow start-up and transportation of waxy crude oil in pipelines-a review. Journal of Non-Newtonian Fluid Mechanics. 2018;**251**: 69-87

[10] Chala GT, Sulaiman SA, Japper-Jaafar A, Abdullah WAKW, Mokhtar MMM. Gas void formation in statically cooled waxy crude oil. International Journal of Thermal Sciences. 2014;**86**:41-47

[11] Guozhong Z, Gang L. Study on the wax deposition of waxy crude in pipelines and its application. Journal of Petroleum Science and Engineering. 2010;**70**(1-2):1-9

[12] Mansour E, Desouky S, Batanoni M, Mahmoud M, Farag A, El-Dars F. Modification proposed for SRK equation of state. Oil and Gas Journal. 2012; **110**(6):78-91

[13] Ahmadpour A, Sadeghy K, Maddah-Sadatieh S-R. The effect of a variable plastic viscosity on the restart problem of pipelines filled with gelled waxy crude oils. Journal of Non-Newtonian Fluid Mechanics. 2014;**205**:16-27

[14] Venkatesan R, Nagarajan N, Paso K, Yi Y-B, Sastry A, Fogler H. The strength of paraffin gels formed under static and flow conditions. Chemical Engineering Science. 2005;**60**:3587-3598

[15] Døble K. Wax Deposition as a Function of Flow Regimes. Trondheim: Norwegian University of Science and Technology; 2018

[16] Shafquet A, Ismail I, Japper-Jaafar A, Sulaiman SA, Chala GT. Estimation of gas void formation in statically cooled waxy crude oil using online capacitance measurement. International Journal of Multiphase Flow. 2015;**75**:257-266

[17] Huang Q, Wang W, Li W, Ren Y, Zhu F. A pigging model for wax removal in pipes. Society of Petroleum Engineers. 2016;**32**(4):1-11

[18] Sarmento RC, Ribbe GAS, Azevedo LFA. Wax blockage removal by inductive heating of subsea pipelines. Heat Transfer Engineering. 2004;**25**(7): 2-12

[19] Paso KG, Fogler HS. Bulk stabilization in wax deposition systems. Energy & Fuels. 2004;**18**:1005-1013

[20] Abdul-Majid S. Determination of wax deposition and corrosion in pipelines by neutron back diffusion collimation and neutron capture gamma rays. Applied Radiation and Isotopes. 2013;**74**:102-108

[21] Piroozian A, Hemmati M, Manan IIMA, Rashidi MM, Mohsin R. An experimental study of flow patterns pertinent to waxy crude oil-water two-phase flows. Chemical Engineering Science. 2017;**164**:313-332

[22] White M, Pierce K, Acharya T. A review of wax-formation/mitigation technologies in the petroleum industry. SPE Production & Operations. 2017; **33**(3):1-10

[23] Alnaimat F, Ziauddin M. Wax deposition and prediction in petroleum pipelines. Journal of Petroleum Science and Engineering. 2019. DOI: 10.1016/j.petrol.2019.106385

[24] Panacharoensawad E, Sarica C. Experimental study of single-phase and two-phase water-in-crude-oil dispersed flow wax deposition in a mini pilot-scale flow loop. Energy & Fuels. 2013;**27**(9): 5036-5053

[25] Agarwal J. Experimental Study of Wax Deposition in Turbulent Flow Conditions by Using Model Oil (MO-14). Tulsa, OK: The University of Tulsa; 2016

[26] Singh A. Experimental and Field Verification Study of Wax Deposition in Turbulent Flow Condition. Tulsa, OK: The University of Tulsa; 2013

[27] Rittirong A, Panacharoensawad and Sarica. Experimental study of paraffin deposition under two-phase gas/oil slug flow in horizontal pipes. In: OnePetro, Offshore Technology Conference, 04-07 May. Houston, Texas, USA; 2014

[28] Chi Y. Investigation of Wax Inhibitors on Wax Deposition Based on Flow Loop Testing. Tulsa, OK: The University of Tulsa; 2015

[29] Singh P, Venkatesan R, Fogler HS, Nagarajan N. Formation and aging of incipient thin film wax-oil gels. AICHE Journal. 2000;**46**(5):1059-1074

[30] Agarwal J. Experimental Study of Wax Deposition in Turbulent Flow Conditions by Using Model Oil (MO-14). Tulsa, OK: The University of Tulsa; 2016

[31] Singh A. Experimental and Field Verification Study of Wax Deposition in Turbulent Flow Condition. Tulsa, OK: The University of Tulsa; 2013

[32] Rittirong A, Panacharoensawad and Sarica. Experimental study of paraffin deposition under two-phase gas/oil slug flow in horizontal pipes. In: Offshore Technology Conference, 04-07 May. Houston, Texas, USA; 2014

[33] Chi Y. Investigation of Wax Inhibitors on Wax Deposition Based on Flow Loop Testing. Tulsa, OK: The University of Tulsa; 2015

[34] Quan Q, Gong J, Wang W, Gao G. Study on the aging and critical carbon number of wax deposition. Journal of PetroleumScienceandEngineering. 2015; **130**:1-5

[35] Cussler EL, Hughes SE, III WJW, Aris R. Barrier membranes.

Journal of Membrane Science. 1988; **38**(2):161-174

[36] Giacchetta G, Marchetti B, Leporini M, Terenzi A, Acqua DD, Capece L, et al. Pipeline wax deposition modeling: A sensitivity study on two commercial software. Petroleum. 2017;**5** (2):1-13

[37] Paso K, Kallevik H, Sjoblom J. Measurement of wax appearance temperature using near-infrared (NIR) scattering. Energy & Fuels. 2009;**23**: 4988-4994

[38] Petoukhov MV, Svergun DI. Applications of small-angle X-ray scattering to biomacromolecular solutions. The International Journal of Biochemistry & Cell Biology. 2013; **45**(2):429-437

[39] Dwiggins CW. A small angle X-ray scattering study of the colloidal nature of petroleum. The Journal of Physical Chemistry. 1965;**69**(10):3500-3506

[40] Bardon C, Barre L, Espinat D, Guille V, Li MH, Lambard J, et al. The colloidal structure of crude oils and suspensions of asphaltenes and resins. Fuel Science and Technology International. 1996;**14**(1–2):203-242

[41] Eyssautier J, Levitz P, Espinat D, Jestin J, Gummel J, Grillo I, et al. Insight into asphaltene nanoaggregate structure inferred by small angle neutron and X-ray scattering. The Journal of Physical Chemistry B. 2011;**115**(21): 6827-6837

[42] Cosultchi A, Bosch P, Lara V. Small-angle X-ray scattering study of oil- and deposit-asphaltene solutions. Colloid and Polymer Science. 2003;**281**(4): 325-330

[43] Morante LR, Poveda JC, Montiel R, Henao JA. Aggregate structure analysis of colombian heavy crude oil-derived asphaltenes using small angle X-ray scattering. Journal of Oil, Gas and Alternative Energy Sources. 2017;**6**(5): 49-58

[44] Martyanov ON, Larichev YV, Morozov EV, Trukhan SN, Kazarian SG. The stability and evolution of oil systems studied via advanced methods in situ. Russian Chemical Reviews. 2017; **86**(11):999

[45] Yen TF, Erdman JG, Pollack SS. Investigation of the structure of petroleum asphaltenes by X-ray diffraction. Analytical Chemistry. 1961; **33**(11):1587-1594

[46] Siljuberg MK. Modelling of Paraffin Wax in Oil Pipelines. Norway: Norwegian University of Science and Technology; 2012

[47] Johnson TL, Cowie L. United States Patent US20070189452A1; 2007

[48] Coutinho JA, Silva JL d, Ferreira A, Soares MR, Daridon J-L. Evidence for the aging of wax deposits in crude oils by Ostwald ripening. Petroleum Science and Technology. 2003;**21**(3–4):381-391

[49] Paso K, Senra M, Yi Y, Sastry AM, Fogler HS. Paraffin polydispersity facilitates mechanical gelation. Industrial and Engineering Chemistry Research. 2005;**44**(18):7242-7254

[50] Chevallier V, Petitjean D, Bouroukba M, Dirand M. Mixtures of numerous different n-alkanes: 2. Studies by X-ray diffraction and differential thermal analyses with increasing temperature. Polymer. 1999;**40**(8): 2129-2137

[51] Wang Z, Lin X, Rui Z, Xu M, Zhan S. The role of shearing energy and interfacial Gibbs free energy in the emulsification mechanism of waxy crude oil. Energies. 2017;**10**(721):2-9

[52] Isono Y, Kawaura H, Komiyatani T, Fujimoto T. Differential dynamic

modulus of polyisobutylene with high molecular weight. 3. Stress development after the onset of steady shear flow. Macromolecules. 1991;**24**:4437-4440

[53] Singh P, Fogler HS, Nagarajan N. Prediction of the wax content of the incipient wax-oil gel in a pipeline: An application of the controlled-stress rheometer. Journal of Rheology. 1999; **43**:1437-1459

[54] Petrus T, Azuraien J. Rheological measurement of waxy crude oil under controlled stress rheometer: Determination of the setting parameters. Journal of Applied Sciences. 2014;**14**:2047-2053

[55] Silva JAL d, Coutinho J o AP. Dynamic rheological analysis of the gelation behaviour of waxy crude oils. Rheologica Acta. 2004;**43**:433-441

[56] Chen S, Øye G, Sjoblom J. Rheological properties of model and crude oil systems when wax precipitate under quiescent and flowing conditions. Journal of Dispersion Science and Technology. 2007;**28**:1020-1029

[57] Fakroun AA. The Development of a Knowledge-Based Wax Deposition, Three Yield Stresses Model and Failure Mechanisms for Re-Starting Petroleum Field Pipelines. United Kingdom: University of Bradford; 2017

[58] Dalla LFR, Soares EJ, Siqueira RN. Start-up of waxy crude oils in pipelines. Journal of Non-Newtonian Fluid Mechanics. 2018;**263**:1-32

[59] Magnin A, Piau J. Cone-and-plate rheometry of yield stress fluids. Study of an aqueous gel. Journal of Non-Newtonian Fluid Mechanics. 1990;**36**: 85-108

[60] Japper-Jaafar A, Bhaskoro PT, Azmi MHH, Sariman MZ, Norpiah R, Shahpin MH. Effects of drilling mud contamination on the properties of waxy crude oil. Platform - A Journal of Engineering. 2018;**2**:1-14

[61] Rodriguez-Fabia S, Fyllingsnes RL, Winter-Hjelm N, Norrman J, Paso KG. Influence of measuring geometry on rheomalaxis of macrocrystalline wax-oil gels: Alteration of breakage mechanism from adhesive to cohesive. Energy & Fuels. 2019;**32**(2):654-664

[62] Kok MV, Létoffé J-M, Claudy P, Martin D, Garcin M, Volle J-L. Comparison of wax appearance temperatures of crude oils by differential scanning calorimetry, thermomicroscopy and viscometry. Fuel. 1996;**75**(7):787-790

[63] Lee HS, Singh P, Thomason WH, Fogler HS. Waxy oil gel breaking mechanisms: Adhesive versus cohesive failure. Energy & Fuels. 2007;**22**(1): 480-487

[64] Coutinhoa JAP, Daridonb J-L. The limitations of the cloud point measurement techniques and the influence of the oil composition on its detection. Petroleum Science and Technology. 2005;**23**(9–10):1113-1128

[65] Alnaimat F, Mathew B, Ziauddin M. Thermal testing of wax layer thickness in crude oil pipeline. Journal of Physics: Conference Series. 2019;**1276**(1):1-9

[66] Alnaimat F, Mathew B, Ziauddin M. The operation of crude oil pipeline: Examination of wax thickness. In: Advances in Artificial Intelligence, Software and Systems Engineering. Springer Nature Switzerland AG: Springer; 2019. pp. 514-522

[67] Hoffmann R, Amundsen L. Single-phase wax deposition experiments. Energy & Fuels. 2010;**24**:1069-1080

[68] Mei ILS, Ismail I, Shafquet A, Abdullah B. Real-time monitoring and measurement of wax deposition in pipelines via non-invasive electrical

capacitance tomography. Measurement Science and Technology. 2016;**27**:1-11

[69] Sardeshpande MV, Harinarayan S, Ranade VV. Void fraction measurement using electrical capacitance tomography and high speed photography. Chemical Engineering Research and Design. 2015; **94**:1-11

[70] Li N, Liu K, Yang X, Cao M. Research on application of wax deposition detection in the nonmetallic pipeline based on electrical capacitance tomography. Journal of Sensors. 2016; **2016**:1-10

[71] Obanijesu EO, Omidiora EO. Artificial neural network's prediction of wax deposition potential of nigerian crude oil for pipeline safety science and technology. Petroleum Science and Technology. 2008;**26**:1977-1991

[72] Qiyu H, Jun M. Prediction of wax deposition of crude using statistical and neural network methods. In: 7th International Pipeline Conference. Alberta, Canada; 2008

[73] Jalalnezhad MJ, Kamali V. Development of an intelligent model for wax deposition in oil pipeline. Journal of Petroleum Exploration and Production Technologies. 2016;**6**:129-133

Section 4

Computational Intelligence Methods

Chapter 13

Comparative Study of Interval Type-2 and Type-1 Fuzzy Genetic and Flower Pollination Algorithms in Optimization of Fuzzy Fractional Order $PI^{\lambda}D^{\mu}$ Controllers

Himanshukumar R. Patel and Vipul A. Shah

Abstract

In this chapter, a comparison between fuzzy genetic optimization algorithm (FGOA) and fuzzy flower pollination optimization algorithm (FFPOA) is bestowed. In extension, the prime parameters of each algorithm adapted using interval type-2 and type-1 fuzzy logic system (FLS) are presented. The key feature of type-2 fuzzy system is alimenting the modeling uncertainty to the algorithms, and hence it is a prime motivation of using interval type-2 fuzzy systems for dynamic parameter adaption. These fuzzy algorithms (type-1 and type-2 fuzzy system versions) are compared with the design of fuzzy control systems used for controlling the dihybrid level control process subject to system component (leak) fault. Simulation results reveal that interval type-2 fuzzy-based FPO algorithm outperforms the results of the type-1 and type-2 fuzzy GO algorithm.

Keywords: interval type-2 fuzzy logic, fuzzy fractional order PID, fuzzy controller, genetic optimization, flower pollination

1. Introduction

Since many years, metaheuristic optimization algorithms have been used to solve numerical optimization problems from the defined search space without any concern of the required parameters. In this chapter, prominent bioinspired optimization algorithm like genetic optimization algorithm (GOA) and flower pollination optimization algorithm (FPOA) is presented for the problem of optimization of membership functions for a fuzzy controller.

In this chapter, two well-known metaheuristic optimization algorithms are used for a comparison in the optimization of a fuzzy system used as a controller of dihybrid level control system. The main reason for preferring GOA and FPOA algorithms is because they use the same methodology for parameter adaptation; however, these two algorithms work with similar inputs in fuzzy system but with dissimilar outputs, considering the outputs of the fuzzy system are parameters of the optimization algorithm which are dynamically adjusted for each iterations of each algorithm.

Genetic algorithms (GA) have been popular since the beginning of the 1960s; from the University of Michigan, Holland started an initial work on GA. His first contribution was based on adaption theory in natural and artificial system [1] in 1975. Genetic algorithms like neural networks are biologically inspired and represent a new computational model having its roots in evolutionary sciences. The core aspect of the GA has been well established with deep theoretical concepts and some practical domain examples in [2]. The GA is very popular in solving complex engineering problems, because of the feasibility and robustness of GA concepts. However, against the prominent advantages of a GA for determining difficult, constrained, and multi-objective functions where other approaches may have failed, the full strength of the GA in engineering application is yet to be exploited [3, 4]. The GA has inadequacy to control parameters which are dynamic in nature. The parameters of GA and a methodology for dynamic parameter adaptation are presented in Section 4.

As a novel metaheuristic algorithm, the flower pollination optimization algorithm (FPOA) is motivated by the pollination philosophy of flowers. In nature, the pollination methods for flowers associate two main types: cross-pollination and self-pollination [5]. In cross-pollination, some birds operate as global pollinators that relocate the pollen to the flowers of higher distant plants. In contradiction, in self-pollination, pollen is disseminated by the wind and only between neighboring flowers on the same plant. From the fundamental understanding of concepts of cross-pollination and self-pollination, the FPOA is created mapping between two types of pollination, core and self, with global and local pollination operators, respectively. The FPOA is gaining more and more attention in recent times due to its advantages, simple in nature, lesser parameters, and user-friendly operation.

The conventional integer proportional integral derivative (PID) controllers are tuned in [6–8] using bioinspired GOA and FPOA for various applications like DC motor, buck converter, and continuous stirred-tank reactor (CSTR) to control the plant. Several approaches had been proposed for GA, for example, in [9] an approach with GA for control vector for loss minimization of induction motor can be seen. Shopova and Vaklieva-Bancheva have introduced in detail a genetic algorithm called BASIC, designed to handle with numerous engineering optimization problems [10]. The self-organizing GA optimization method has been used for deriving PID controller parameters to escape incomplete convergence and to accomplish good optimization performance [11]. Krohling and Rey have examined a strategy to design an optimal disturbance rejection PID controller based on genetic algorithms for solving the constrained optimization problem in a servomotor system [12]. Kumar et al. have illustrated the design of GA-based controller for a bioreactor model that outperformed Ziegler-Nichols and Skogestad's tuned controller in terms of overshoot and undershoot as well as disturbance rejection and set point tracking [13].

A hybridization of the algorithms was performed in [14]; the authors publish a hybridization between the particle swarm optimization (PSO) and GO algorithms to minimize the cost and materials required for the elaboration of a metal cylinder. To increase reliability and performance of the optimization algorithm, ambiguous data, and the extension of the type-1 fuzzy sets (T1FSs) which is intuitionistic fuzzy sets (IFSs), it is used to find the optimal parameters in the algorithms. In recent scenarios, intuitionistic fuzzy logic system is an effective technique in bioinspired algorithms. A new model for decision-making is advised in [14], which is based on the IFSs; the objective of the new model is to eliminate parameter uncertainty in the data to help in the right decision-making. This techniques have been significantly enhanced timing and pressure by judgmenter. In [15], with the intention of improving the accuracy in optimization algorithms, the author proposed the interval type-2 fuzzy set (IT2FS), which helps to find the level of membership of an object to something else in analytical terms, and this method undoubtedly improves the accuracy of membership of a data set. There are certain works by Garg et al. in which fuzzy logic has been used for the same metaheuristics [16, 17].

This is why we examine the IT2FS for parameter adaptation in bioinspired algorithms. As the major contribution of this research work, two bioinspired optimization algorithms are proposed and their fuzzy variants with dynamical parameter adjustment using type-1 fuzzy logic system and interval type-2 fuzzy logic systems as tools for modeling nonlinear complex problems, exclusively for the stability of the dihybrid level control system subject to system component (leak) faults. Despite using these algorithms as tools to optimize the fuzzy fractional order PID controller membership function, a comparison is performed with all variants of GOA and FPOA algorithms.

The rest of the chapter is organized as follows. Section 2 describes the state of the art with related works for each bioinspired optimization algorithm. Section 3 contains a more exhaustive information of the internal work of the two algorithms. Section 4 describes the methodology used to dynamically adjust the parameters of each method. Section 5 contains the versions of the algorithms with dynamic parameter adaptation using the proposed strategy with a type-1 and an interval type-2 fuzzy system. Section 6 describes briefly the problem statement in which the two bioinspired optimization algorithms were tested. Section 7 shows the results of applying the bioinspired optimization algorithms to the optimization of a fuzzy system used in control. Finally, in Section 8, the conclusion and future work of this paper are presented.

2. Related works

The genetic algorithm is an optimization technique based on indiscriminate search method, which can be used to optimize the complex problem and even solve the nonlinear equations. In GOA the first parameter is chromosomes (genotype or individuals); it is a set of parameters which contains the potential solution to the problem that the GA is trying to solve and derive iteratively. The second parameter is "generation"; it defines each iteration of the algorithm. The progression of the solutions is simulated over a fitness function and other genetic operators like reproduction, mutation, and crossover [18]. In the literature there are many variants and improvements of this algorithm, for example, in [19], the authors present a work, where the algorithm is combined with techniques of fuzzy logic and the algorithm parameters are tuned using fuzzy logic system, and found improved results. In addition, in [20], an empirical study of the GA is presented in that the author tuned the parameter of the GA by GA itself.

Homayouni and Tang [21, 22] propose the use of fuzzy logic-based GA for the scheduling of handling/storage equipment in an automated manner. In recent times type-1 fuzzy logic is used to optimize the crossover and mutation rate for GA; in [23, 24] the authors proposed the methods and solve the real-time problem like rail-freight crew-scheduling problem. In 2013, Maldonado et al. [25] proposed fuzzy-based system for PSO and GA optimization, and it is used for FPGA applications, and the fuzzy system is used to control the speed of DC motor.

Flower pollination optimization algorithm (FPOA) is a new high-performance heuristic optimization algorithm which is always welcome to solve real-world problems [26]. References [27–29] represent that FPOA has the promote solution and robustness than other published methods and also it has shown reasonable superiority over GA. FPOA has only one parameter p (switch probability) which causes the simple algorithm to implement and quickly reach optimum solution [30]. The other special competence of FPOA is a wide-ranging domain search with quality and the same texture solution. And hence it is used alone with DE for multi-objective optimal dispatch problem [31]. FPOA is used to solve many complex problems and also compared with various optimization algorithms [32], and its performance assure to implement for present problem.

3. Bioinspired optimization methods

3.1 Genetic algorithms for optimization

The simple genetic algorithm can be expressed in pseudo code with the following cycle.

Algorithm 1. Genetic Algorithm Optimization.

```
Generate the initial population of individuals aleatorily P (0).
While (Number Generations <= Maximum Numbers Generations)
Do:
{
Evaluation;
Selection;
Reproduction;
Generation ++;
}
Show results
End
```

The modified GOA with parameter adaptation is modified, where the main difference with respect to original GOA is the calculation of the new crossover K_1 and mutation rate K_2.

3.2 Characteristics of bioinspired flower pollination algorithm

The flower pollination is a nature-inspired metaheuristic optimization algorithm; it is based on the concepts of "flower optimal breeding" and genetic algorithm "survival of the fittest." The primary types of the flower pollination are divided into two types: biotic and abiotic [26]. The majority flowering plants is a biotic pollination; it is around 85–90%. The transportation of the pollen is by natural resources such as birds, bees, insects, and animals. However, the remaining pollination of 15–10% takes the help of abiotic sources, for example, wind and diffusion in water. The pollination activity accomplished by self-pollination or cross-pollination is presented in **Figure 1** [26]. The term self-pollination can be defined as the fertilization of one flower from pollen of the same flower (autogamy) or the nearby flowers of the same plant (geitonogamy) [26]. It can develop when a flower consists of both male and female eggs. The basic characteristic of the self-pollination is that it takes place generally at short distance and without pollinators. It is a proof for the local pollination. In the contradictory, allogamy (cross-pollination) materialize in the case of cereal are moved to a flower from another plant. These process can be done with the help of biotic or abiotic operators as pollinators. It is observed as the global pollination. Bees and birds as biotic pollinators operate Lèvy flight behavior [27] with jump or fly distance steps which adhere to a Lèvy distribution. The FPO algorithm proposed by Yang [26] can be summed up by pseudo code for FPO Algorithm 2.

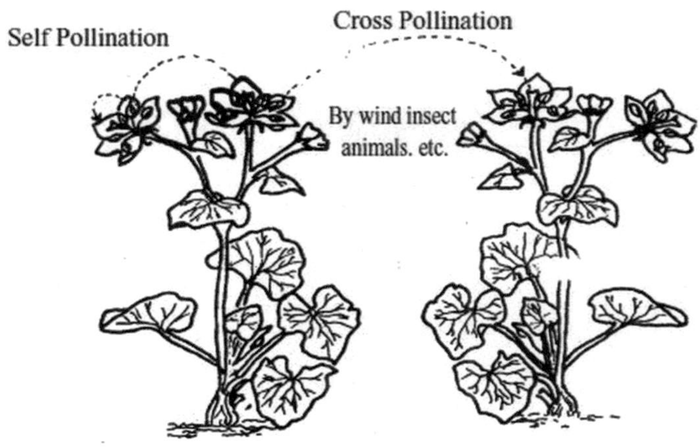

Figure 1.
The behavior of FPOA in terms of fitness function and iteration count [26].

3.3 Mathematical modeling of FPOA

By confirming the characteristic of the flower pollination procedure, we can define the following rules by pollinators [27]: When pollinators relocated pollen by operating Lèvy flight with procedure of global pollination, then it is treated as cross-pollination and biotic. The classes of biotic and self-pollination are foreseen from the local pollination. Now for the same attributes of flower pair, it is proportional to the probability of the breeding, and we can say it is pollinator constancy. Transformation possibility $p \in [0, 1]$ is used to control the procedure of local and global pollination. In the entire pollination process due to physical closeness and another circumstance such as wind. From the investigation of attributes of the pollination procedures, the FPOA methodology has developed; therefore, we first convert the above four rules into the mathematical equations. In the beginning, the basic steps of global pollination, pollen eggs are transferred using pollinators, for example, flying insects can fly over a long distance. This process establishes that pollination and breeding of the competent results are characterized as f^*. Rule 1 and flower dependability perhaps are shown as Eq. (1) [28]:

$$y_i^{t+1} = y_i^t + P\left(y_i^t + f^*\right) \tag{1}$$

where y_i^t describes the pollen i or vector of solution y_i at generation t. The pattern f^* presents the current optimal solution at the current no. of iterations that are constructed among all the solution. Durability of the pollination is expressed by element P, which is essentially a step size.

The Lèvy flight defined as the characteristics of flying insects can migrate over a large distance during transferring the pollen. That is, $0 > P$ embellished from a Lèvy distribution [28]:

$$P \sim \frac{\lambda \Gamma(\lambda) \sin(\pi\lambda/2)}{\pi} \frac{1}{s^{1+\lambda}}, (s >> s_0 > 0). \tag{2}$$

In Eq. (2) classic gamma function is expressed by $\Gamma(\lambda)$, and this type of distribution is applicable for large steps $0 > s$. Considered value of λ is 1.5. Eq. (3) clarifies the local pollination, and rule 2 plus flower dependability can be mathematically modeled as [28]:

$$y_i^{t+1} = y_t^i + \in \left(y_j^t - y_k^t\right) \tag{3}$$

where pollen of different flowers of the same plant is shown by y_j^t and y_k^t. This essentially mimics the flower constancy in a limited neighborhood. If y_j^t and y_k^t belong to the same category and same population, this becomes a local random walk if we express $\in$ from a uniform distribution in the range [0,1] [28].

Algorithm 2. Pseudo code for FPO Algorithm

```
0: Initialize objective as minimization.
   Define the population for n flowers.
   Find current best solution f* in the initial population.
   Describe the switch probability p ∈ [0, 1].
   while (t < MaxIteration) do
     for i=1: n do
       if rand < p then
         Define step size P which follow Lèvy distribution. Use Eq. (1) to perform global
         pollination.
       else
         Define ∈ for uniform distribution [0,1].
         Randomly select a j and k among all the solution.
         Perform local pollination by Eq. (3).
       end if
       Calculate new solution.
       If calculate solution is better, then update it in population.
     end for
     Get the optimal solution f*.
   end while
```

The modified FPOA with parameter adaptation is modified, where the main difference with respect to FPO algorithm is the calculation of switch probability P and the use of a fuzzy system to calculate new switch probability P.

4. Bioinspired methods with parameter adaptation

In this work, two bioinspired optimization algorithms are used: GA and FPOA. Both the methods used in these works use the same procedure for parameter adaptation; however, small adjustment is enforced to each optimization method, because of, for parameter adaptation fuzzy logic system is used to revise the one or more parameter's value during the execution of each iteration of the algorithm. To find out the new parameter values, the fuzzy system uses as input the percentage of transpire iterations and the degree of diversity $Q(t)$ of individuals from each bioinspired method, and now from these metrics, these parameters are used as an input for the fuzzy system as defined by Eqs. (4) and (5), respectively:

$$Iteration = \frac{Current \;\; iteration}{Maximum \;\; of \;\; iterations} \tag{4}$$

$$Diversity(Q(t)) = \frac{1}{n_s} \sum_{i=1}^{n_s} \sqrt{\sum\nolimits_{j=1}^{n_x} \left(X_{ij}(t) - \overline{X}_j(t)\right)^2 n_s} \tag{5}$$

To implement the methodology in context, the iteration number is given in terms of percentage of what iteration we are presently in, and hence we can model the fuzzy rules for updating the parameters rely on early-mid-final iterations of algorithm and, consequent to full knowledge of this, changing the parameters of bioinspired optimization algorithm, respectively.

The second input parameters is diversity $Q(t)$ which gives us a degree of closeness to the global best individuals with respect to individuals. From this parameter we can control the speed of the algorithm and control the local (individuals get close together) and global (individual gets removed for the rest) search.

By bringing together these two matrices such as **iteration** and **diversity**, we can easily control the performance of a bioinspired optimization algorithm by manipulating its parameters; this is of course possible; however, it requires previous knowledge of the algorithm.

5. Parameter adaption of bioinspired algorithm using type-1 and interval type-2 fuzzy logic system

The GO algorithm has proven to be a good technique to optimize parameters [18, 19]; that is why we perform a computerized search that grants good performance of the genetic algorithm. For this research work in the case of fuzzy controller, first we define the fitness function of the GOA, and it is a mean square error (MSE) which is shown in Eq. (6). For each mutation and for N iterations, type-1 fuzzy system design for the GOA is calculated and trying to minimize the objective function (MSE error). Accordingly, for designing the fuzzy systems, which dynamically adapt the K_1 and K_2 parameters, the iteration and diversity metrics are considered as inputs:

$$MSE = \frac{1}{n}\sum_{i=1}^{n}\left(Y_i - \hat{Y}_i\right)^2 \tag{6}$$

where n are the predictions generated from a sample of n data points on all variables, Y_i is the vector of observed values of the variable being predicted, and $\hat{Y}_i$ is the predicted values of the variable.

The bifurcation of the membership functions for inputs and outputs is completed in a symmetrical appearance. The layout of input variables in terms of linguistic variables is depicted in **Figure 2** for the type-1 fuzzy logic system.

The type-1 fuzzy system is depicted in **Figure 2**, having two inputs—one is iteration, and the second one is diversity $Q(t)$—and having two outputs K_1 (crossover) and K_2 (mutation rates). In this case, each inputs were divided into three linguistic variables with triangle membership functions, and the outputs were granulated into four linguistic variables with triangle membership functions. The proposed type-1 fuzzy system consists of nine fuzzy *if-then* rules to control the performance of GOA.

Now, in **Figure 3** appreciated interval type-2 fuzzy system, it has similar *if -then* fuzzy rules and linguistic variables for inputs as well as for outputs. However the rule base is the same, meaning expert knowledge is the same but only type of membership functions is changing from the type-1 fuzzy system.

Consequently, **Table 1** presents the *if-then* fuzzy rules for both the fuzzy systems type-1 and type-2 in **Figures 2** and **3**, respectively. The rule base is designed based on past simulation results to generate the expert knowledge base from parameters of GOA and how to control the performance.

At the time of fuzzy system designing, symmetric triangle membership functions for inputs and outputs for type-2 FLS were taken. Type-1 FLS and interval type-2 FLS have the same number of fuzzy *if-then* rules and use Mamdani style.

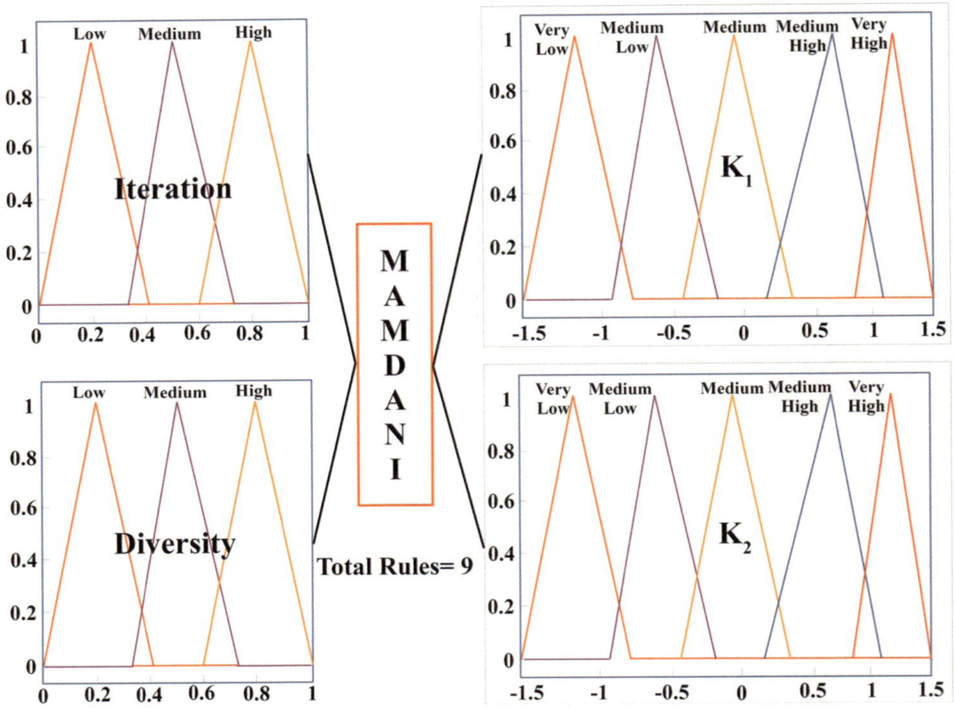

Figure 2.
Type-1 fuzzy system for dynamic parameter adaptation in genetic algorithm optimization (GAO).

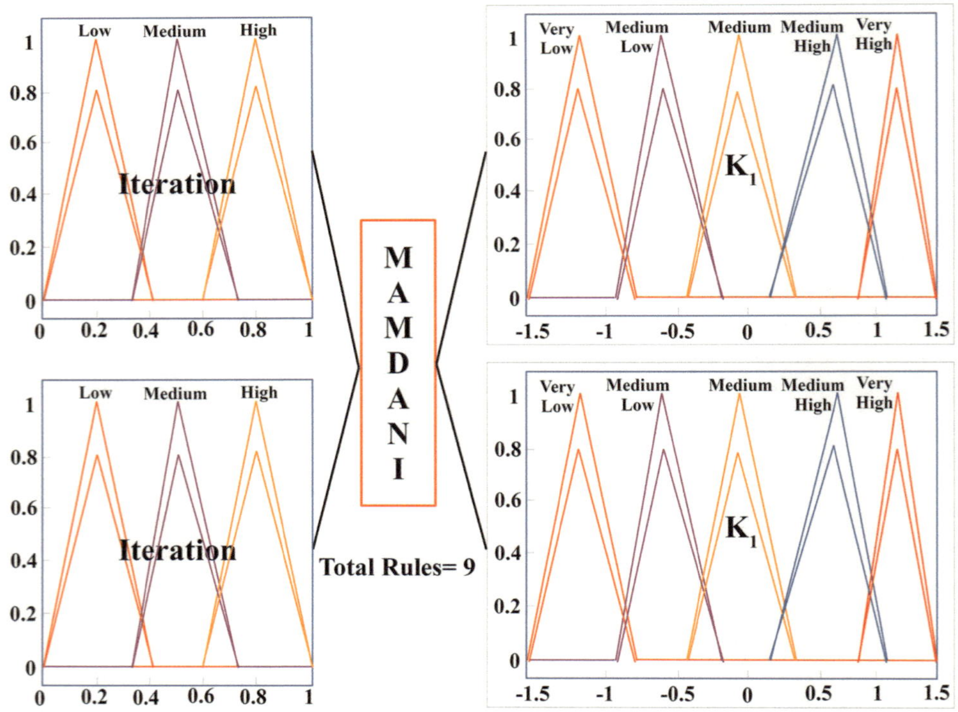

Figure 3.
Interval type-2 fuzzy system for dynamic parameter adaptation in GAO.

From the illustrated strategy for parameter adaptation of bioinspired optimization algorithm in Section 4, only one parameter can play the vital role in performance of the FPOA, and it is the best parameter, which is the switch probability P from Eq. (2). So controlling this parameters can control the entire FPO algorithm behavior. For the parameter adaptation in FPOA, the same like GOA type-1 and interval type-2 FLC is designed.

No.	Inputs		Outputs	
	Iteration	Diversity	K_1	K_2
1.	Low	Low	Very high	Very low
2.	Low	Medium	Medium high	Medium
3.	Low	High	Medium high	Medium low
4.	Medium	Low	Medium high	Medium low
5.	Medium	Medium	Medium	Medium
6.	Medium	High	Medium low	Medium high
7.	High	Low	Medium	Very high
8.	High	Medium	Medium low	Medium high
9.	High	High	Very low	Very high

Table 1.
Fuzzy rules for dynamic K_1 and K_2.

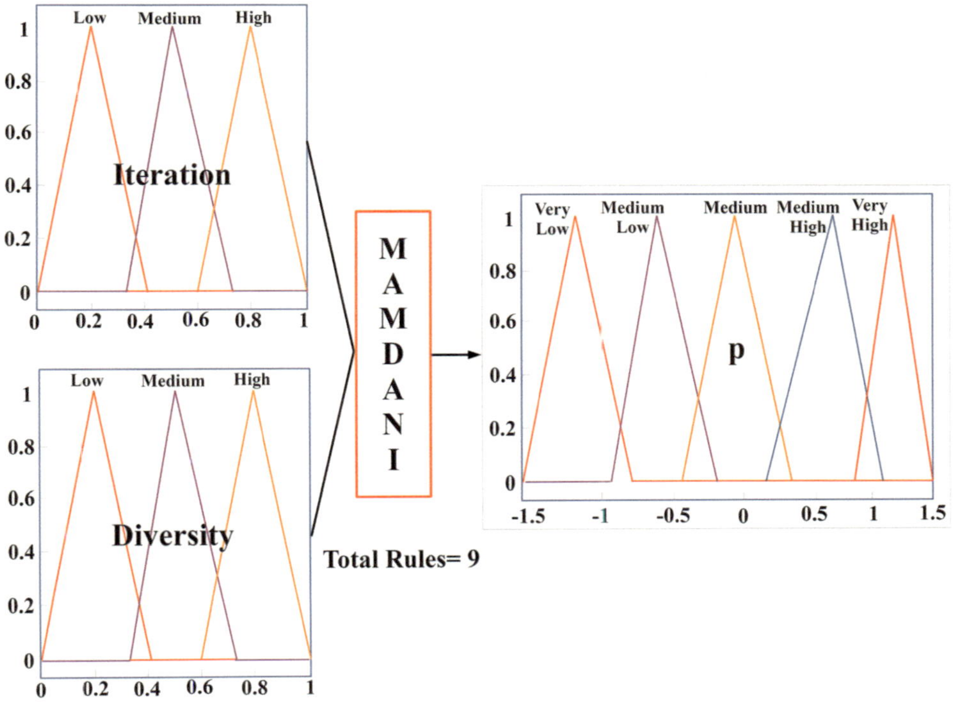

Figure 4.
Type-1 fuzzy system for dynamic parameter adaptation in flower pollination optimization algorithm (FPOA).

Figures 4 and **5** present the type-1 fuzzy system and the interval type-2 fuzzy system used to dynamic parameter adaptation of the parameters of flower pollination optimization algorithm (FPOA), using as inputs metric iteration and diversity.

The fuzzy system illustrated in **Figure 4** is type-1, has iteration and diversity as inputs, and has the parameter *P* as output. For this case, the inputs were divided into three different linguistic triangle membership functions and the outputs into five triangle membership functions. The proposed Mamdani fuzzy system contains the nine *if -then* fuzzy rules that control the behavior of FPOA.

The fuzzy system in **Figure 5** is an interval type-2 and has the same number of membership per input and output, but now as type-2 triangular membership functions, the fuzzy rule base is the same as in the type-1 because the knowledge is not changing, only the type of membership functions.

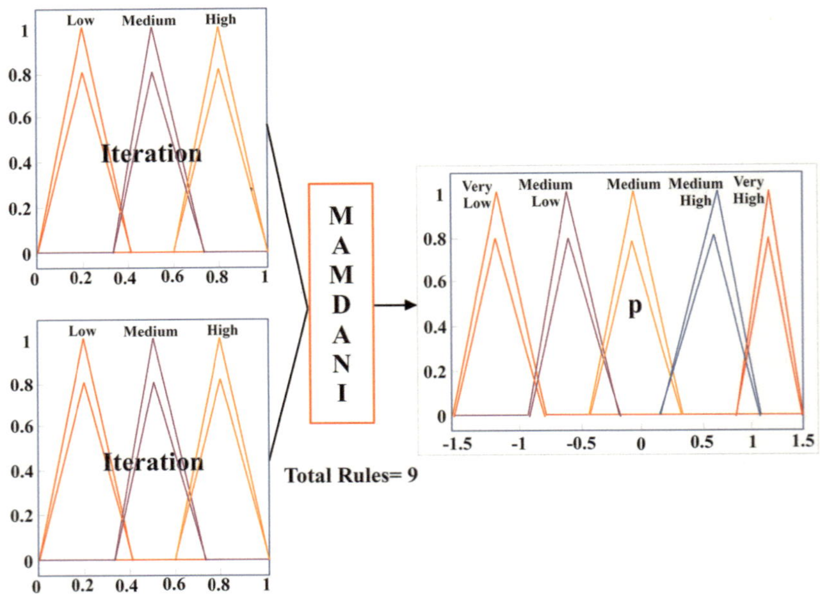

Figure 5.
Interval type-2 fuzzy system for dynamic parameter adaptation in FPOA.

No.	Inputs		Outputs
	Iteration	**Diversity**	***P***
1.	Low	Low	Very high
2.	Low	Medium	Medium high
3.	Low	High	Medium high
4.	Medium	Low	Medium high
5.	Medium	Medium	Medium
6.	Medium	High	Medium low
7.	High	Low	Medium
8.	High	Medium	Medium low
9.	High	High	Very low

Table 2.
Fuzzy rules for dynamic parameter adaption in FPOA.

Table 2 contains the fuzzy rules used for the fuzzy systems in **Figures 4** and **5**; these rules were designed based on several experiments to create knowledge of the parameters of FPOA and how to control its behavior.

6. Formulation of the problem

To test the proposed methods with dynamic parameter adaptation, a complex problem was selected which is commonly used in various industries like petrochemical, pharmaceutical, food processing, chemical, etc. In this case, the optimization of a fuzzy system design used for controlling a dihybrid level control system subject to system component (leak) fault, the task of the fuzzy controller is to provide a way to control the two inlet flow rates of the dihybrid tank 1 and tank 3 in order to minimize the steady-state error. The type-1 fuzzy logic is used to design the fractional order PID controller. The prototype model of the dihybrid system is illustrated in **Figure 6** and has three tanks; out of these two side-by-side tanks are identical. The system has two identical pumps which provide the inlet flow rate to the tank 1 and tank 3. The intermediate tank is unique in terms of dynamics of the tank, which added nonlinear response to the outer tanks. The dynamics of dihybrid system subject to leak faults represented by the following set of equations from [33]

$$S\frac{dh_1(\mathbf{t})}{dt} = f_1(t) - sign[h_1(t) - h_2(t)]\, f_{12}(t) - f_{l1}(t) \tag{7}$$

$$S'\frac{dh_2}{dt} = sign[h_1(t) - h_2(t)] f_{12}(t) - sign[h_2(t) - h_3(t)] f_{23}(t) \tag{8}$$

$$S\frac{dh_3(\mathbf{t})}{dt} = f_2(t) + sign[h_2(t) - h_3(t)] f_{23}(t) - f_{l3}(t) \tag{9}$$

where $S' = \frac{\pi}{3}\left[3r_{b2}^2 + 6r_{b2}\left(\frac{R-r_{b2}}{H_2}\right)h_2 + 3\left(\frac{R-r_{b2}}{H_2}\right)^2 {h_2}^2\right]$

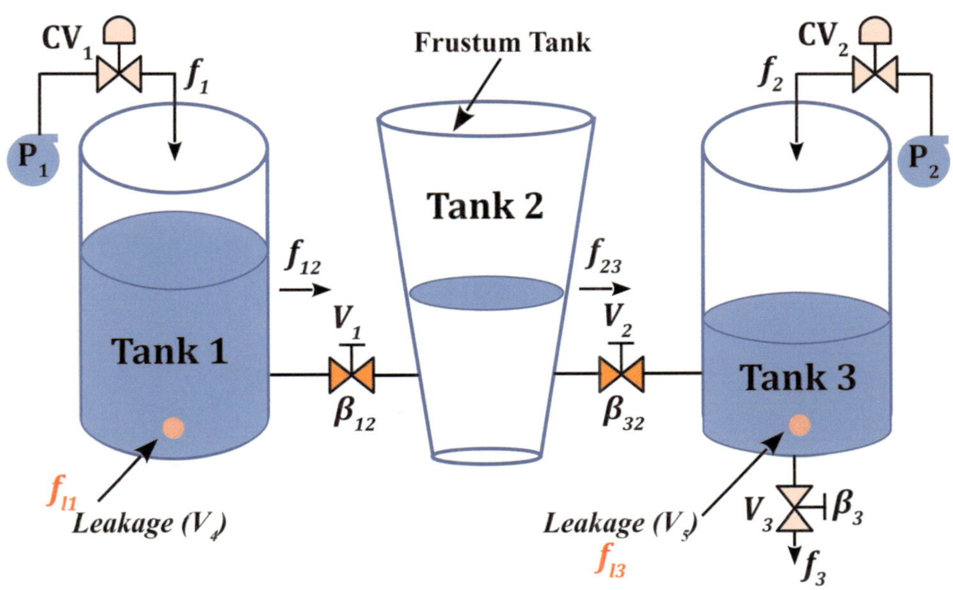

Figure 6.
Prototype model of the dihybrid system [33, 34].

The S' equation given above is the area of conical frustum tank at any height of tank given by equation [33, 35, 36].

The proposed control strategy is depicted in **Figure 7**.

The fuzzy system is shown in **Figure 8** and is used for the complex plant to control the level of tank 1 and tank 3 of the dihybrid level control system. This is the fuzzy system that the bioinspired optimization algorithms will optimize; in this case, only the parameters of the membership functions are optimized. The fuzzy controller of **Figure 8** has two inputs, the error $e(s)$ and derivative of error $de(s)/dt$,

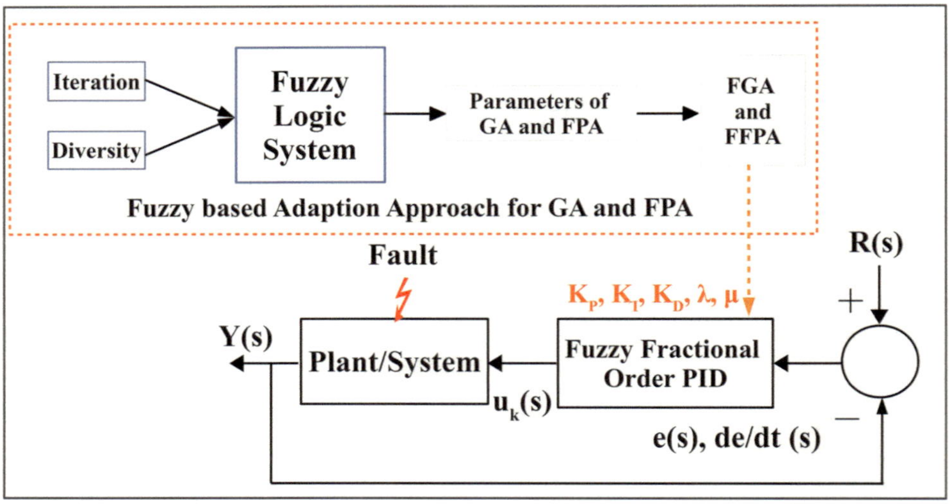

Figure 7.
Proposed control scheme.

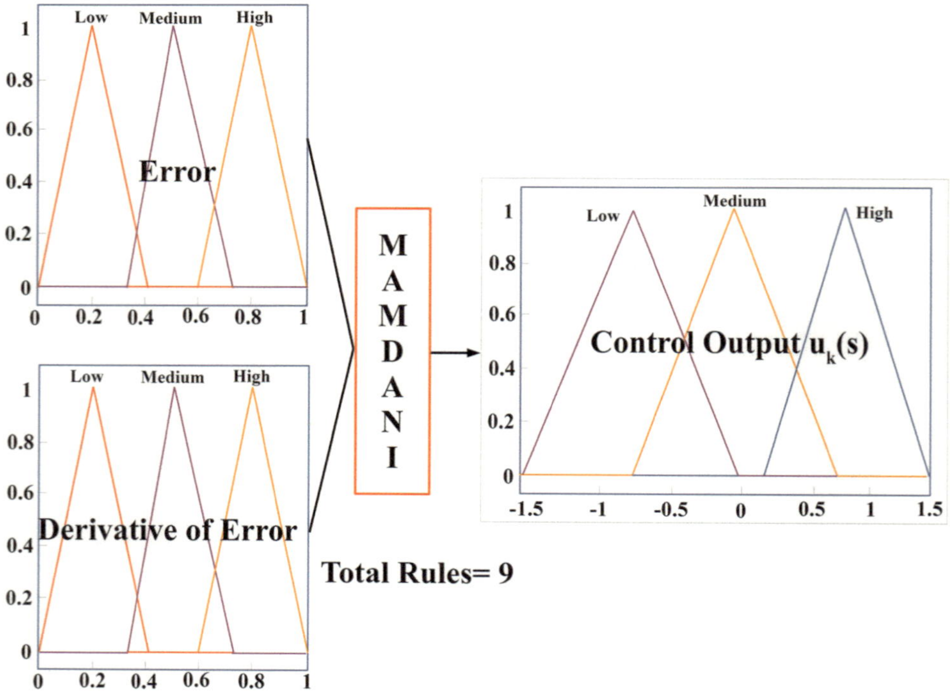

Figure 8.
Membership functions for input and output of type-1 FLC.

and one output, control signal $u_k(s)$, for each control valve of the dihybrid level control system. The inputs are granulated into three triangle membership functions, and the outputs are granulated into three triangular membership functions. Also this fuzzy controller uses the fuzzy rule set from **Table 3**. The desired reference trajectory for level is illustrated in result figures, where it starts from 0 to 80 cm and system component (leak) fault is added into tank 1 and tank 3, in order to create a complex control problem [33, 35]. The main difference between the type-1 and interval type-2 fuzzy systems is the ability of the latter to handle uncertainty; thus, to simulate this problem and to provide better tools to perform the comparison, a modification of the original plant is added, where leak fault is added to the desired reference trajectory.

No.	**Inputs**		**Outputs**
	$e(s)$	$\dot{e}(s)$	$u_k(s)$
1.	Low	Low	Low
2.	Low	Medium	Medium
3.	Low	High	Medium
4.	Medium	Low	Medium
5.	Medium	Medium	Medium
6.	Medium	High	High
7.	High	Low	Medium
8.	High	Medium	High
9.	High	High	High

Table 3.
Fuzzy rules for fuzzy controller.

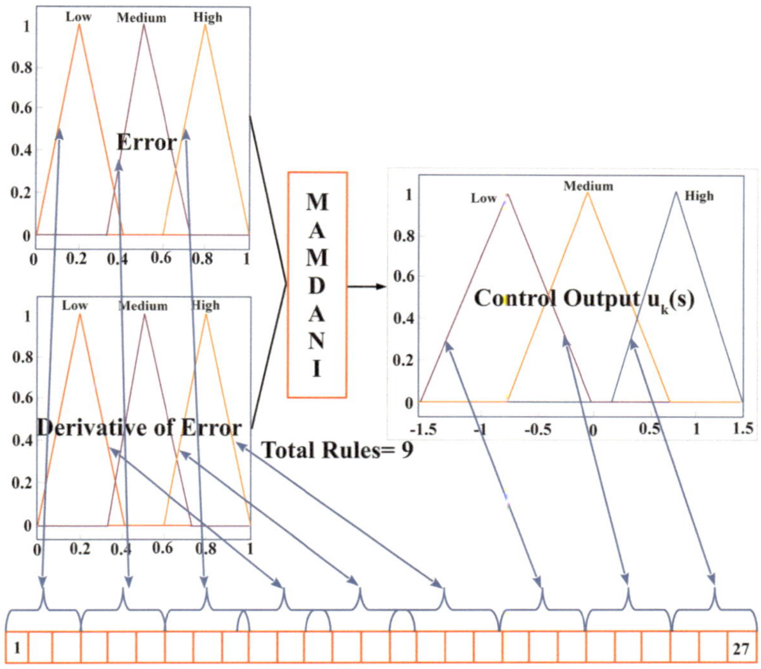

Figure 9.
Optimization problem (points of the membership functions).

The optimization problem can be stated as follows: to optimize all the points of the membership functions of the fuzzy system used for control from **Figure 8**, this is illustrated in **Figure 9** where for each membership function, the bioinspired methods will try to find the best values for each point. In this case, the fuzzy controller has two inputs with three triangle membership functions and one input each and one output with three triangular membership functions; for each triangle membership function, the bioinspired methods need to find four values and three values for each triangular membership function, with a total of 27 points (values) for these particular fuzzy system for control as optimization problem. In this case the fuzzy rule set from **Table 3** was not modified, and only the membership functions were optimized.

The objective function is to minimize the trajectory error created by the optimized fuzzy FOPID controller using Eq. (6); this means that each bioinspired method will try to find the best values for each point of each membership function, and with this the optimized fuzzy FOPID controller creates a trajectory with the lowest possible error.

7. Simulation results

The proposed bioinspired optimization algorithm with parameter adaptation is now tested on problem of dihybrid level control system subject to system component (leak) faults. This system taken for the testing is novel in terms of its dynamics; the system is a highly nonlinear and interacting process. Also the system has two system component (leak) faults; one is f_{sys1} leak in tank 1, and the second is f_{sys2} leak in tank 3. So we proposed the novel system like dihybrid level process and tested proposed fuzzy bioinspired optimization algorithms such as GOA and FPOA. The result section presents the comparisons of different variants of GOA and FPOA based on fuzzy logic system.

The control scheme of the overall design is presented in **Figure 7**. The bioinspired optimization algorithms were applied to the optimization of the fuzzy system (fractional order PID controller) (**Figures 8** and **9**) for control of the dihybrid level control system, using the same parameters, such as population, iterations, and number of experiments described in **Tables 4** and **5** for the GOA and FPOA.

These parameters were selected based on several experiments with all the methods applied to the optimization of some benchmark mathematical function, such as Rosenbrock, just to search for the best parameters, while also trying to use almost the same parameters for all the bioinspired methods.

The metric used to evaluate the performance of all methods is the mean square error described in Eq. (6), calculated from the desired reference trajectory and the

Parameter	Original GOA	Proposed fuzzy GOA
Population	100	100
Iterations	40	40
Crossover (K_1)	Single point crossover	Dynamic
Mutation rates (K_2)	Uniform	Dynamic
Encoding	Binary	Binary
Selection	Uniform	Uniform

Table 4.
Parameters for the GOA method original/proposed.

trajectory created by the optimized fuzzy controller. In addition, each method is applied 40 times to each problem and presents the average, best, worst, and standard deviation of those experiments. The fractional order PID controller parameters are chosen as $K_P = 5.9801$, $K_I = 2.0781$, $K_D = 2.09681$, $\lambda = 0.5491$, and $\mu = 0.2349$.

For comparison purposes, there are two variations of GOA, which are described below. All of these variations use the parameters described in **Table 4**.

GOA + T1FS is the GOA method with parameter adaptation using the type-1 fuzzy system illustrated in **Figure 2**.

GOA + IT2FS is the GOA method with parameter adaptation using the interval type-2 fuzzy system illustrated in **Figure 3**.

Table 6 contains the results of applying the variations of FPOA to the optimization of the membership functions from the fuzzy FOPID controller illustrated in **Figure 9**, using the plant without fault shown in **Figure 7**. In this case, results in bold are the best from all methods on each category.

From the results in **Table 6**, the proposed FPOA + IT2FS method obtains the best results on average, best, worst, and standard deviation, when compared with the FPOA + T1FS.

For comparison, there are two variants of the genetic optimization algorithm, GOA + T1FS, which is the GOA algorithm with type-1 fuzzy system for dynamical parameter adaptation, and GOA + IT2FS, which is the GOA algorithm with interval type-2 fuzzy system for dynamical parameter adaptation.

Table 7 contains the results of applying the variations of the GOA to the optimization of the membership functions from the fuzzy controller illustrated in

Parameter	Original FPOA	Proposed fuzzy FPOA
Population size	100	100
Iterations	40	40
Probability P	0.8	Dynamic
Dimension	3	3

Table 5.
Parameters for the FPOA method original/proposed.

MSE	FPOA +T1FS	FPOA +IT2FS
Average	4.013×10^{-2}	4.63×10^{-3}
Best	6.284×10^{-2}	2.291×10^{-3}
Worst	5.7816	4.701×10^{-2}
Standard deviation	4.9218×10^{-2}	8.1219×10^{-3}

Table 6.
Comparison of results of the plant without system fault for the FFPOA.

MSE	GOA + T1FS	GOA + IT2FS
Average	10.79	**9.091**
Best	1.7×10^{-3}	2.09×10^{-3}
Worst	68.91	**64.12**
Standard deviation	**14.93**	15.61

Table 7.
Comparison of results of the plant without system fault for the FGOA.

Figure 9, using the plant without system component (leak) fault. **Table 8** contains the results of applying the variations of the GOA algorithm to the optimization of the membership functions from the fuzzy controller illustrated in **Figure 9**, using the plant with noise shown in **Figure 6**. Results highlighted in bold are the best from all methods on each category.

Results in **Table 6** show that the FPOA, which uses an interval type-2 fuzzy system for parameter adaptation, can obtain on average better results than FPOA as well as the lowest MSE in both the cases without system component (leak) fault. This is the best of all controllers; its worst results are lower on MSE than on the FPOA methods and finally also obtain the lowest standard deviation.

Table 9 contains a comparison of results with the best methods using the plant with system component (leak) fault. In this case, from the FPO algorithm, it is FPOA + T1FS, and from FPO Algorithm, it is FPOA + IT2FS.

The next result figures illustrate the best trajectories from each method, for visual comparison purpose; note that, in each figure, all trajectories from the optimized fuzzy system used for control are very similar to desired trajectory.

Table 10 contains the results with the Rosenbrock function for proposed bioinspired optimization method with parameter adaptation using an interval type-2 and type-1 fuzzy system.

Figure 10 contains the comparison of the best trajectory from IT2FS variations of GOA and FPOA for tank 1 and tank 3 of dihybrid level control system, in this case using original IT2FS GOA with a MSE 2.191×10^{-3}.

Figure 11 contains the comparison of the trajectory from T1FS variations of GOA and FPOA for tank 1 and tank 3 of dihybrid level control system, in this case using original T1FS GOA with a MSE 3.8212.

The results shows that an interval type-2 fuzzy system used for parameter adaptation can help FPOA to obtain better quality results than GOA, even when GOA methods use the same methodology for dynamic parameter adaptation and also use a type-1 or interval type-2 fuzzy system for the same task.

The second simulations are carried out with the system component (leak) faults introduced into tank 1 and tank 3 of dihybrid level control system. The proposed

MSE	GOA + T1FS	GOA + IT2FS
Average	**15.61**	15.78
Best	2.89	**2.36**
Worst	**78.15**	77.91
Standard deviation	17.83	17.81

Table 8.
Comparison of results of the plant with system fault for the FGOA.

MSE	FPOA + T1FS	FPOA + IT2FS
Average	4.8253	**3.3771**
Best	3.8212	**2.5614**
Worst	5.8916	**4.0941**
Standard deviation	5.0628×10^{-2}	4.1494×10^{-2}

Table 9.
Comparison of results of the plant with system fault for the FFPOA.

Rosenbrock function						
Population size	**Dimensions**	**Iterations**	**FPOA + IT2FS**	**FPOA + T1FS**	**GOA + IT2FS**	**GOA + T1FS**
20	10	500	10.3481	15.3791	13.5471	17.2019
	20	1000	24.5901	29.7513	26.9416	32.4569
	30	1500	42.5822	49.4692	46.87651	58.5821
40	10	500	5.6792	8.9389	6.9021	9.8093
	20	1000	11.4911	15.0921	12.9916	17.8137
	30	1500	19.0921	23.5821	21.9091	25.9810

Table 10.
Comparison against all variations of FGOA and FFPOA.

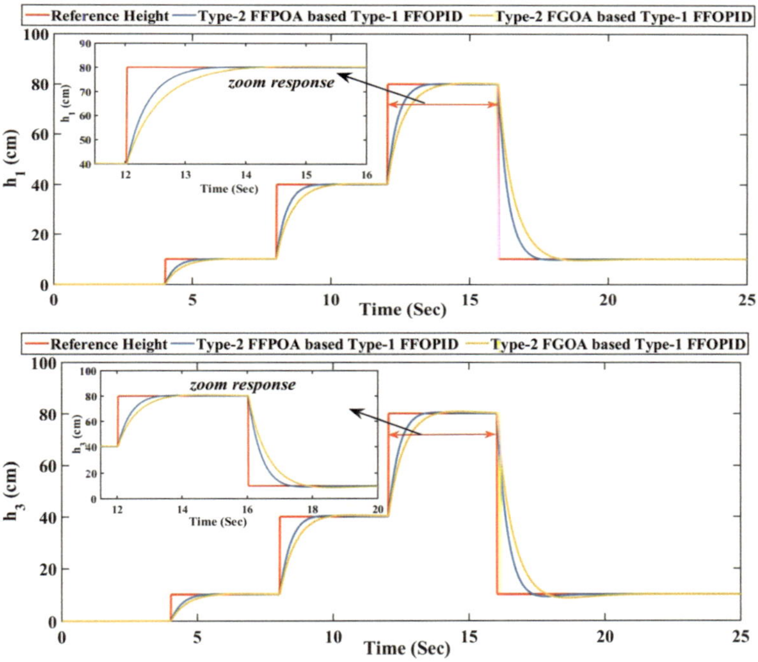

Figure 10.
Best simulation of GO and FPO algorithm with IT2FS variations for the dihybrid level control system without system component (leak) fault.

fuzzy-based parameter adaption techniques are applied, and results are produced with very good accuracy. There are two simultaneous leak faults $f sys1$ and f_{sys2} introduced into system at time t = 10 sec and t = 14 sec, respectively.

Figure 12 shows the best results of GO and FPO algorithm from IT2FS variations, and result clearly shows that FPOA + IT2FS outperforms all the variants of GOA and FPOA + T1FS even though system component (leak) faults are present in the dihybrid level control system.

In the same manner, **Figure 13** shows the results of GO and FPO algorithm from T1FS variations, and result clearly shows that FPOA + IT2FS outperforms the GOA + IT2FS even though system component (leak) faults are present in the dihybrid level control system.

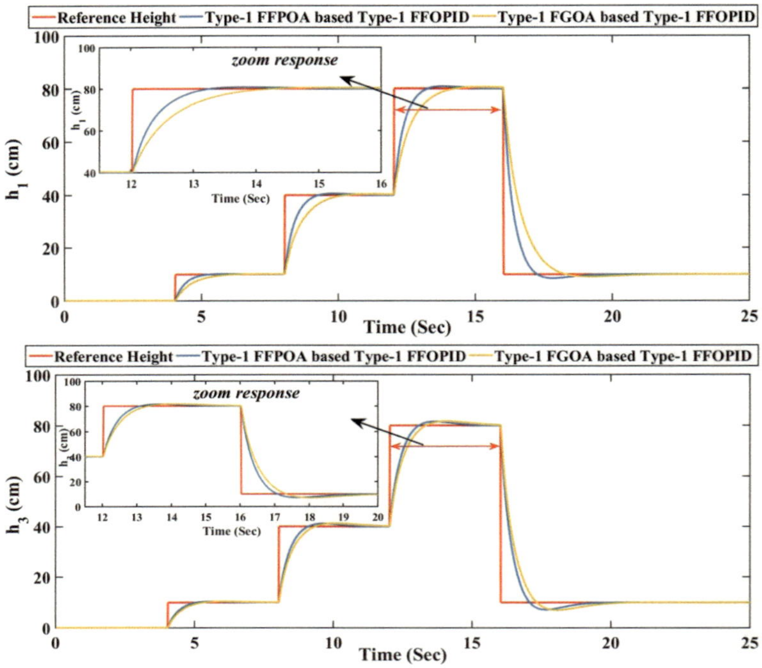

Figure 11.
Best simulation of GO and FPO algorithm with T1FS variations for the dihybrid level control system without system component (leak) fault.

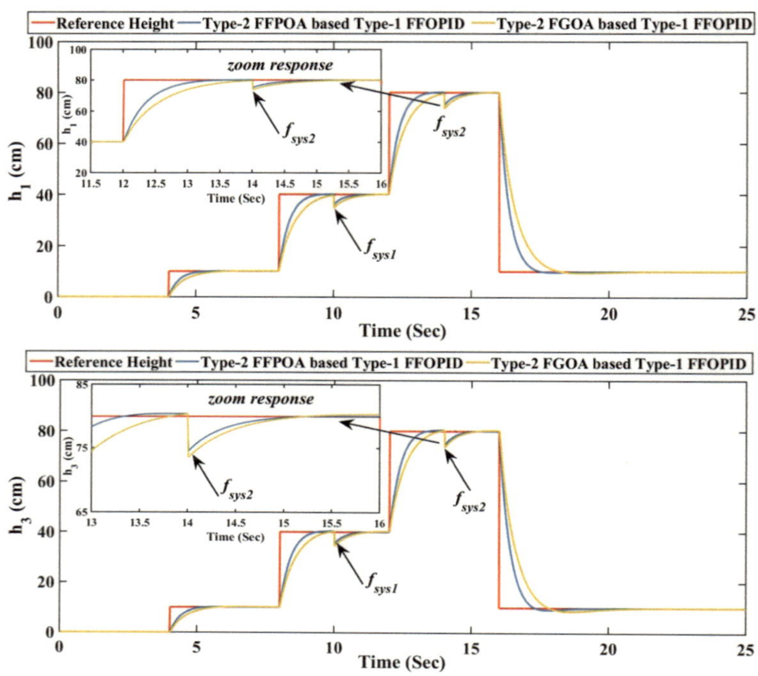

Figure 12.
Best simulation of GO and FPO algorithm with IT2FS variations for the dihybrid level control system with system component (leak) fault.

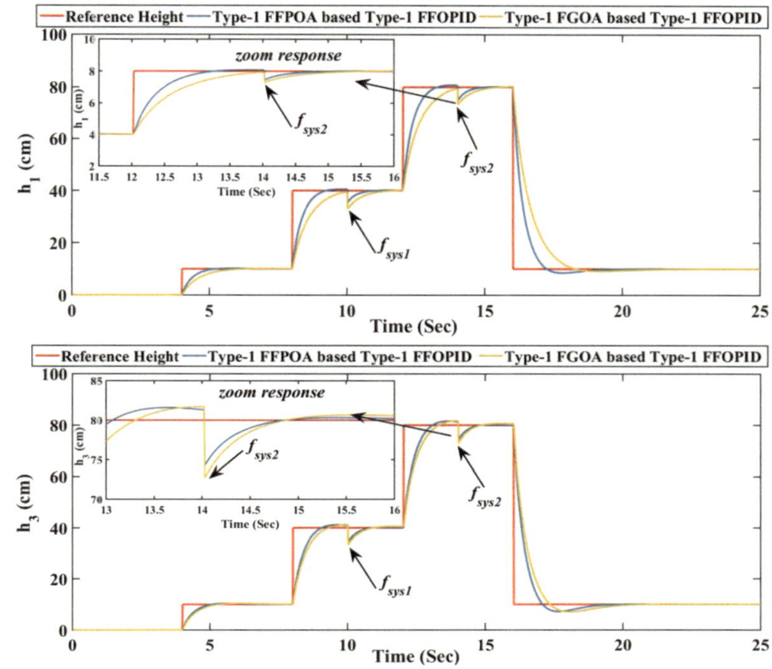

Figure 13.
Best simulation of GO and FPO algorithm with T1FS variations for the dihybrid level control system with system component (leak) fault.

From observing the simulation results, FPOA + IT2FS gives better results than GOA + IT2FS, GOA + T1FS, and FPOA + T1FS with a MSE of 2.5614.

Aside from the results with the dihybrid level control system, a comparison against a fuzzy GOA and FPOA is also presented. The following results (contained in **Table 10**) were obtained using the same parameters. In addition, results highlighted in bold are the best results. **Table 10** contains the results with the Rosenbrock function for FPOA + IT2FS, FPOA + T1FS, GOA + IT2FS, and GOA + T1FS. From the results in **Table 10**, it is clear that our IT2FS + GOA obtains on average better results than the other methods.

8. Conclusions

Flower pollination and genetic algorithms are an exceptional good bioinspired optimization algorithms; it is adequate to handle complex engineering problems and achieve impressive results. In this particular case, we optimize the membership functions from a fuzzy FPID controller. From the results, in the observation of two variations of FPOA, the FPOA + IT2FS version, which uses an interval type-2 fuzzy logic system for dynamic parameter adaptation, it can give exceptional results than FPOA + T1FS version.

The flower pollination optimization algorithm is used to optimize membership functions of a fuzzy logic controller applied to track the trajectory of dihybrid level control process subject to system component (leak) fault, with the motive of diminishing an error. Subsequently from examining and interpreting the obtained results, we can summarize that GOA is competent in optimizing the problems. In this case, two versions of GOA, GOA + IT2FS and GOA + T1FS, can achieve decent results; however, the algorithm presents unsatisfactory results when the fault occurred into the system, but it will maintain the system stability, i.e., some simulations are good, and some are bad.

The motivation for the advancement of this research work was to authenticate the improvement with our proposed methodology for parameter adaptation through fuzzy logic system, applied to different bioinspired methods. The main contribution of this work is a comparative study based on two bioinspired algorithms for the design and implementation of fuzzy fractional order PID controllers. In addition, a comparative study proposed methods with parameter adaptation using type-1 and interval type-2 fuzzy logic systems as tools for modeling complex problems in control engineering.

For future work, we want to broaden the proposed technique for parameter adaptation to other bioinspired methods (i.e., differential evolutionary algorithm) as well as use other potential engineering applications, such as the optimization of neural networks, fuzzy systems applied to other problems, or even a hybridization of two or more bioinspired algorithms.

Acknowledgements

The project outcome is a PhD work of the corresponding (first) author of this article. This research received no external funding. The authors take this opportunity to express their sincere thanks to Editor Prof. Yang Yi and his entire team and IntechOpen staff members for their valuable guidance, excellent co-operation and timely help extended. Useful contributions and cooperation received from all the cited sources of references are also gratefully acknowledged.

Funding

This research received no specific grant from any funding agency in the public, commercial, or not-for-profit sectors.

Conflict of interest

The authors declare no conflict of interest.

Notes/Thanks/Other declarations

Compliance with ethical standards.

Nomenclature

IT2FS	interval type-2 fuzzy system
T1FS	type-1 fuzzy system
DEA	differential evolutionary algorithm
FOPID	fractional order proportional integral derivative
FFOPID	fuzzy fractional order proportional integral derivative
GOA	genetic optimization algorithm
FPOA	flower pollination optimization algorithm

FGOA	fuzzy genetic optimization algorithm
FFPOA	fuzzy flower pollination optimization algorithm
FPGA	field programmable gate array
PSO	particle swarm optimization
CSTR	continuous stirred-tank reactor

Abbreviations

f_{sys1}	system component (leak) fault 1
f_{sys2}	system component (leak) fault 2
K_P	proportional gain
K_I	integral gain
K_D	derivative gain
λ	fractional order integral parameter
μ	fractional order derivative parameter
K_1	crossover
K_2	mutation rate
P	pollination switch probability

Author details

Himanshukumar R. Patel* and Vipul A. Shah†
Department of Instrumentation and Control Engineering, Faculty of Technology, Dharmsinh Desai University, Nadiad, Gujarat, India

*Address all correspondence to: himanshupatelp32@gmail.com; vashahin2010@gmail.com

† This author contributed as a PhD. supervisor.

References

[1] Fogel DB. An introduction to simulated evolutionary optimization. IEEE Transactions on Neural Networks. 1994;**5**(1):3-14. [Accessed: 14 November 2018]

[2] Man KF, Tang KS, Kwong S. Genetic Algorithms: Concepts and Designs. Heidelberg: Springer; 1999. [Accessed: 18 January 2018]

[3] Back T, Fogel DB, Michalewicz Z, editors. Handbook of Evolutionary Computation. Oxford: Oxford University Press; 1997. [Accessed: 01 November 2017]

[4] Castillo O, Valdez F, Melin P. Hierarchical genetic algorithms for topology optimization in fuzzy control systems. International Journal of General Systems. 2007;**36**(5):575-591. [Accessed: 02 July 2018]

[5] Wiangtong T, Sirapatcharangkul J. PID design optimization using flower pollination algorithm for a buck converter. In: 17th International Symposium on Communications and Information Technologies (ISCIT); Cairns, QLD; 2017. pp. 1-4. DOI: 10.1109/ISCIT.2017.8261202 [Accessed: 12 June 2018]

[6] Dwi L, Muhammad RD, Widodo IR. Optimization of PID controller design for DC motor based on flower pollination algorithm. In: 2015 International Conference on Electrical, Telecommunication and Computer Engineering (ELTICOM 2015); Aryaduta Hotel, Medan; 2015. DOI: 10.13140/RG.2.1.3028.6963. [Accessed: 12 June 2018]

[7] Jayachitra A, Vinodha R. Genetic algorithm based PID controller tuning approach for continuous stirred tank reactor. Advances in Artificial Intelligence. 2014;**2014**:791230. DOI: 10.1155/2014/791230. [Accessed: 12 June 2018]

[8] Kim D, Hirota K. Vector control for loss minimization of induction motor using GAPSO. Applied Soft Computing. 2008;**8**:1692-1702. [Accessed: 12 June 2018]

[9] Shopova EG, Vaklieva-Bancheva NG. BASIC: A genetic algorithm for engineering problems solution. Computers and Chemical Engineering. 2006;**30**(8):1293-1309. [Accessed: 28 June 2018]

[10] Zhang J, Zhuang J, Du H, Wang S. Self-organizing genetic algorithm based tuning of PID controllers. Information Sciences. 2009;**179**(7):1007-1018. [Accessed: 28 June 2018]

[11] Krohling RA, Rey JP. Design of optimal disturbance rejection PID controllers using genetic algorithms. IEEE Transactions on Evolutionary Computation. 2001;**5**(1):78-82. [Accessed: 28 June 2018]

[12] Kumar SMG, Jain R, Anantharaman N, Dharmalingam V, Begum KMMS. Genetic algorithm based PID controller tuning for a model bioreactor. Indian Chemical Engineering. 2008;**50**(3):214-226. [Accessed: 28 June 2018]

[13] Garg H. A hybrid PSO-GA algorithm for constrained optimization problems. Applied Mathematics and Computation. 2016;**274**:292-305. [Accessed: 28 June 2018]

[14] Garg H. Performance analysis of an industrial system using soft computing based hybridized technique. Journal of the Brazilian Society of Mechanical Sciences and Engineering. 2017;**39**:1441-1451. [Accessed: 28 July 2019]

[15] Singh S, Garg H. Distance measures between type-2 intuitionistic fuzzy sets and their application to multicriteria decision-making process. Applied Intelligence. 2017;**46**:788-799. [Accessed: 28 July 2019]

[16] Garg H. Reliability, availability and maintainability analysis of industrial systems using PSO and fuzzy methodology. Mapan. 2014;**29**:115-129. [Accessed: 28 July 2018]

[17] Garg H. An approach for solving constrained reliability-redundancy allocation problems using cuckoo search algorithm. Beni-Suef University Journal of Basic and Applied Sciences. 2015;**4**: 14-25. [Accessed: 28 July 2018]

[18] Man KF, Tang KS, Kwong S. Genetic algorithms: Concepts and applications. IEEE Transactions on Industrial Electronics. 1996;**43**(5):519-534. [Accessed: 02 August 2018]

[19] Fevrier V, Patricia M, Oscar C. Fuzzy Logic for Parameter Tuning in Evolutionary Computation and Bio-inspired Methods. LNAI 6438. Berlin/ Heidelberg: Springer-Verlag; 2010. pp. 465-474. [Accessed: 22 August 2018]

[20] Bernd F, Michael H. Optimization of Genetic Algorithms by Genetic Algorithms. Artificial Neural Nets and Genetic Algorithms. Vienna: Springer; 1993. [Accessed: 28 August 2018]

[21] Homayouni S, Tang S. A fuzzy genetic algorithm for scheduling of handling/storage equipment in automated container terminals. International Journal of Engineering and Technology. 2015;**7**(6):497-501. [Accessed: 28 August 2018]

[22] Lau HCW, Nakandala D, Zhao L. Development of a hybrid fuzzy genetic algorithm model for solving transportation scheduling problem. Journal of Information Systems and Technology Management. 2015;**12**(3):505-524. [Accessed: 28 August 2018]

[23] Plerou A, Vlamou E, Papadopoulos V. Fuzzy genetic algorithms: Fuzzy logic controllers and genetics algorithms. Global Journal For Research Analysis. 2016;**5**:497-500. [Accessed: 28 August 2018]

[24] Khmeleva E, Hopgood AA, Tipi L. Fuzzy-logic controlled genetic algorithm for the rail-freight crew-scheduling problem. Künstliche Intelligenz. 2018;**32**(1):61-75. [Accessed: 02 September 2019]

[25] Maldonado Y, Castillo O, Melin P. Particle swarm optimization of interval type-2 fuzzy systems for FPGA applications. Applied Soft Computing. 2013;**13**(1):496-508. [Accessed: 02 September 2018]

[26] Yang X-S. Flower Pollination Algorithm for Global Optimization. Berlin Heidelberg: Springer; 2012. [Accessed: 02 September 2018]

[27] Huang S-J, Gu P-H, Su W-F, Liu X-Z, Tai T-Y. Application of flower pollination algorithm for placement of distribution transformers in a low-voltage grid. In: IEEE International Conference on Industrial Technology (ICIT); 2015. pp. 1280-1285 [Accessed: 02 September 2018]

[28] Oda ES, Abdelsalam AA, Abdel-Wahab MN, El-Saadawi MM. Distributed generations planning using flower pollination algorithm for enhancing distribution system voltage stability. Ain Shams Engineering Journal. 2017;**8**(4):593-603. [Accessed: 02 September 2018]

[29] Draa A. On the performances of the flower pollination algorithm qualitative and quantitative analyses. Applied Soft Computing. 2015;**34**:349-371. [Accessed: 02 September 2018]

[30] Abdelaziz EA, Abd Elazim S. Optimal sizing and locations of capacitors in radial distribution systems via flower pollination optimization algorithm and power loss index. Engineering Science and Technology. 2016;**19**(1):610-618. [Accessed: 02 September 2018]

[31] Dubey HM, Pandit M, Panigrahi BK. A hybrid flower pollination algorithm with time-varying fuzzy selection mechanism for wind integrated multi-objective dynamic economic dispatch. Renewable Energy. 2015;**83**:188-202. [Accessed: 02 September 2018]

[32] Dubey HM, Pandit M, Panigrahi BK. A biologically inspired modified flower pollination algorithm for solving economic dispatch problems in modern power systems. Cognitive Computation. 2015;**7**(5):594-608. [Accessed: 02 September 2018]

[33] Patel HR, Shah VA. Stable fuzzy controllers via LMI approach for nonlinear systems described by type-2 T-S fuzzy model. In: 15th European Workshop on Advanced Control and Diagnosis; Bologna, Italy; November 21–22, 2019

[34] Patel HR, Shah VA. Actuator and system component fault tolerant control using interval type-2 Takagi-Sugeno fuzzy controller for hybrid nonlinear process. International Journal of Hybrid Intelligent Systems. 2019;**15**(3):143-153

[35] Lakshmanaprabu SK, Wahid Nasir A, Sabura Banu U. Design of Centralized Fractional order PI controller for two interacting conical frustum tank level process. Journal of Applied Fluid Mechanics. 2017;**10**:23-32

[36] Patel HR, Shah VA. Stable fault tolerant controller design for Takagi-Sugeno fuzzy model based control systems via linear matrix inequalities: Three conical tank case study. Energies. 2019;**12**(11):2221

Chapter 14

Uncertainty in Measurement

Carlo Ferri

Abstract

Measurements of physical quantities are the corner stone upon which we humans have built the scientific perception of the world. Measurements are the distinctive means to tell the scientific truth apart from any other approach to knowledge. The fundamental concepts of measurement and uncertainty in measurement have been analysed with reference to authoritative documents produced by the International Bureau of Weights and Measures (BIPM). The need for the introduction of the concept of uncertainty and its theoretical implications are analysed. The practical consequences in the development of industrial products have been illustrated for a specific measurement in proton exchange-membrane fuel cell assembly. A short critical analysis of the relationship between the evaluation of uncertainty in measurement and intelligent systems led then to a few open questions.

Keywords: fuel cell, GUM, metrology, PEM, PUMA, uncertainty, VIM

1. Introduction

The fact that intelligent systems have been specifically introduced to overcome the difficulties in handling vagueness and qualitative knowledge in computational environments has generated a misconception quite widespread: once a quantity has been measured, then all the vagueness have vanished because the quantity is described by a number. By contrast, any measurement result is inherently uncertain. Quantifying this uncertainty is the theme of this chapter.

The chapter is based on three of the most authoritative sources of reference: 'The Guide to the Expression of Uncertainty in Measurement' (GUM) [1], 'The International Vocabulary of Metrology' (VIM) [2] and 'The International System of Units' (SI), also known as the SI Brochure [3]. The concepts defined in these documents are analysed and illustrated with an example drawn from the assembly of a proton exchange membrane fuel cell (PEMFC).

The main components of a proton exchange membrane fuel cell (PEMFC) are described in the next section. In this way, examples from PEMFC manufacturing can be introduced in the following sections. In Section 3, the concept of measurement is explored. Reference is made to its fundamental components: the measurand, the reference and the measurement model. In Section 4, the definition of uncertainty in the GUM and in the VIM is discussed. Then, a method of analysis is introduced and illustrated with an example referring to a PEMFC critical measurement. In Section 5, some ideas about the relationship between uncertainty in measurement and intelligent systems are briefly presented. Conclusions are drawn thereafter.

2. Fuel cell components

The main components of a fuel cell are illustrated in the schema of **Figure 1**. A cell provides a voltage of less than 1 V in typical working conditions. Multiple cells are usually connected in series in a stack to increase the voltage to a level suitable for the load that is intended to power.

The bipolar plates (BPPs) represented in **Figure 1** are typically made of metal, graphite or composites. Their primary function is to support the cell and to provide electrical contact with neighbouring cells in the stack. A set of channels may be present on a BPP to convey the reactant gasses onto the gas diffusion layer (GDL). This set of channels is often called a flow field.

GDLs are thin porous sheets inserted to provide a pathway for the gaseous reactants to diffuse evenly from the plates to the membrane electrode assembly (MEA). GDL sheets also take away the water produced in the electrochemical reaction at the cathodic MEA surface and residue of the MEA hydration from the reaction area. A GDL must offer little resistance to the passage of electrons, to enable them to reach the BPP's from the electrochemical reaction sites. These sheets are usually made either of carbon paper or carbon cloth. The first are hard and brittle, with negligible compressibility and generally thinner than the second, which are flexible and can sustain higher levels of compression when assembled in the stack. Therefore, paper-based GLD's need care in handling to avoid chipping and tighter tolerances in the stack due to the poor compressibility. Instead, the compressibility of cloth-based GDLs enables them to be elastically deformed. GDLs may also have microporous layers (MPLs) and polytetrafluoroethylene (PTFE) hydrophobic coatings to balance the requirement of retaining some water to hydrate the MEA in order to keep it conductive with the requirement of mantaining the micropores open for the gaseous reactants to diffuse.

Two variants of MEAs are generally used: three-layer MEAs, also called catalyst-coated membranes (CCMs), and five-layer MEAs. Three-layer MEAs are composed by a proton exchange membrane, also known as polymer electrolyte membrane (PEM), and two catalyst layers. Common membranes are made of an ion-conducting polymer (ionomer) that, when conveniently hydrated, are selectively permeable only to cations while having high electronic resistivity. PEM has also the function of keeping the fuel (hydrogen) and the oxidant (oxygen in the air) separate. The thickness of PEMs usually varies between 10 and 100 micrometres. A catalyst layer is typically made of a mixture of very thin powders of platinum and carbon powder blended with a ionomer. Catalyst layers are applied on the PEM

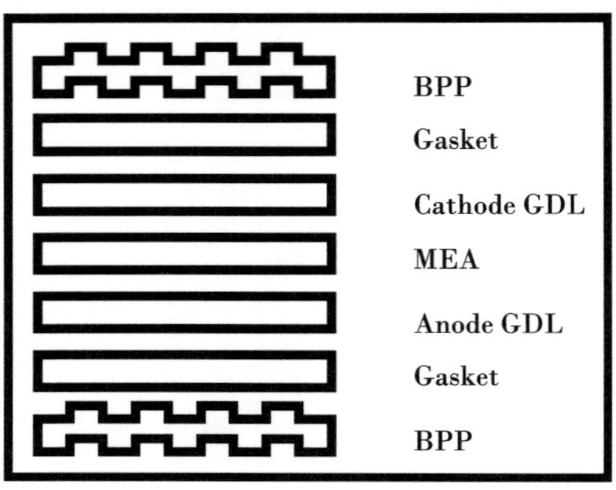

Figure 1.
Schema of a proton exchange membrane fuel cell (BPP, bipolar plate; GDL, gas diffusion layer; MEA, membrane electrode assembly).

surfaces and are the sites where the electrochemical reactions occur. The layer where the oxidation of hydrogen occurs is the anode electrode. The layer where the reduction of oxygen occurs is the cathode electrode. The electrodes, i.e. the catalyst layers, must have low electronic resistivity, to enable the reaction-generated electrons to reach the BPP via the GDL at the anode or to be reached by them at the cathode. Five-layer MEAs include also the two GDLs and may differ from CCMs for the sites of application of the catalyst layers, which can be the PEM-facing surface of the GDLs. In this chapter, only CCMs are considered.

The gaskets prevent unexpected leakage of the fluids, i.e. gaseous reactants and water, to the environment and to the other side of the MEA. The first is often referred to as overboard leakage and the second as cross-over leakage. There is always, however, an expected flow rate of the gaseous reagents across the MEA. Such an expected flow rate can be calculated on the basis of a number of variables. Among these are, for example, the MEA thickness and the temperature [4]. The choice of the gasket material depends widely on the operating conditions of the PEMFC (e.g. temperature and pressure of the reactants). Among other materials, gaskets can be made of PTFE or of elastic polymers, e.g. silicone.

With this chapter, a video has been provided which displays a graphical simulation of a PEMFC automated assembly system. The simulation shows the stations where the anode GDL and the gasket are placed onto the BPP and the station where the gap between anode GDL and gasket is inspected. The inspection is based on a vision system. The video is part of the simulation studies carried out by Mr. David Urquhart of the WMG Automation Systems research group of the University of Warwick (UK), within the scope of the EU-funded research project DigiMan. The video is available at the following link: https://bit.ly/2JalW8Z

3. Fundamentals

Measurement is any experimental process aimed at obtaining one or more number and reference pairs that can be attributed to a property of a body, a phenomenon or a substance (cf. Sections 1.1, 1.19, 2.1 in [2]). This property is called a quantity and its magnitude is defined as the number and the reference considered together. The reference typically is a measurement unit (e.g. the kilogramme, when measuring a mass), but it can be a measurement procedure (e.g. Rockwell C, when measuring hardness) or a reference material (e.g. the concentration of luteinizing hormone in a specimen of human blood plasma, cf. Sections 1.1 and 1.19 in [2]). The measurement unit is a quantity selected conventionally to which any other quantity of the same kind can be compared. The result of this comparison is called the ratio of the two quantities and is expressed as a number (Section 1.9 in [2]).

The above definitions suggest that the following circumstances are necessary for a measurement result not to be intrinsically uncertain:

1. The quantity intended to be measured should be defined without any indeterminacy. This quantity is called the measurand (Section 2.3 in [2]).

2. The reference, in particular the unit of measurement, should have an unambiguous magnitude.

3. The knowledge of which quantities are influencing the measurement and the effects of these quantities on the measurement result should be complete. For example, it should always be possible to compare the measurand and the measurement unit so that no indeterminacy is present in the numerical quantity value (cf. Section 1.20 in [2]).

Unfortunately, none of these conditions holds, as it is described in Sections 3.1, 3.2 and 3.3, respectively.

3.1 Measurand

To define unambiguously a measurand, an infinite amount of information is needed. This unavoidable intrinsic vagueness in the definition of a measurand is called definitional uncertainty (cf. Section 2.27 in [2]). An example can clarify. In PEMFC, if the GDL-gasket gap width is defined as 'the length of a segment with extremes P_g^* and P_{GDL}^* respectively on the gasket and GDL straight edges', then there are infinite ways in which each of the two extremes can be chosen. The example is illustrated in **Figures 2** and **3**, where a distance satisfying the measurand definition above is shown.

A range of different choices can be made when associating geometric entities, which are abstract concepts of the rational world, to entities of the sensory world.

In the schema of **Figure 3**, the representation of two lines is associated with the GDL and the gasket boundaries, respectively. A line, by definition, is an entity without any width. So it does not exist in the sensory world, because it cannot be perceived. It can only be represented in an approximation.

For the same reason, in **Figure 3**, there are only two representations of two straight lines in the visible world and not two real straight lines, which are abstract concepts existing only in the human mind. If a straight line representation is on a plane, as in this case, then it needs to occupy a surface portion to be perceived; if it is in space, then, for the same reason, it needs to occupy a volume region.

How then to identify unambiguously in the sensory world a point P_g^* on the straight line l_g^* associated with the gasket boundaries? It is not possible, because

Figure 2.
GDL (A), gap (B), gasket (C), edges (d), straight line fitted to each edge (l_g and l_{GDL}).

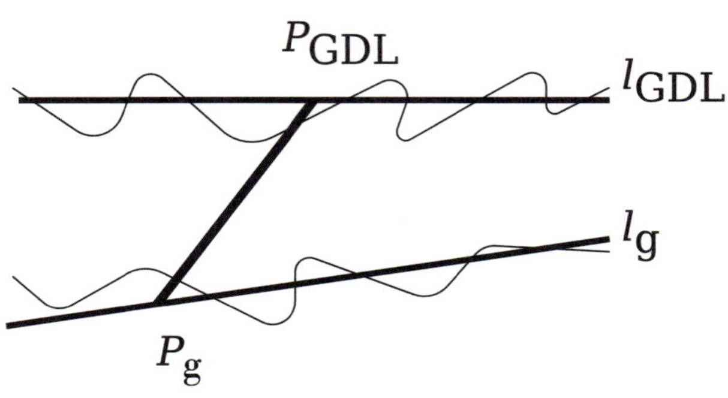

Figure 3.
A schema of a gap where the 'length' P_gP_{GDL} realises the given definition of gap width. The thin lines represent the edges.

neither the point nor the line exist in the sensory world. The surface portions P_g and l_g, respectively, representing P_g^* and l_g^* can however be identified. Yet each of these surface portions has an extension, which can be thought of as made up of infinite points. Which point to select between this infinity cannot be determined. The same reasoning applies to P_{GDL}^* and l_{GDL}^*. It then follows that the distance between P_{GDL}^* and l_{GDL}^*, i.e. the length of the segment joining the two points, cannot be determined unambiguously in the sensory world. The gap width cannot be defined without indeterminacy.

In **Figure 3**, $P_{GDL}P_g$ represents an infinity of lengths. If the figure is magnified enough, the surface area representing the segment is apparent and so is therefore the infinity of lengths. For this reason, quotation marks have been used in the caption for the term length.

The representations P_{GDL} and P_g can then be chosen in infinite ways on the lines while still complying with the given definition of gap width. This indeterminacy can, however, be removed by specifying further the measurand. For example, the measurand definition can also prescribe that the points P_g^* and P_{GDL}^* are chosen so that their distance is minimal.

Additional detail may be included in the measurand definition to make its realisation in the sensory world less vague. But how much detail? When to stop adding?

A strategy is to define a parameter that expresses the vagueness of the measurement result of the measurand that is about to be defined, for example, a standard deviation, and then to define jointly the measurand and an upper limit for the just defined parameter. If this upper limit is not exceeded, it makes the measurement result fit for purpose. When expressed as a standard deviation, this parameter is called standard uncertainty $u()$ (Section 2.30 in [2]). Its upper limit is referred to as the target uncertainty u^* (Section 2.34 in [2]).

The kind of information contained in a measurement definition is ideally a balance between the need of representing the intended purpose of the measurement result and the need of performing the measurement.

For example, in product design, the specification of the geometrical product characteristics are established by designers for a product to fulfil an intended purpose (e.g. functional, aesthetic, safety-related, regulatory). Often this purpose is referred to as the design intent. Designer product specifications should however account for the limits of available manufacturing and verification processes (sometimes they do not). Providing objective evidence of conformity of a product to its specification is called verification (cf. Section 2.44 in [2]). A verification where product requirement specifications are proved to be adequate for the intended purpose is called validation (cf. Section 2.45 in [2]). The objective evidence relied upon in verification of geometrical product characteristics is provided by measurement.

In the gap width case, for example, a designer is given the information that if the GDL overlaps with the gasket, an overboard leakage of some significance is likely to occur. He or she therefore specifies a lower limit to the minimum gap width. The metrologist who is called to verify the conformity of a gap width to this specification needs to know how to associate the boundaries of the GDL and the gasket (sensory world) to two edges. He or she then needs also to associate a straight line to each of the two edges and to find on them two points at minimum distance. The detail of how they accomplish this depends on the specific characteristics of the measurement they select.

As the example shows, a translation is needed into a practical measurand definition that however does not void the verification of the geometric characteristic.

To facilitate this translation process, the geometrical product specification system of standards (GPS) has been established by the ISO. More formally, the aim of the ISO GPS system is 'to describe certain workpiece characteristics through some of the different stages of its life cycle (design, manufacture, inspection, etc.)' (cf. Section 2 in [5]).

3.2 References

Units are the most typical reference used in a quantity value. Hence, they are discussed in this section. Adaptation may be needed if a measurement procedure or a reference material is used as reference of a quantity value. The units of measurement do not have an unambiguous unique magnitude. To support this statement, the concept of 'definition of a unit' must be distinguished and kept apart from the concept of 'realization of a unit', as explained in the SI Brochure (cf. Sections 1 and 2.2 in [3]). A measurement unit is a quantity that is conventionally defined so as it has solid theoretical foundations and it enables measurements as reproducible as possible. The realisation of a unit is instead a process where a quantity value is associated to a quantity of the same kind as the unit and that fulfils the definition given for the unit. A realised unit is a quantity existing in the sensory world and not just on the paper as the unit definition. The process of unit realisation is made clear by referring to the primary methods of realisation. For a method of unit realisation to be called primary, it needs to allow 'a quantity to be measured in a particular unit directly from its definition by using only quantities and constants that themselves do not contain that unit.' (Appendix 4 in [3]). This means that bringing the definition of a unit in existence into the sensory world requires a measurement where the measurand is the unit to be realised.

If a measurement is involved in realising a unit, then the vagueness of a realised unit is the same as that of the measurement that realises it. To limit this vagueness, the realising measurement for the definition of a base unit is specified in a document called a *mise en pratique*. (see Appendix 2 and Section 2.31 in [3]). *Mises en pratique* are only published in electronic form to facilitate frequent revision.

3.3 Measurement model

A measuring system is any set of devices that is designed to generate measured quantity values (cf. Sections 3.2 and 2.10 in [2]). Typically, the nature of the interaction measurand-measuring system cannot be isolated from other quantities characterising the conditions in which the measurement takes place.

For example, when measuring the gap width between gasket and GDL with a vision system, the measurement result is not only a function of the gap but also of other quantities and conditions. Among these, there may be the temperature and humidity of the air affecting the PEMFC component size, the colours of the PEMFC components, the light conditions, the camera (e.g. field of view, magnification and resolution) and the algorithms used to process the acquired images.

This interdependence between quantities is captured by 'a mathematical relation among all quantities known to be *involved* in a measurement' which is called in the VIM the measurement model (cf. Section 2.48 in [2]). Namely, it holds:

$$h(Y, X_1, X_2, \ldots, X_n) = 0. \tag{1}$$

In Eq. 1, Y is the measurand that is also called output quantity of the model to highlight that the value of Y is calculated with the measurement model 1 when the quantity value of $X_1, X_2, \ldots, X_n$ is known. The quantities $X_1, X_2, \ldots, X_n$ are correspondingly called the input quantities of the measurement model (cf. 2.50 in [2]). The value of the input quantities is known either by measurements or by other means like calibrated measurement standards, certified reference materials, reference data obtained manufacturer's specifications, handbooks and certificates. Often Eq. 1 can be explicitly defined as follows:

$$Y = f(X_1, X_2, \ldots, X_n). \tag{2}$$

In Eq. 2, the function $f(X_1, X_2, \ldots, X_n)$ is referred to as measurement function (cf. Section 2.49 in [2]).

This situation generates indeterminacy of the measurement in at least two different ways: in the selection of the input quantities and in their effect on the measurement result.

The input quantities in a measurement model are not uniquely known. Different people may consider different quantities to be relevant in a measurement on the basis of their knowledge of the phenomena involved in the process of measurement. The expression 'all the quantities known to be *involved* in a measurement' that appears in the VIM measurement model definition does not identify a unique set of input quantities in its implementation.

The VIM definition appears to suggest that an effort should be made to include in the model all the quantities believed to influence the measurement result, not just those influencing it the most.

The GUM refers to the inclusion in the model of 'every quantity that can contribute a *significant* component of uncertainty to the measurement result' (4.1.2 in [1]).

But how can a component of uncertainty contribution to the measurement result be considered *significant* if it is not first included in the model (GUM case)? On the other hand, how to know if a quantity is *involved* in a measurement if it is not first included in the model (VIM case)?

This difficulty may be overcome by considering the selection of input quantities on the basis of knowledge available prior to the formulation of the model. There is an element of subjectivity in this selection that ultimately relies on honesty and professional skill of those who are building the model (3.4.8 in [1]).

Some of the input quantities may then in their turn 'be viewed as measurands and may themselves depend on other quantities' (4.1.2 in [1]). That is to say, some of the quantities are the output of another measurement model. A measurement model is then a hierarchy of different models nested one in the other. How many levels of hierarchy do exist depends on the specific measurement. The levels, however, must be finite, because they ultimately reach the realisation of the definitions of the base units (see the SI Brochure for the recently changed definition of base unit [3]).

The measuring model $h(\ldots)$ (or function $f(\ldots)$) is typically not known analytically, barring those cases when some physical law describing the measurement is available. Ideally, the effect of an input quantity on the output is determined by an experimental investigation. The extension of the investigation is a balance between its cost and the intended use of the measurement result.

For example, the effect of the air humidity on the GDL-gasket gap width could be estimated by conducting an experiment if the cost of the experiment was affordable and the width measurement result needed to be confidently relied upon. But what if that cost was not affordable? Should the model builders give up to their prior knowledge that led them to include air humidity in the model because they cannot estimate its effect on the gap width?

Perhaps, an option is to rely on the skill and scientific knowledge of the model builders in estimating the sensitivity of the gap width to the humidity on the basis of their prior knowledge that led them to include the humidity in the first place. But then, where would it be the experimental validation that is typical of any scientific investigation? The issue is controversial.

When using prior knowledge to determine the effect of the humidity on the gap width, the specialists building the model may for example subjectively adjudge that this effect is negligible. They may then be so confident in their judgement that they model the air humidity effect with a constant in the model. Differently, if they are less confident, they can model the effect as a random variable with a mean of small value and a standard deviation subjectively attributed (for example, by inferring it from a survey in handbooks or other reputable sources). If the constant or the mean of the random variable was zero, someone may wonder why they have included the humidity in the model. The sole reason would be to show due diligence in their analysis.

4. Uncertainty

The argumentation presented so far is aimed to make the intrinsic indeterminacy of measurement results apparent.

In the GUM, this indeterminacy or vagueness that expresses a doubt about the result of a measurement is referred to as uncertainty (cf. Section 2.2.1 in [1]). In the same document, the term uncertainty is, however, also used in a more specific way to designate a parameter providing a quantitative measurement of this generic concept of doubt; namely in the GUM, uncertainty is defined as a 'parameter, associated with the result of a measurement, that characterizes the dispersion of the values that could reasonably be attributed to the measurand' (cf. Section 2.2.3 in [1]).

In the VIM instead, measurement uncertainty is defined as a 'non-negative parameter characterizing the dispersion of the quantity values being attributed to a measurand, based on the information used' (cf. Section 2.26 in [2]).

Typically this parameter is the standard deviation of a probability density function that models the incomplete or partial knowledge of the measurand achievable with measurements. This partial knowledge is described respectively by the expressions 'reasonably attributed' and 'based on information used' in the two definitions.

In the error approach, as the VIM puts it, 'a true quantity value is considered unique and, in practice, unknowable' (cf. Section 2.11, note 1 in [2]).

The uncertainty approach is to recognise that owing to the inherently incomplete amount of detail in the definition of a quantity, there is not a single true quantity value but rather a set of true quantity values consistent with the definition. However, in principle and in practice, this set of values cannot be known (see Appendix D in the GUM [1], D.3 in particular). Then, even if an imaginary measurement is capable of producing measurement results of a measurand without any indeterminacy, that measurand would still be known with vagueness.

The introduction of a probability density function of a measurand Y and of an input quantity X_i is the same to say that these quantities are represented by a random variable, whose realisations y_i and x_{ij} are all their possible observations.

4.1 Uncertainty analysis

This section has similarities with the BS EN ISO 14253-2:2011 'Procedure for Uncertainty MAnagement (PUMA)' [6]. However, the analysis conducted here adheres more strictly to the GUM: the evaluated uncertainty is not overestimated as it is instead in the iterative PUMA procedure (cf. Section 5 in [6]). A measurement may be seen as an iterative process consisting of seven steps. These steps are listed below and illustrated in the diagram of **Figure 4**.

- **The measurand definition** has been discussed in Section 3.1. The amount of detail to include in the definition is determined. The detail may include 'physical states and conditions' (D.1 in [1]). If the target uncertainty test fails in the following step, attempts to satisfy it by modifying the in-between steps within the limits of the resources available are made. If these attempts are ineffective, then further detail may be added to the definition. However, adding detail to the measurand should be consistent with the purpose for which the measurement result is intended to be used.

- **The measurement principle** is the physical, chemical or biological phenomenon on which the measurement is based, as defined in Section 2.4 of the VIM [2]. Once a principle has been chosen, additional characteristics concerning the measurement principle may be added to the steps that follow.

- **The measurement method** is a 'generic description of a logical organization of operations used in a measurement'(2.5 in [2]). The category of operations encompassed by the same measurement method is typically large.

- **The measurement procedure** is a description of the measurement with enough detail to allow an operator to perform it. This description is typically a sequence of instructions for the operator. The sequence is sometimes called standard operating procedure (SOP, cf. 2.6 in [2]) and includes a statement of target uncertainty.

- **The measurement model** is established on the basis of the measurand. This concept has been defined in Section 3.3. **Figure 5** displays a fish-bone diagram which is an elaboration of the content from Section 7 of BS EN ISO 14253-2:2011 [6]. It displays a grouping of candidate input quantities. The figure suggests a systematic procedure of input quantity investigation for inclusion in a measurement model for uncertainty evaluation. The model refers to the measurement of a geometric characteristic, which is the case of the gap width. The figure has been produced using qcc, an r software package [7].

- **A target uncertainty test** is performed. The measurement uncertainty ($u(Y)$) is first evaluated and then compared with the target uncertainty (u^*).

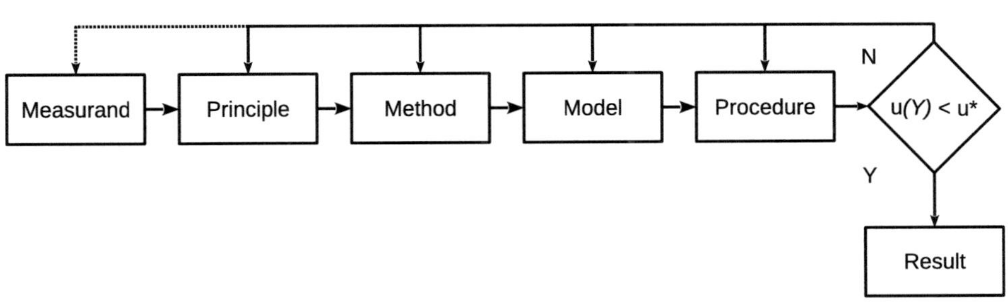

Figure 4.
Diagram illustrating the steps of a measurement ($u(Y)$ and u^* are the evaluated and the target measurement uncertainty, respectively).

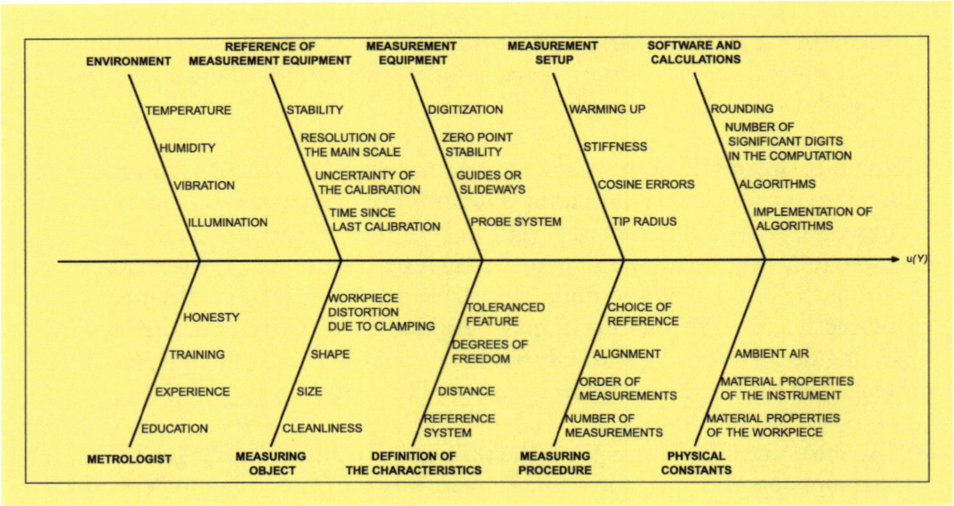

Figure 5.
An Ishikawa (fish-bone) diagram displaying a categorical hierarchy of potential input quantities typically considered for inclusion in measurement models for uncertainty evaluation $u(Y)$ of geometric quantities (elaboration from Section 7 of BS EN ISO 14253-2:2011 [6]).

If $u(Y) > u^*$, then all the previous steps are put under scrutiny. The source of uncertainty that is found to contribute more to the violation of the target uncertainty constraint or that is known the least is respectively mitigated or further analysed, if possible. If not, the second most severe or less known uncertainty source is considered, and so on. Once one source of uncertainty has been chosen, it is acted upon and the test $u(Y) < u^*$ is run again. The sequence is repeated until $u(Y) < u^*$ or the measurement is recognised incompatible with the given target uncertainty.

- **The measurement result** is presented in a form that also contains an expression of the evaluated uncertainty. As much detail as possible about how the evaluation was performed is recommended by the GUM. Specific guidance is given in its Section 7 [1].

The uncertainty of the measurand Y that appears in the target uncertainty test is 'evaluated' and not 'estimated'. The verb 'to evaluate' is used to highlight that the input quantities X_i are typically grouped into two categories. The standard uncertainty of those in the first group is determined by their repeated observation (Type A evaluation, Section 4.2 in [1]). The standard uncertainty of those in the second is instead determined by the 'scientific judgement based on all of the available information' (Type B evaluation, cf. Section 4.3 in [1]). In the first case, uncertainty evaluation is based on probability density functions estimated from frequency distribution of observations. In the second, the evaluation is based on probability density functions postulated on the basis of reputable sources of information like handbooks or calibration certificates. In both Type A and Type B evaluations, the complete characterisation of the probability density functions $p(X_i)$ of the input quantities is needed. Complete characterisation means that the mean $E(X_i) = \mu_i$, the standard uncertainty $u(X_i)$ and the distribution type (e.g. normal, triangular, rectangular/uniform) must be made available for each X_i. Then, the measurement function is expanded into a partial sum of the Taylor series around the input quantity means. By applying the definition of variance to this partial sum, the variance of the output quantity is expressed as a sum of the variances and covariances of the input quantities, where each term is multiplied by constants (law of propagation of uncertainty, E.3 in [1]). These constants are the coefficient of sensitivity.

According to the GUM definition, measurement uncertainty refers to the result of a measurement. If the conditions in which the measurement is carried out are believed to be influencing the result, they have also to be specified: input quantities are introduced in the measurement model to represent them. For example, if multiple measurement runs are necessary to calculate the result, as it is the case when the result is an average, then the repetition settings may enter the model, if believed influential. Typical repetition settings are repeatability and reproducibility conditions (2.20 and 2.24 in [2]). Repetition settings may not be quantities as defined in Section 3, because they represent conditions and not properties expressed as number and reference pairs. They enter the model as random variables. When they are meant to contribute only to the uncertainty of the measurement result, these random variables have location parameter (e.g. mean, median) set at zero and unknown standard deviations.

The statement of the uncertainty $u(Y)$ and of the input quantities that most contribute to $u(Y)$ together with the evaluation of their uncertainty and their combination to give $u(Y)$ is called uncertainty budget (2.33 in [1]). The next section illustrates the uncertainty analysis of the gap width.

4.2 Gap width uncertainty

The steps of the gap width measurement are listed here below.

- **The gap width definition** has been given in Section 3.1 and illustrated in **Figure 3**. The requirement of minimum distance between P_g and P_{GDL} is a part of the definition considered here. The symbol w is used for the measurand. If satisfying the target uncertainty requirement is not practicable with this definition, then a more detailed measurand definition is needed. An example of a new definition is the following: the gap edge is the minimum distance between P_g and P_{GDL} when the straightness of the GDL and gasket edges is $t = 0.05\,\text{mm}$ (cf. 17.2 in [8] for a definition of straightness). This new definition may provide a measurand with less variation in the gap width. The process of verifying the edge straightness specification may be described by a dedicated input quantity in the measurement model. This variable is in its turn the output of a measurement.

- **The measurement principle** on which the gap width measurement is based is the selective reflection of visible light by the GDL, the gasket and the gap background. For the anode GDL and the cathode GDL, the gap background is respectively the portion in view of the BPP and the MEA. In **Figure 2**, the background was replaced with a white paper sheet. Characteristics like wavelength, intensity, number and locations of the light sources may enter the measurement procedure and measurement model, if needed. But how to ascertain whether there is a need to include them? If they are included, their contribution to the uncertainty of the measured gap width is estimated. Generating the data for an estimation requires an experimental investigation whose the cost may not be affordable. In these cases, a Type B evaluation is performed.

- **The measurement method** of the gap width is the digital image processing of an image where the gap is visible. The gap width definition is realised in this image. The position and orientation of the camera relative to the gap, the characteristics of the optics and of the CCD sensor all contribute to the captured view of the gap.

- **The measurement procedure** of the gap width depends heavily on the vision system used and its degree of automation. The operator can be a person, a robotic system holding a camera, a computer that runs the image processing algorithms or a combination of all of these. The sequence of instructions in the procedure has therefore to account for these different kinds of operators. To clarify, the case of a fixed camera orthogonally placed above the gap is considered. The acquisition process is completely automated. The light conditions and camera set-up are fixed. The instructions in this case consist of commands written in a computer program. The commands are grouped in modules, whose sequence is illustrated in **Figure 6**. In each single module represented in this diagram, discretionary decisions may be taken in the selection of an algorithm and its parameters. Examples of these decisions are the following: a Canny edge detection algorithm is selected among the many available algorithms; the edges are considered acceptable if they have no more than a predetermined number of pixels groups disconnected from the largest edge; the image filtering is done with a Gaussian filter to reduce the number of disconnected group of pixels, i.e. the occurrence of edge false detection; and straight lines are fitted to the edges using an orthogonal non-linear least squares algorithm (ONLS). As it can be deducted from these examples, the decisions taken in the modules affect the measured gap width. The subjective behaviour of the specialists taking these decisions is reflected in a contribution to the gap width uncertainty.

- **The measurement model** chosen to describe the gap width measurement is given by the following equation:

$$w = \sqrt{c^2\left(u_{\mathrm{GDL}} - u_{\mathrm{g}}\right)^2 + a^2c^2\left(v_{\mathrm{GDL}} - v_{\mathrm{g}}\right)^2} + e. \tag{3}$$

To enhance readability, the convention used in the example of Section H.1 of the GUM was adopted [1]: random variables are in lowercase italic shape and constants are in the normal lowercase shape. In the vision system, the image coordinate system with coordinates expressed in number of pixels is such as $P_{\mathrm{g}} = P_{\mathrm{GDL}}(u_{\mathrm{GDL}}, v_{\mathrm{GDL}})$ and $P_{\mathrm{g}} = P_{\mathrm{g}}(u_{\mathrm{g}}, v_{\mathrm{g}})$, c is the calibration factor which is the ratio of an imaged calibration length to the number of pixels contained in it and a is the aspect ratio of a pixel (width over height). The measurement function of Eq. 3 is based on the following assumption: the imaged artefact realising the calibration

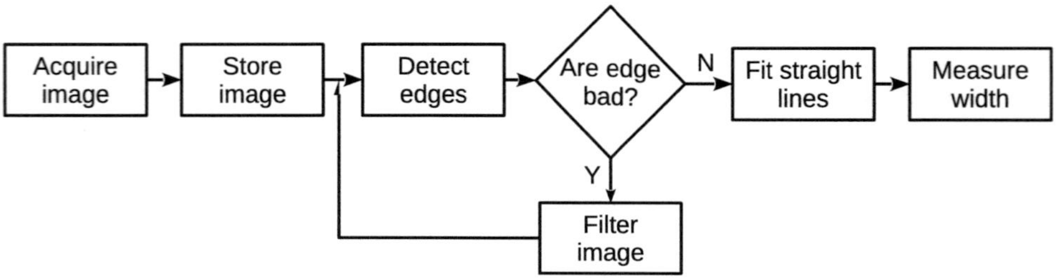

Figure 6.
A diagram of the sequence of modules that constitute the measurement procedure.

length needs to be placed so that the calibration length is aligned with the x axis of the image coordinate system.

- **A target uncertainty test** is performed. If the target uncertainty test fails, then one input quantity is selected for a more detailed uncertainty evaluation. The selection criteria are described in Section 4.2. For example, in the vision system the magnification of the lens is the ratio of the length on the image to the length in the scene, i.e. $m = l_{image}/l_{scene}$, where $l_{image} = sn_{\mathrm{CCD}}$, with s indicating the pixel size on a CCD sensor and n_{CCD} representing the number of pixels in the imaged length. Then, the following equations holds:

$$c = l_{scene}/n_{\mathrm{CCD}} = s/m. \tag{4}$$

To clarify, the case of a CCD sensor with pixel size 6.45 μm and a lens magnification 100 would result in a calibration factor 0.0645. A new level of detail is introduced in the original measurement model of Eq. (3) by the nested model of Eq. (4). The uncertainty of the calibration factor c is then evaluated by combining the uncertainty evaluations of s and m.[1]

- **The measurement result** is presented in a statement like, for example, the following: 'the gap width is 0.08 mm with a standard uncertainty $\mathrm{u}(w) = 0.014\,\mathrm{mm}$'. A report illustrating how the measurement and the evaluation took place would also be attached.

5. Intelligent systems and uncertainty

The relationship between intelligent systems and the uncertainty in measurement is analysed in this section. The concept of uncertainty in measurement has been explored in the previous sections. The concept of an intelligent system is more elusive. It requires first to understand what intelligence means. With intelligence, it is typically intended a set of human rational abilities that include reasoning, learning and adaptation to changing conditions. Yet, which of these abilities are the distinguishing characteristics needed for a system to be called intelligent is controversial. Failing to recognise the differences in the behaviour between an artificial system and a human is often advocated as a criterion for considering the artificial system as intelligent (Turing test [9]).

In uncertainty evaluation, circumstances arise where discretionary decisions of human specialists are necessary. Realising an intelligent system whose behaviour constitutes a reference for the specialists in making these decisions would greatly reduce their discretion. For example, in the definition of the gap width measurand and of its measurement model, a trusted intelligent system would facilitate the task. To be considered intelligent, such a system would need to demonstrate to take decisions independently from human input and not just 'executing' a predetermined behaviour that the system is programmed to have. Conversely, an uncertainty evaluation is purposely carried out to quantify how much trust there is in a measurement result. The question then arises whether an uncertainty analysis where an artificial intelligent system is 'the specialist' taking discretionary decision unsupervised by humans can be trusted as a human (or more).

---

[1] The symbols introduced in this example have not been included in the end-of-chapter symbol list to keep it clear.

A framework to evaluate the uncertainty of an intelligent system providing a measurement uncertainty analysis is needed. No evidence of that has been found in the literature. Research surveyed about the adoption of intelligent systems in measurement uncertainty evaluations dates back to about 10 years ago. An expert system for uncertainty evaluation in analytical chemistry has been developed by Rösslein and Wampfler [10]. Their motivation was to provide a means of performing complex uncertainty evaluations that otherwise would have been too simple for their purpose or too costly. An analysis of how much trust to put into their system has not however been found.

Intelligent systems may then contribute to reduce uncertainty in any measurement phase where parameter-dependant software modules are involved. A parameter, by definition, needs to be given a value that is determined by a specialist. He or she does so drawing from their knowledge. This may leave room for subjective decisions that contribute to the uncertainty. With an intelligent system replacing the judgement of the specialist in the determination of the parameter value, it may seem possible to eliminate that uncertainty. This possibility would require the intelligent system not to be dependant in its turn on other parameters, which is hardly the case.

The situation is illustrated with the edge detection module in the gap width case. If a Canny edge algorithm is used, then two threshold parameters need to be determined. An intelligent system (e.g. a neural network) may be trained on a set of images for this purpose. But this would still have at least one component of uncertainty which is how the choice of the training set of images (equivalently, the parameters identifying the training set of images).

6. Conclusions

Uncertainty as a technical term introduced by authoritative international bodies as a means of representing the intrinsic indeterminacy in measurements has been discussed. The analysis presented in this chapter provides an interpretation of the documents elaborated by these international bodies. The framework defined in these documents is interpreted as a method that makes it possible to organise consistently the knowledge about measurement uncertainty within the limitations of the resources available.

Acknowledging that measurements are always unavoidably uncertain has some profound implications on how humans construct their knowledge about the physical world in science and technology. Measurements are the ultimate source of knowledge in science: any statement to be scientifically accepted must be substantiated by experiments or observations that are expressed in terms of measurement results. If measurement results are inherently uncertain, all that can be inferred from them can be only uncertain. One consequence is, for example, that talking of 'exact science' when referring to Physics can be quite prone to misinterpretations. Science may be considered exact only in its methods of dealing with approximations and uncertain or partial knowledge. Stretching this view to its extreme may lead to consider science as an activity with very useful practical effects but with little use in the unambiguous identification of the truth.

Acknowledgements

The author acknowledges the support of the research project 'DIGItal MANufacturing and Proof-of-Process for Automotive Fuel Cells' (DigiMan) funded by the programme 'The Fuel Cells and Hydrogen Joint Undertaking' (FCH 2 JU, grant agreement ID: 736290) in the EU framework Horizon 2020 (H2020).

Thanks

The author is grateful to Mr. David Urquhart for his help.

Nomenclature

Symbols

Y	output quantity in a measurement model
y_i	the ith observation or realisation of an output quantity
X_i	the ith input quantity in a measurement model
x_{ij}	the jth observation or realisation of the input quantity X_i
$E(X_i) = \mu_i$	mean or expected value of X_i
$p(\ldots)$	probability density function
$h(\ldots) = 0$	measurement model
$Y = f(\ldots)$	measurement function
$u()$	standard uncertainty
u^*	target uncertainty
$l^*_{\mathrm{GDL}}, l^*_{\mathrm{g}}$	GDL and gasket straight edges
$l_{\mathrm{GDL}}, l_{\mathrm{g}}$	representations of the straight edges of GDL and gasket in the sensory world
$P^*_{\mathrm{GDL}}, P^*_{\mathrm{g}}$	points defining the gap width
$P_{\mathrm{GDL}}, P_{\mathrm{g}}$	representations in the sensory world of the points defining the gap width
$(u_{\mathrm{GDL}}, v_{\mathrm{GDL}})$	coordinates of P_{GDL} in the image reference system
$(u_{\mathrm{g}}, v_{\mathrm{g}})$	coordinates of P_{g} in the image reference system
c	calibration factor, i.e. length of a pixel width in units of length
k	image aspect ratio
m	lens magnification
s	physical size of a pixel on a CCD chip

Abbreviations

BIPM	Bureau International des Poids et Mesures (International Bureau of Weights and Measures)
BPP	bipolar plate
CCM	catalyst coated membrane
GDL	gas diffusion layer
GPS	geometrical product specification
GUM	Guide to the Expression of Uncertainty in Measurement
ISO	International Organisation for Standardisation
MEA	membrane electrode assembly
MPL	microporous layer
ONLS	orthogonal non-linear least squares
PEM	proton exchange membrane (polymer electrolyte membrane)
PEMFC	proton exchange membrane fuel cell

PTFE	polytetrafluoroethylene
PUMA	procedure for uncertainty management
SI	Le Système International d'unités (The International System of Units)
SOP	standard operation procedure
VIM	International Vocabulary of Metrology

Author details

Carlo Ferri
WMG, University of Warwick, Coventry, UK

*Address all correspondence to: c.ferri@warwick.ac.uk

References

[1] Joint Committee for Guides in Metrology (JCGM). JCGM 100:2008. GUM. 1995 with minor corrections. Evaluation of Measurement Data—Guide to the Expression of Uncertainty in Measurement. 1st ed. 2008. Corrected Version 2010 edition, 2008

[2] Joint Committee for Guides in Metrology (JCGM). JCGM 200:2012. International Vocabulary of Metrology—Basic and General Concepts and Associated Terms (VIM). third, Version with Minor Corrections Edition; 2008

[3] Le système international d'unitès (SI). The International System of Units (SI). Bureau International des Poids et Mesures (BIPM). Organisation Intergouvernementale de la Convention du Mètre. Available from: http://www.bipm.org/en/publications/si-brochure/. Visited in August 2019, 9th ed. ISBN 978-92-822-2272-0; 2019

[4] Husar A, Serra M, Kunusch C. Description of gasket failure in a 7 cell pemfc stack. Journal of Power Sources. 2007;**169**(1):85-91. CONAPPICE 2006

[5] BS EN ISO 17450-1:2011 geometrical product specifications (GPS)—General concepts. Part 1: Model for geometrical specification and verification; 2011

[6] BS EN ISO 14253-2:2011 geometrical product specifications (GPS)—Inspection by measurement of workpieces and measuring equipment. Part 2: Guidance for the estimation of uncertainty in GPS measurement, in calibration of measuring equipment and in product verification; 2011

[7] Scrucca L. qcc: An r package for quality control charting and statistical process control. R News. 4/1:11–17; 2004

[8] BS EN ISO 1011:2017 geometrical product specifications (GPS)—Geometrical tolerancing—Tolerances of form, orientation, location and run-out (ISO 1101:2017); 2017

[9] Chen J. Chapter Artificial Intelligence. In: Encyclopedia of Computer Science and Technology. Taylor & Francis; 2017. pp. 144-153

[10] Rösslein B, Wampfler M. Expert system for the evaluation of measurement uncertainty: Making use of the software tool uncertaintymanager®. CHIMIA International Journal for Chemistry. 2009;**63**(10):624-628

Chapter 15

Quantized Neural Networks and Neuromorphic Computing for Embedded Systems

Shiya Liu and Yang Yi

Abstract

Deep learning techniques have made great success in areas such as computer vision, speech recognition and natural language processing. Those breakthroughs made by deep learning techniques are changing every aspect of our lives. However, deep learning techniques have not realized their full potential in embedded systems such as mobiles, vehicles etc. because the high performance of deep learning techniques comes at the cost of high computation resource and energy consumption. Therefore, it is very challenging to deploy deep learning models in embedded systems because such systems have very limited computation resources and power constraints. Extensive research on deploying deep learning techniques in embedded systems has been conducted and considerable progress has been made. In this book chapter, we are going to introduce two approaches. The first approach is model compression, which is one of the very popular approaches proposed in recent years. Another approach is neuromorphic computing, which is a novel computing system that mimicks the human brain.

Keywords: machine learning, deep learning, model compression, algorithms, pattern recognition, neuromorphic computing

1. Introduction

Deep learning is a branch of machine learning that is inspired by the biological processes of human brain and it is not a new concept. The reason it was not popular earlier is because there were not enough computational power and data available many years ago. With the development of the semiconductor industry and Internet, the stronger computational power and tremendous data generated by the Internet make the use of deep learning techniques possible [1–10].

Even though deep learning techniques have made great success in many fields, we still have not realized their full potential, especially in embedded systems because such systems do not have enough computation power. In the era of mobile computing, enabling deep learning techniques running on mobile devices is very important and a lot of researchers have been working on this area [4, 6]. Researches have been conducted in two directions. The first direction aims to reduce model size and computation of deep learning models. The second direction is to design new hardware architectures that have much stronger computation power. In this chapter, we are going to introduce two approaches. The first technique is quantization, which is used to reduce the computation and model size of deep learning models. The second technique is neuromorphic computing, which is a new hardware architecture to enhance the computation power.

2. Neural network

Artificial neural network is a computing system that is capable of mimicking the human brain. The purpose of an artificial neural network is to identify patterns in input data and learn an approximate function that maps inputs to outputs. The most basic building units of neural network are neurons, which have inputs, outputs and a processing unit. To better learn complicated patterns in input data, a neural network consists of a huge number of neurons, which are organized into layers of neurons [11, 12]. These layers of neurons are stacked on each other so that the output of a layer is the input of the following layer. A neuron in a layer is connected to multiple neurons in previous layer in order to receive data from those neurons. Data received from neurons in previous layer are multiplied by corresponding weights and the product results are accumulated together to generate an output.

2.1 Single neuron

The most basic building unit of a neural network is neuron. A neuron receives data from multiple neurons from previous layer and each of these data is multiplied by a weight. Then, these weighted data are accumulated together to generate an output. A non-linear function is applied to the output before the output is sent out to other neurons. More details about why we need a non-linear function are presented in Section 2.3. The working mechanism of a single neuron can be expressed as,

$$Out = \sum_i w_i x_i + b \tag{1}$$

where x_i is the i_{th} input, w_i is the weight corresponding to i_{th} input, b is the bias value and Out is the accumulated output.

Figure 1 illustrates a single neuron with three inputs where x_1, x_2 and x_3 are the three inputs and w_1, w_2 and w_3 are weights corresponding to inputs x_1, x_2 and x_3. The output of this neuron can be computed using Eq. (1).

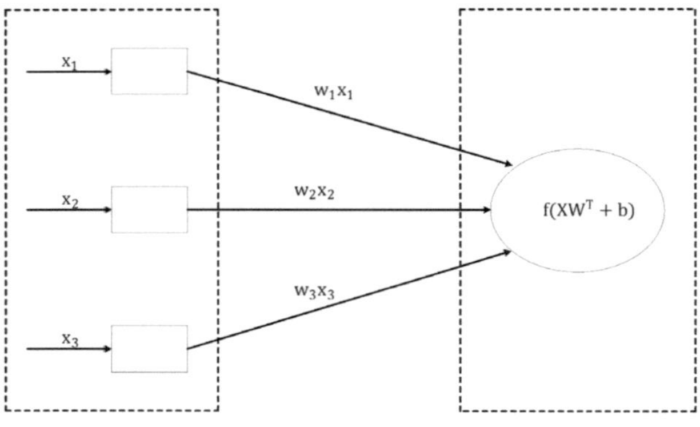

Figure 1.
Single neuron with three inputs.

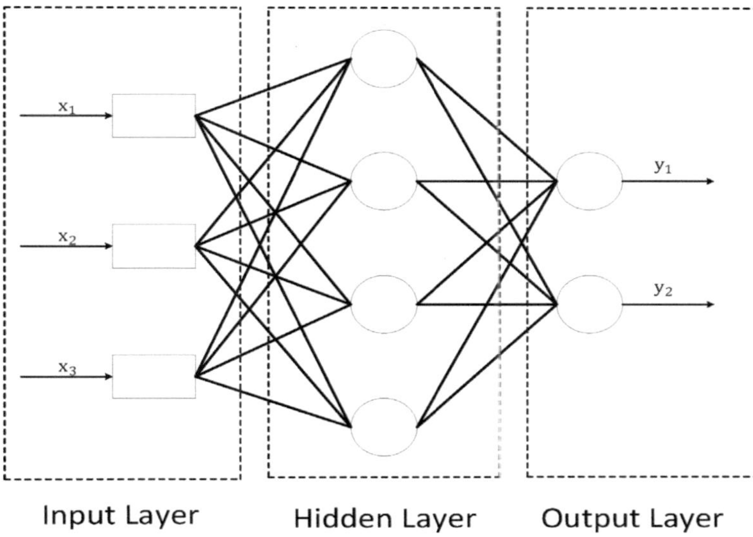

Figure 2.
Neural network with hidden layer.

2.2 Multilayer perceptron

Multilayer perceptron (MLP) is a kind of neural network that has at least one layer of neurons between the input and output layer. Hidden layers are layers between the input and output layer. The reason why multilayer perceptron is introduced is that it makes a neural network much more powerful and is able to learn very complicated patterns in input data. If there is no layer between the input and output layer, the input is transformed to output by a linear transformation function and the neural network can only work on linearly separable data. To enable neural network to handle data that are not linearly separable, we need to have at least one layer between the input and output layer. Meanwhile, non-linear functions are applied to the outputs of hidden layers. Let us take the MLP neural network in **Figure 2** as an example. This neural network has three inputs, two outputs and one hidden layer with four neurons. Lines represent connections between neurons. Each connection has an associated weight and we use weight matrices to represent the connections between the input, hidden and output layer.

The behaviour of the neural network in **Figure 2** can be expressed mathematically using Eq. (2).

$$\begin{aligned} h_1 &= \theta(W_1 \cdot x + b_1) \\ y &= W_2 \cdot h_1 + b_2 \end{aligned} \tag{2}$$

where θ is the non-linear activation function; *symbol* $\cdot$ is the dot product between two matrices; W_1 is the weight matrix between the input and hidden layer; W_2 is the weight matrix between the hidden and output layer; h_1 is the output of the hidden layer; x is the neural network input and y is the neural network output.

Input x is mapped to four neurons by a weight matrix W_1 first and then each neuron is applied a non-linear activation function. h_1 is the output of the hidden layer and this output is multiplied by another weight matrix W_2 to obtain final result y.

2.3 Non-linear activation function

Non-linear activation function [13, 14] is very important for MLP. Without a non-linear activation function, neural network does a linear transformation from input to output no matter how many hidden layers exist between the input and output layer. It is because the linear transformation of a linear transformation is still a linear transformation and thus any number of hidden layers can be deducted to a single linear transformation. Let us take the MLP neural network in **Figure 2** as an example. Without non-linear activation, the neural network can be expressed mathematically as,

$$\begin{aligned} h_1 &= W_1 \cdot x + b_1 \\ y &= W_2 \cdot h_1 + b_2 \end{aligned} \tag{3}$$

Substituting h_1 with $W_1 \cdot x + b_1$, we can have,

$$\begin{aligned} y &= W_2 \cdot (W_1 \cdot x + b_1) + b_2 \\ => y &= W_2 \cdot W_1 \cdot x + W_2 \cdot b_1 + b_2 \end{aligned} \tag{4}$$

Assume, $W = W_1 \cdot W_2$ and $b = W_2 \cdot b_1 + b_2$,

$$y = W \cdot x + b \tag{5}$$

Therefore, the MLP neural network in **Figure 2** can be expressed mathematically as a single linear transformation from input to output.

2.4 Types of hidden layers

In MLP neural networks [15], there are hidden layers between the input and output layer and these hidden layers play a very important role in performances of MLP neural networks. There are many different types of hidden layers such as convolutional layers, fully-connected layers, pooling layers and so on. In this section, we are going to present more details about convolutional layers and fully-connected layers.

2.4.1 Fully-connected layers

In fully-connected layers, each neuron is connected to all neurons in previous layers and each connection has an associated weight. Each output of a neuron from previous layers is multiplied by a weight associated with the connection. Then, the product result is accumulated together.

Let us take the hidden layer of MLP neural network in **Figure 2** as an example. The hidden layer in **Figure 2** is a fully-connected layer. Each neuron in the hidden layer connects all three inputs in the input layer and generates one output. The weight matrix is represented in Eq. (6). In the weight matrix, each row represents the weights of a neuron and thus the matrix size is 4 × 3 since there are four outputs and three inputs.

$$W = \begin{bmatrix} w_{11} & w_{12} & w_{13} \\ w_{21} & w_{22} & w_{23} \\ w_{31} & w_{32} & w_{33} \\ w_{41} & w_{42} & w_{43} \end{bmatrix} \tag{6}$$

The input can be represented as matrix that has the size of 3 × 1.

$$X = [x_1 \;\; x_2 \;\; x_3] \tag{7}$$

Mathematically, fully-connected layer is computed as a matrix multiplication,

$$Out = W \cdot X \tag{8}$$

2.4.2 Convolutional layer

The convolutional layer is a layer used in many deep learning applications, especially in computer vision [1, 2, 16–18]. In computer vision, processing and understanding an image is a major task. An image has three dimensions, which are width, height and channel. Meanwhile, an image is highly structured and has strong spatial dependency [11].

The convolution layer has a group of kernels and each of these kernels has three dimensions, which are width, height and channel. The width and height of a kernel are hyper-parameters defined by designers. The size of a channel is equal to the channel size of previous layer. Unlike a fully-connected layer, each neuron in a convolutional layer is only connected to a small spatial region of neurons but all channels in the previous layer. The size of this spatial region depends on the width and height of each kernel. Each kernel slides over the whole image with a specific stride to extract features such as edge feature from the image. Therefore, each kernel extracts a specific feature we want to obtain from each local region.

Let us use **Figure 3** above as an example to demonstrate how convolution layer works. In **Figure 3**, the image only has one channel with size $6 \times 6 \times 1$ and there is one kernel with size $3 \times 3 \times 1$. Assume the weight of this kernel is

$$W = \begin{bmatrix} 1 & 0 & 1 \\ 0 & 1 & 0 \\ 1 & 0 & 1 \end{bmatrix} \tag{9}$$

Then, the weight matrix W is multiplied by the pixels of a small region in the image element-wise and then these product results are accumulated together. Assume we are applying our kernel on the yellow region of the image in **Figure 3**. Then, we can get the output Out using Eq. (10),

$$\begin{aligned} Out &= W \odot X \\ Out &= 1 \times 0 + 0 \times 1 + 1 \times 0 + \\ &\quad 0 \times 0 + 1 \times 1 + 0 \times 1 + \\ &\quad 1 \times 1 + 0 \times 1 + 1 \times 1 = 3 \end{aligned} \tag{10}$$

where $\odot$ represents element-wise product between two matrixes.

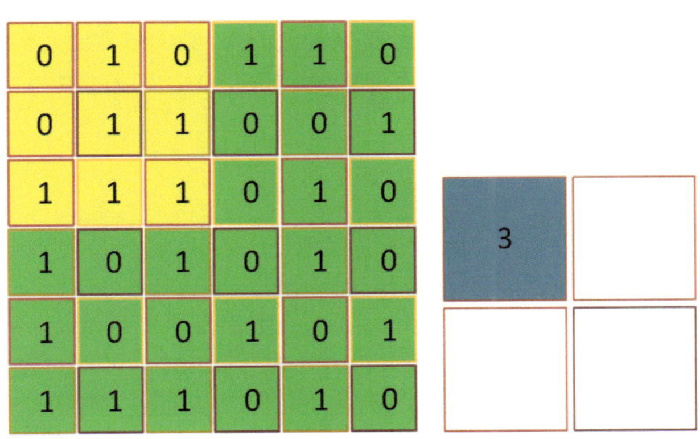

Figure 3.
Image and convoluted output.

Convolution layer has a couple of advantages compared to other layers when dealing with images. These features make the convolution layer very popular in the area of computer vision. First of all, convolutional layers need much less weights compared to fully-connected layers. In a fully-connected layer, a neuron is connected to all neurons in previous layers. If the dimension of previous layer is very large, the number of weights required by the fully-connected layer is very large since the total number of weights is equal to the number of neurons in previous layer times the number of neurons in fully-connected layer. Secondly, the convolution layer focuses on local spatial regions instead of the whole image and many applications benefit from this characteristic. For example, when dealing with object detection in an image, we only need to focus on regions where the object appears and other regions such as background are not needed when we are trying to detect the object in an image. Thirdly, the convolution layer is translation invariant. It means that the responses of a kernel to an object are the same regardless the location of the object in an image.

3. Model compression

In the era of mobile computing, enabling deep learning techniques running on mobile devices is very important. In general, large and complicated models tend to have high performance, but it increases the computation requirement dramatically. Embedded systems do not have sufficient computation and memory resource to support the model complexity of a high-performance deep learning model. Therefore, deploying deep learning models in embedded systems without sacrificing much performance has been a hot research topic [6, 19–22].

3.1 Model quantization

Deep learning models use floating-point arithmetic in both training and inference phases. Floating-point arithmetic needs many computation resources, which is one of the reasons why deploying deep learning models in embedded systems is difficult. To address this issue, researchers have proposed many approaches [4–6, 23–29] to replace the floating-point arithmetic during the inference phase. Low bit-precision arithmetic [30–32] is one of those approaches.

In the low-bit arithmetic approach, floating-point numbers and floating-point arithmetic are still used in the training phase. After training is done, model weights and activation layers are quantized using low-bit integer numbers such as 8-bit or 16-bit integers. During inference, integer arithmetic is used instead of floating-point arithmetic and thus the computation resource requirement is reduced dramatically.

3.1.1 Quantization scheme

In Ref. [4], the authors proposed a quantization scheme that is successfully adopted in TensorFlow [33]. During the inference phase, the proposed quantization scheme uses integer-only arithmetic while floating-point arithmetic is still used in the training phase. Since this approach uses different data type in training and inference phases, creating a one-to-one mapping between floating-point number and integer number is needed. The authors use Eq. (11) [4] to describe the mapping between a floating-point number and integer number.

$$r = S(q - Z) \tag{11}$$

In this equation, S and Z are quantization parameters, which are constant for each layer. There is only one set of quantization parameters associated for each activation layer and weights layer.

The constant S is a floating-point number, which is a scale constant to represent the size of each quantization level. For example, assume we are going to quantize a floating point r in a layer to an 8-bit integer number. To calculate the scale constant S, we obtain the maximum and minimum floating-point number of the layer first. We use r_{max} and r_{min} to represent the maximum and minimum floating-point number respectively. Since we are using an 8-bit integer number, there are $n = 2^8 = 256$ quantization levels. Then, the constant scale S can be computed as,

$$S = \frac{r_{max} - r_{min}}{n - 1} \tag{12}$$

In terms of Z, it represents real number 0 using quantized integer. The reason why we need an exact number to represent number 0 is that number 0 is wildly used in deep learning such as zero-padding in convolutional neural network. Representing number 0 exactly improves the performance of deep learning models. The number 0 can be calculated using Eq. (13) [4].

$$Z = q_{min} - rounding\left(\frac{r_{min}}{S}\right) \tag{13}$$

In Eq. (13), q_{min} is the minimum quantization level of our quantized integer. For example, if we use an unsigned 8-bit integer, q_{min} is equal to 0. r_{min} and S are two floating-point numbers representing the minimum values of a layer and the scale constant of this layer, respectively. Because Z is an integer, we need to round $\frac{r_{min}}{S}$ to the nearest integer. However, Eq. (13) only works for the case where r_{min} is smaller than 0 and r_{max} is larger than 0. To make it work for all cases, we use the following approaches to handle those cases. If r_{min} is larger than 0, we set r_{min} to 0 and calculate scale constant S using the new r_{min}. If r_{max} is smaller than 0, then we set r_{max} to 0 and calculate scale constant S using the new r_{max}. After we obtain the scale constant S, then we can calculate the zero-point.

3.1.2 Integer-only multiplication

After floating-point numbers are quantized into integers using the quantization scheme described in Eq. (11), the authors in [4] describe their approaches of how to compute multiplication between two floating-point numbers using quantized integers.

Assume we are going to compute the multiplication between two floating-point numbers r_1 and r_2. The multiplication result is stored in floating-point number r_3. We first quantize the floating-point numbers r_1, r_2 and r_3 into quantized integer numbers q_1, q_2 and q_3 respectively using Eq. (11). We have scale constants S_1, S_2 and S_3 corresponding to r_1, r_2 and r_3. Meanwhile, we have quantized zero-points Z_1, Z_2 and Z_3 corresponding to r_1, r_2 and r_3.

$$\begin{aligned} r_1 &= S_1(q_1 - Z_1) \\ r_2 &= S_2(q_2 - Z_2) \\ r_3 &= S_3(q_3 - Z_3) \end{aligned} \tag{14}$$

Then, we want to compute the product between r_1 and r_2. We have,

$$\begin{aligned} r_3 &= r_1 r_2 \\ &=> S_3(q_3 - Z_3) = S_1(q_1 - Z_1)S_2(q_2 - Z_2) \\ &=> q_3 = Z_3 + \frac{S_1 S_2}{S_3}(q_1 - Z_1)(q_2 - Z_2) \end{aligned} \tag{15}$$

In Eq. (15) [4], every arithmetic is between two integers except $\frac{S_1S_2}{S_3}$, which is a floating-point number. To make the whole computation integer-only, the authors in [4] proposed an approach to quantize the floating-point number $\frac{S_1S_2}{S_3}$.

Firstly, the authors found that $M = \frac{S_1S_2}{S_3}$ is always in the interval (0, 1) and used Eq. (16) [4] to describe the relationship between M and M_0,

$$M = 2^{-n} M_0 \tag{16}$$

In Eq. (16), the authors in [4] set M_0 to a number between 0.5 and 1. n is a positive integer number. Using Eq. (16), the authors in [4] make M to be a fixed-point multiplier. If a 16-bit integer is used in the multiplication, M_0 can be represented as a 16-bit integer, which is $2^{15}M_0$ and bit-shift operation is used to compute the multiplication of $2^{-\mathbf{n}}$. Then, the whole expression can be computed using integer-only arithmetic.

3.2 Quantization-aware training

There are two common approaches to train quantized neural networks. The first approach is to train neural networks using floating-point numbers and then quantize weights and activation layers after training. However, this approach might not work for some models. In [4], the authors found that this approach does not work for small models because small models tend to have significant accuracy drops. The authors in [4] listed two reasons for accuracy drops. The first one is that weight distribution is large for different output channels. The large weight distribution makes channels with small weights range have large quantization errors. The second reason is that outlier weight values cause the quantization of weights much less accurate.

Because of the reasons mentioned above, the authors in [4] proposed a training approach that includes the quantization effects in the forward pass of training. Backward pass of training works as traditional training method and floating-point numbers are still used for weight and activation layers. During forward pass of training, the authors in [4] use Eq. (17) to quantize each layer and these equations are applied to each floating-point number element-wise.

$$\begin{aligned} &Clamp(r; a, b) := min\,(max\,(x, a), b) \\ &s(a, b, n) := \frac{b - a}{n - 1} \\ &q(r; a, b, n) := round\left(\frac{clamp(r; a, b) - a}{s(a, b, n)}\right)s(a, b, n) + a \end{aligned} \tag{17}$$

where r is a floating-point number; a, b are the maximum and minimum values of a layer and n is quantization level. For example, n = $2^8 = 256$ quantization levels if a 8-bit integer is used.

The function *round* is to rounding the number to its nearest integer.

For weights, the authors' proposed to set a and b to the minimum and maximum floating-point number of a weight layer respectively. In terms of activation layer, the authors used exponential moving averages to track the minimum and maximum floating-point numbers of an activation layer. After we have the range parameter a and b, we can compute other parameters easily. This approach has been implemented in Tensorflow [33, 34].

3.3 Comparison between different quantization approaches

Binarized neural network: Binarized neural network is an aggressive quantization approach that quantizes each weight to a binary value. In binary neural networks, dot product between two matrices can be completed by bit count operation, which is an operation to count the number of 1 s in a vector. The binary neural network in [5] achieves 32× reduction in model size and 58× speed up without losing much accuracy compared to equivalent neural network using single-precision values.

DoReFa-Net: DoReFa-Net is one of the most popular quantization approaches. This quantization approach not only applies quantization on weight and activation layers, but also on gradients. Through applying quantization on gradients, the training speed could be increased significantly. In [25], the proposed DoReFa-Net achieved 46.1% top-1 accuracy on ImageNet using 2-bit activations, 1-bit weights and trained with 6-bit gradients.

Log-based quantization: In [35], the authors proposed a multiplication-free hardware accelerator for deep neural networks. The proposed approach quantizes each weight to the nearest powers of two using logarithmic and rounding functions. In terms of the activation layer, the authors quantize each output to an 8-bit integer. By quantizing each weight to the nearest powers of two, multiplication between two integers could be replaced by bit-shift operations, which could reduce the resource utilization significantly. In [35], the authors demonstrate that the proposed quantization approach achieves almost the same accuracy as floating-point version but reduces energy consumption significantly.

3.4 Progress in model compression

Besides quantization approaches, many other model compression approaches are proposed. Pruning approach is one of the most popular approaches for model compression [6]. Pruning approach reduces the size of weights by removing some weights if these weights meet certain criteria. Besides weights, pruning approach could be also applied to activations and biases.

Knowledge distillation is another very popular approach for model compression [36]. There are two models, namely teacher model and student model, in knowledge distillation approach. Teacher model is a trained model. In addition, it has much larger model size than the student model. The main idea of knowledge distillation is to transfer the knowledge of teacher model to student model so that student model could have comparable performance to that of teacher model.

4. Neuromorphic computing

Neuromorphic computing [10, 37–50] is an emerging computing system that mimics the architecture of human brain. Carver Mead proposed the concept of neuromorphic computing in the late 1980s [43, 51–53]. Neuromorphic computing systems exploit spiking neural network to process information. Compared to conventional neural networks, spiking neural networks are more analogous to human brains and consume much less power. Recently, neuromorphic computing has been successfully applied to many applications [54–59].

4.1 Spiking neurons

The basic building block of spiking neuron networks is spiking neurons. The working mechanism of spiking neuron is different from that of neurons introduced in Section 2.1. Spiking neurons exchange information through electrical pulses, which are also called spikes. Spikes are discrete, time-dependent data and are represented as binary signals. In [8], the authors introduced several properties of spiking neurons. In the first place, spiking neurons receive information from many inputs and generate one output. Secondly, generating a spike depends on the amount of excitatory and inhibitory inputs. Thirdly, a spiking neuron's received spikes from other spiking neurons are integrated over time and will fire spikes if the integrated result is over a certain threshold.

4.1.1 Neuron models

There are several commonly used spiking neuron models such as leaky integrate-and-fire, Hodgkin-Huxley [60] and Fitzhugh Nagumo neuron models.

Leaky integrate-and-fire: According to [61], leaky integrate-and-fire neuron model is the simplest model to implement and the operation of leaky integrate-and-fire neuron can be completed using few floating-point operations such as additions and multiplications. However, there is no phasic spiking in leaky integrate-and-fire model since the model only has one variable. Meanwhile, spiking latencies do not exist in spikes because the threshold is fixed. The behaviour of leaky integrate-and-fire neuron model can be expressed using Eq. (18) [61]. If voltage V reaches a certain threshold level V_{th}, then a spike is fired and voltage V is reset to c.

$$\frac{dV}{dt} = I + a - bV, ifV \geq V_{th}, thenVresettoc \tag{18}$$

In Eq. (18), a, b, c and V_{th} are the parameters.

FitzHugh-Nagumo: FitzHugh-Nagumo neuron model [61] is more complicated compared to the leaky integrate-and-fire model and needs slightly more floating-point operations. The model has multiple variables and thus it has phasic spiking. Meanwhile, spikes of FitzHugh-Nagumo neuron model have spiking latencies because the threshold is not fixed. The behaviour of the FitzHugh-Nagumo neuron model can be expressed using Eq. (19) [61].

$$\begin{aligned} \frac{\mathrm{dV}}{\mathrm{dt}} &= a + bV + cV^2 + dV^3 - u \\ \frac{\mathrm{du}}{\mathrm{dt}} &= \varepsilon(eV - u) \end{aligned} \tag{19}$$

Hodgkin-Huxley: Hodgkin-Huxley [60] is a much more complicated neuron model compared to leaky integrate-and-fire and Fitzhugh-Nagumo neuron models. It is described by multiple equations and many parameters. In [61], the authors state that the parameters of the Hodgkin-Huxley neuron model are biophysically meaningful. More importantly, the Hodgkin-Huxley neuron model is very helpful for researchers to investigate single-cell dynamics. However, this model is hard to implement since it requires over 100 floating-point operations. More details about this model can be found in [60].

4.1.2 Leaky integrate-and-fire spiking neuron model

In this section, we are going to present more details about leaky integrate-and-fire spiking neuron models. The behaviour of integrate-and-fire spiking neuron model can be described using Eq. (18). If voltage V is above a certain voltage threshold V_{th}, it will fire spikes and voltage V is reset to 0. The behaviour of leaky integrate-and-fire model can be described by the circuit shown in **Figure 4**.

4.2 Neuromorphic computing for embedded systems

As we stated above, embedded systems have very limited computation resources and power constraints. Compared to conventional neural networks, spiking neural networks are more analogous to human brains and consume much less power. Because of these features, neuromorphic computing is suitable for embedded systems. A lot of researchers [62–66] have implemented neuromorphic computing in embedded systems such as FPGA. In [63], the authors implement liquid state machine on FPGA for speech recognition. The overall architecture achieves 88× speed up compared to CPU implementation. Meanwhile, the proposed approach reduces 30% power consumption.

4.3 Hardware implementation of spiking neural networks

A lot of researchers have been working on the hardware implementation of spiking neural networks and many neuromorphic chips have been developed. For example, in Stanford University, Neurogrid [67] and TrueNorth [68] have been developed by IBM.

Neurogrid: Neurogrid [67] is a mixed-signal hardware system for simulating biological brains. This system exploits analog circuits to implement all circuits except axonal arbors to improve energy efficiency and axonal arbors are implemented using digital circuits. The whole system consists of 16 Neurocores and each Neurocore has a 256 × 256 silicon-neuron array, a receiver, a transmitter and two RAMs. Neurogrid is able to simulate a million neurons by only consuming few watts.

TrueNorth: TrueNorth [68] is a brain-inspired neurosynaptic processor and uses non-von Neumann architecture. The whole system has 4096 cores, 1 million digital neurons and 256 million synapses. TruhNorth achieves 58 giga-synaptic operations per second (GSOPS) and 400 GSOPS per watt. More importantly, the authors have successfully implemented several applications such as object

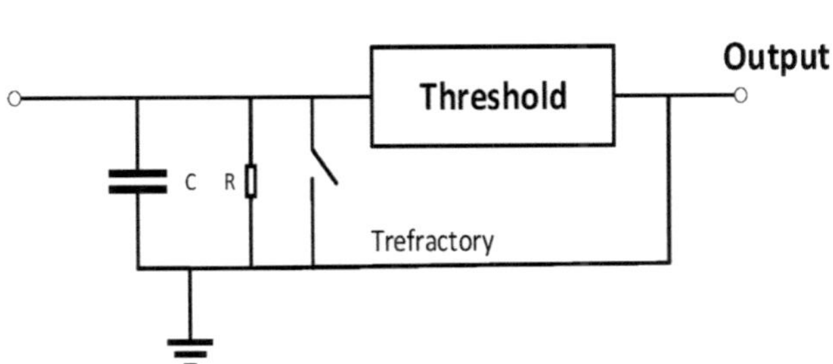

Figure 4.
Leaky integrate-and-fire model.

recognition on TrueNorth, and it has much lower power consumption compared to conventional processors.

4.4 Recent progress in neuromorphic computing

Spiking neural networks exploit spikes to represent information and thus effective and efficient approaches of representing information using spikes are very important. In spiking neural networks, input is encoded into spikes and each spike is represented as a single binary bit. There are two types of encoding approaches [7, 10, 45, 47, 69–72]. The first type of encoding approach is rate encoding. In rate encoding, input is encoded as the rate of spikes over an encoding window [7, 69]. Temporal encoding is also an encoding approach. Inter-spike interval encoding is a method of doing temporal encoding. In inter-spike encoding, the information is encoded by the time difference between two adjacent spikes [7, 69, 70].

Researchers have successfully implemented neural encoder using hardware [10, 62, 69]. In [62], the authors proposed a spike time-dependent encoder on FPGA. In [69], the authors implemented an inter-spike interval-based encoder for neuromorphic processors using analog integrated circuits. The proposed analog implementation of inter-spike interval encoder gets rid of ADCs and Op-amp and thus consumes less power.

In recent years, an increasing number of researchers have started to implement neuromorphic computing using analog integrated circuits [46, 47, 49, 50, 71, 73–79]. Compared to digital implementation, analog implementation of neuromorphic computing is more energy efficient. Meanwhile, analog implementation consumes less chip area.

Three-dimensional integrated circuits (3D IC) technique [80–82] is an emerging technique to improve the performance of integrated circuits. Compared to conventional fabrication techniques, three-dimensional integrated circuits technique consumes less power and uses small footprint. Recently, 3D IC technique has been applied to neuromorphic computing [79, 83–93]. Through the 3D IC technique, power consumption and chip area are reduced dramatically [88].

5. Conclusion

In the era of mobile computing and internet of things, embedded systems are everywhere. It can be found in consumer electronics, automobile, industrial and many other applications. Without embedded systems, our daily life would become extremely inconvenient. Deep learning is a technology, which is as important as embedded systems to our daily life. In recent years, deep learning is becoming a fundamental technology that impacts every aspect of our daily life. Therefore, deploying deep learning in embedded systems draws a lot of attention nowadays. Researchers have been conducting researches in many directions. For example, researchers are designing new layers and applying quantization techniques to reduce computation. Meanwhile, new architectures such neuromorphic computing are proposed. Through these techniques, many deep learning models are implemented in embedded systems successfully.

Author details

Shiya Liu* and Yang Yi
The Bradley Department of Electrical and Computer Engineering, Virginia Polytechnic Institute and State University (Virginia Tech), Virginia, USA

*Address all correspondence to: shiyal@vt.edu

References

[1] He K, Zhang X, Ren S, Sun J. Deep residual learning for image recognition. In: Proceedings of the IEEE Conference on Computer Vision and Pattern Recognition. 2016. pp. 770-778

[2] Ren S, He K, Girshick R, Sun J. Faster r-cnn: Towards real-time object detection with region proposal networks. In: Advances in Neural Information Processing Systems. 2015. pp. 91-99

[3] Graves A, Mohamed A-R, Hinton G. Speech recognition with deep recurrent neural networks. In: 2013 IEEE International Conference on Acoustics, Speech and Signal Processing; IEEE. 2013. pp. 6645-6649

[4] Jacob B et al. Quantization and training of neural networks for efficient integer-arithmetic-only inference. In: Proceedings of the IEEE Conference on Computer Vision and Pattern Recognition. 2018. pp. 2704-2713

[5] Rastegari M, Ordonez V, Redmon J, Farhadi A. Xnor-net: Imagenet classification using binary convolutional neural networks. In: European Conference on Computer Vision; Springer. 2016. pp. 525-542

[6] Han S, Mao H, Dally WJ. Deep compression: Compressing deep neural networks with pruning, trained quantization and huffman coding. Computer Vision and Pattern Recognition. 2015

[7] Bouvier M et al. Spiking neural networks hardware implementations and challenges: A survey. ACM Journal on Emerging Technologies in Computing Systems (JETC). 2019;**15**(2):22

[8] Ponulak F, Kasiński A. Introduction to spiking neural networks: Information processing, learning and applications. Acta Neurobiologiae Experimentalis. 2011;**71**(4):409-433

[9] Rathi N, Panda P, Roy K. STDP-based pruning of connections and weight quantization in spiking neural networks for energy-efficient recognition. Neural and Evolutionary Computing. 2018;**38**(4):668-677

[10] Zhao C et al. Energy efficient spiking temporal encoder design for neuromorphic computing systems. IEEE Transactions on Multi-Scale Computing Systems. 2016;**2**(4):265-276

[11] Zhang A, et al. Dive into deep learning; Unpublished draft. 2019. p. 319

[12] Goodfellow I, Bengio Y, Courville A. Deep Learning. MIT Press; 2016

[13] Maas AL, Hannun AY, Ng AY. Rectifier nonlinearities improve neural network acoustic models. Proceedings ICML. 2013;**30**(1):3

[14] Leshno M, Lin VY, Pinkus A, Schocken SJ. Multilayer feedforward networks with a nonpolynomial activation function can approximate any function. 1993;**6**(6):861-867

[15] Pinkus A. Approximation theory of the MLP model in neural networks. 1999;**8**:143-195

[16] Qi CR, Su H, Mo K, Guibas LJ. Pointnet: Deep learning on point sets for 3d classification and segmentation. In: Proceedings of the IEEE Conference on Computer Vision and Pattern Recognition. 2017. pp. 652-660

[17] Krizhevsky A, Sutskever I, Hinton GE. Imagenet classification with deep convolutional neural networks. In: Advances in Neural Information Processing Systems. 2012. pp. 1097-1105

[18] Ji S, Xu W, Yang M, Yu K. 3D convolutional neural networks for human action recognition. IEEE

Transactions on Pattern Analysis and Machine Intelligence. 2012;**35**(1): 221-231

[19] Han S, Pool J, Tran J, Dally W. Learning both weights and connections for efficient neural network. In: Advances in Neural Information Processing Systems. 2015. pp. 1135-1143

[20] Desoli G et al. 14.1 a 2.9 tops/w deep convolutional neural network soc in fd-soi 28 nm for intelligent embedded systems. In: 2017 IEEE International Solid-State Circuits Conference (ISSCC); IEEE. 2017. pp. 238-239

[21] Howard AG et al. Mobilenets: Efficient convolutional neural networks for mobile vision applications. 2017. arXiv preprint arXiv:1704.04861

[22] Sandler M, Howard A, Zhu M, Zhmoginov A, Chen L-C. Mobilenetv2: Inverted residuals and linear bottlenecks. In: Proceedings of the IEEE Conference on Computer Vision and Pattern Recognition. 2018. pp. 4510-4520

[23] Hubara I, Courbariaux M, Soudry D, El-Yaniv R, Bengio Y. Quantized neural networks: Training neural networks with low precision weights and activations. 2017;**18**(1):6869-6898

[24] Wu J, Leng C, Wang Y, Hu Q, Cheng J. Quantized convolutional neural networks for mobile devices. In: Proceedings of the IEEE Conference on Computer Vision and Pattern Recognition. 2016. pp. 4820-4828

[25] Zhou S, Wu Y, Ni Z, Zhou X, Wen H, Zou Y. Dorefa-net: Training low bitwidth convolutional neural networks with low bitwidth gradients. 2016. arXiv preprint arXiv:1606.06160

[26] Zhu C, et al. Trained ternary quantization. 2016. arXiv preprint arXiv:1612.01064

[27] Lin X, Zhao C, Pan W. Towards accurate binary convolutional neural network. In: Advances in Neural Information Processing Systems. 2017. pp. 345-353

[28] Han S, Mao H, Dally WJ. A deep neural network compression pipeline: Pruning, quantization, huffman encoding. 2015;**10**

[29] Zhang X, Zhou X, Lin M, Sun J. Shufflenet: An extremely efficient convolutional neural network for mobile devices. In: Proceedings of the IEEE Conference on Computer Vision and Pattern Recognition. 2018. pp. 6848-6856

[30] Lin D, Talathi S, Annapureddy S. Fixed point quantization of deep convolutional networks. In: International Conference on Machine Learning. 2016. pp. 2849-2858

[31] Anwar S, Hwang K, Sung W. Fixed point optimization of deep convolutional neural networks for object recognition. In: 2015 IEEE International Conference on Acoustics, Speech and Signal Processing (ICASSP); IEEE. 2015. pp. 1131-1135

[32] Zhou A, Yao A, Guo Y, Xu L, Chen YJ. Incremental network quantization: Towards lossless cnns with low-precision weights. 2017. arXiv preprint arXiv:1702.03044

[33] Abadi M et al. Tensorflow: A system for large-scale machine learning. In: 12th {USENIX} Symposium on Operating Systems Design and Implementation ({OSDI} 16). 2016. pp. 265-283

[34] Abadi M et al. Tensorflow: Large-scale machine learning on heterogeneous distributed systems. 2016. arXiv preprint arXiv:1603.04467

[35] Tann H, Hashemi S, Bahar RI, Reda S. Hardware-software codesign of accurate, multiplier-free deep neural

networks. In: 2017 54th ACM/EDAC/IEEE Design Automation Conference (DAC); IEEE. 2017. pp. 1-6

[36] Hinton G, Vinyals O, Dean J. Distilling the knowledge in a neural network. Machine Learning. 2015

[37] Burr GW et al. Neuromorphic computing using non-volatile memory. Advances in Physics: X. 2017;**2**(1):89-124

[38] Furber S. Large-scale neuromorphic computing systems. Journal of Neural Engineering. 2016;**13**(5):051001

[39] Kim D, Kung J, Chai S, Yalamanchili S, Mukhopadhyay S. Neurocube: A programmable digital neuromorphic architecture with high-density 3D memory. In: 2016 ACM/IEEE 43rd Annual International Symposium on Computer Architecture (ISCA); IEEE. 2016. pp. 380-392

[40] Liu X et al. RENO: A high-efficient reconfigurable neuromorphic computing accelerator design. In: 2015 52nd ACM/EDAC/IEEE Design Automation Conference (DAC); IEEE. 2015. pp. 1-6

[41] Monroe D. Neuromorphic computing gets ready for the (really) big time. Communications of the ACM. 2014;**57**(6):13-15. DOI: 10.1145/2601069

[42] Schuman CD et al. A survey of neuromorphic computing and neural networks in hardware. 2017. arXiv preprint arXiv:1705.06963

[43] Mead C, Ismail M. Analog VLSI Implementation of Neural Systems. Vol. 80. Springer Science & Business Media; 2012

[44] Calimera A, Macii E, Poncino M. The human brain project and neuromorphic computing. Functional Neurology. 2013;**28**(3):191

[45] Zhao C, et al. Spike-time-dependent encoding for neuromorphic processors. ACM Journal on Emerging Technologies in Computing Systems (JETC). 2015; **12**(3):23

[46] Zhao C, Danesh W, Wysocki BT, Yi Y. Neuromorphic encoding system design with chaos based CMOS analog neuron. In: 2015 IEEE Symposium on Computational Intelligence for Security and Defense Applications (CISDA); IEEE. 2015. pp. 1-6

[47] Zhao C, Li J, Liu L, Koutha LS, Liu J, Yi Y. Novel spike based reservoir node design with high performance spike delay loop. In: Proceedings of the 3rd ACM International Conference on Nanoscale Computing and Communication; ACM. 2016. p. 14

[48] Mosleh S, Sahin C, Liu L, Zheng R, Yi Y. An energy efficient decoding scheme for nonlinear MIMO-OFDM network using reservoir computing. In: 2016 International Joint Conference on Neural Networks (IJCNN); IEEE. 2016. pp. 1166-1173

[49] Bai K, Yi Y. DFR: An energy-efficient analog delay feedback reservoir computing system for brain-inspired computing. ACM Journal on Emerging Technologies in Computing Systems (JETC). 2018;**14**(4):1-22

[50] Bai K, Li J, Hamedani K, Yi Y. Enabling an new era of brain-inspired computing: Energy-efficient spiking neural network with ring topology. In: 2018 55th ACM/ESDA/IEEE Design Automation Conference (DAC); IEEE. 2018. pp. 1-6

[51] Mead C. Neuromorphic electronic systems. Proceedings of the IEEE. 1990; **78**(10):1629-1636

[52] Douglas R, Mahowald M, Mead C. Neuromorphic analogue VLSI. Review of Neuroscience. 1995;**18**(1):255-281

[53] Mead C. Analog VLSI and neutral systems. NASA STI/Recon Technical Report A. 1989;**90**

[54] Mosleh S, Liu L, Sahin C, Zheng YR, Yi Y. Brain-inspired wireless communications: Where reservoir computing meets MIMO-OFDM. IEEE Transactions on Neural Networks and Learning Systems. 2017;**99**:1-15

[55] Hamedani K, Liu L, Atat R, Wu J, Yi Y. Reservoir computing meets smart grids: Attack detection using delayed feedback networks. IEEE Transactions on Industrial Informatics. 2017;**14**(2): 734-743

[56] Danesh W, Zhao C, Wysocki BT, Medley MJ, Thawdar NN, Yi Y. Channel estimation in wireless OFDM systems using reservoir computing. In: 2015 IEEE Symposium on Computational Intelligence for Security and Defense Applications (CISDA); IEEE. 2015. pp. 1-5

[57] Li J, Liu L, Zhao C, Hamedani K, Atat R, Yi Y. Enabling sustainable cyber physical security systems through neuromorphic computing. IEEE Transactions on Sustainable Computing. 2017;**3**(2):112-125

[58] Shafin R et al. Realizing green symbol detection via reservoir computing: An energy-efficiency perspective. In: 2018 IEEE International Conference on Communications (ICC); IEEE. 2018. pp. 1-6

[59] Yi Y. Neuron Design in Neuromorphic Computing Systems and Its Application in Wireless Communications. Lawrence: The University of Kansas Center for Research, Inc.; 2017

[60] Hodgkin AL, Huxley AF. A quantitative description of membrane current and its application to conduction and excitation in nerve. The Journal of Physiology. 1952;**117**(4):500-544

[61] Izhikevich EM. Which model to use for cortical spiking neurons? 2004; **15**(5):1063-1070

[62] Yi Y et al. FPGA based spike-time dependent encoder and reservoir design in neuromorphic computing processors. Microprocessors and Microsystems. 2016;**46**:175-183

[63] Wang Q, Li Y, Li P. Liquid state machine based pattern recognition on FPGA with firing-activity dependent power gating and approximate computing. In: 2016 IEEE International Symposium on Circuits and Systems (ISCAS); IEEE. 2016. pp. 361-364

[64] Gomar S, Ahmadi A. Digital multiplierless implementation of biological adaptive-exponential neuron model. IEEE Transactions on Circuits and Systems I: Regular Papers. 2013; **61**(4):1206-1219

[65] Rostro-Gonzalez H, Cessac B, Girau B, Torres-Huitzil C. The role of the asymptotic dynamics in the design of FPGA-based hardware implementations of gIF-type neural networks. Journal of Physiology-Paris. 2011;**105**(1–3):91-97

[66] Neil D, Liu S-C. Minitaur, an event-driven FPGA-based spiking network accelerator. 2014;**22**(12):2621-2628

[67] Benjamin BV et al. Neurogrid: A mixed-analog-digital multichip system for large-scale neural simulations. 2014; **102**(5):699-716

[68] Akopyan F et al. Truenorth: Design and tool flow of a 65 MW 1 million neuron programmable neurosynaptic chip. 2015;**34**(10):1537-1557

[69] Zhao C et al. Interspike-interval-based analog spike-time-dependent encoder for neuromorphic processors. IEEE Transactions on Very Large Scale Integration (VLSI) Systems. 2017;**25**(8): 2193-2205

[70] Zhao C, Li J, Yi Y. Making neural encoding robust and energy efficient:

an advanced analog temporal encoder for brain-inspired computing systems. In: Proceedings of the 35th International Conference on Computer-Aided Design; ACM. 2016. p. 115

[71] Li J, Zhao C, Hamedani K, Yi Y. Analog hardware implementation of spike-based delayed feedback reservoir computing system. In: 2017 International Joint Conference on Neural Networks (IJCNN); IEEE. 2017. pp. 3439-3446

[72] Zhao C, Li J, An H, Yi Y. Energy efficient analog spiking temporal encoder with verification and recovery scheme for neuromorphic computing systems. In: 2017 18th International Symposium on Quality Electronic Design (ISQED); IEEE. 2017. pp. 138-143

[73] Jiang H et al. Cyclical sensing integrate-and-fire circuit for memristor array based neuromorphic computing. In: 2016 IEEE International Symposium on Circuits and Systems (ISCAS); IEEE. 2016. pp. 930-933

[74] Li J, Bai K, Liu L, Yi Y. A deep learning based approach for analog hardware implementation of delayed feedback reservoir computing system. In: 2018 19th International Symposium on Quality Electronic Design (ISQED); IEEE. 2018. pp. 308-313

[75] Bai K, An Q, Yi Y. Deep-DFR: A memristive deep delayed feedback reservoir computing system with hybrid neural network topology. In: Proceedings of the 56th Annual Design Automation Conference; 2019. ACM. 2019. p. 54

[76] Bai K, Bradley YY. A path to energy-efficient spiking delayed feedback reservoir computing system for brain-inspired neuromorphic processors. In: 2018 19th International Symposium on Quality Electronic Design (ISQED); IEEE. 2018. pp. 322-328

[77] Yi Y. Analog Integrated Circuit Design for Spike Time Dependent Encoder and Reservoir in Reservoir Computing Processors. Lawrence, United States: University of Kansas Center for Research, Inc.; 2018

[78] Zhao C, Hamedani K, Li J, Yi Y. Analog spike-timing-dependent resistive crossbar design for brain inspired computing. IEEE Journal on Emerging and Selected Topics in Circuits, and Systems. 2017;**8**(1):38-50

[79] Ehsan MA, An H, Zhou Z, Yi Y. Design challenges and methodologies in 3D integration for neuromorphic computing systems. In: 2016 17th International Symposium on Quality Electronic Design (ISQED); IEEE. 2016. pp. 24-28

[80] Garrou P, Bower C, Ramm P. Handbook of 3D Integration, Volume 1: Technology and Applications of 3D Integrated Circuits. John Wiley & Sons; 2011

[81] Knickerbocker JU et al. 3D silicon integration. In: 2008 58th Electronic Components and Technology Conference; IEEE. 2008. pp. 538-543

[82] Topol AW et al. Three-dimensional integrated circuits. IBM Journal of Research and Development. 2006;**50** (4.5):491-506

[83] An H, Zhou Z, Yi Y. 3D memristor-based adjustable deep recurrent neural network with programmable attention mechanism. In: Proceedings of the Neuromorphic Computing Symposium; ACM. 2017. p. 11

[84] Ehsan MA, An H, Zhou Z, Yi Y. Adaptation of enhanced TSV capacitance as membrane property in 3D brain-inspired computing system. In: 2017 54th ACM/EDAC/IEEE Design Automation Conference (DAC); IEEE. 2017. pp. 1-6

[85] An H, Ehsan MA, Zhou Z, Yi Y. Electrical modeling and analysis of 3D neuromorphic IC with monolithic inter-tier vias. In: 2016 IEEE 25th Conference on Electrical Performance of Electronic Packaging and Systems (EPEPS); IEEE. 2016. pp. 87-90

[86] An H, Ehsan MA, Zhou Z, Yi Y. Electrical modeling and analysis of 3D synaptic array using vertical RRAM structure. In: 2017 18th International Symposium on Quality Electronic Design (ISQED); IEEE. 2017. pp. 1-6

[87] Ehsan MA, Zhou Z, Yi Y. Modeling and analysis of neuronal membrane electrical activities in 3d neuromorphic computing system. In: 2017 IEEE International Symposium on Electromagnetic Compatibility & Signal/Power Integrity (EMCSI); IEEE. 2017. pp. 745-750

[88] An H, Ehsan MA, Zhou Z, Shen F, Yi YJI. Monolithic 3D neuromorphic computing system with hybrid CMOS and memristor-based synapses and neurons. Integration. 2019;**65**:273-281

[89] Ehsan MA, Zhou Z, Yi Y. Neuromorphic 3D integrated circuit: A hybrid, reliable and energy efficient approach for next generation computing. In: Proceedings of the on Great Lakes Symposium on VLSI 2017; ACM. 2017. pp. 221-226

[90] An H, Zhou Z, Yi Y. Opportunities and challenges on nanoscale 3D neuromorphic computing system. In: 2017 IEEE International Symposium on Electromagnetic Compatibility & Signal/Power Integrity (EMCSI); IEEE. 2017. pp. 416-421

[91] Ehsan MA, Zhou Z, Yi Y. Hybrid three-dimensional integrated circuits: A viable solution for high efficiency neuromorphic computing. In: 2017 International Symposium on VLSI Design, Automation and Test (VLSI-DAT); IEEE. 2017. pp. 1-2

[92] Ehsan MA, An H, Zhou Z, Yi Y. A novel approach for using TSVs as membrane capacitance in neuromorphic 3-D IC. IEEE Transactions on Computer-Aided Design of Integrated Circuits and Systems. 2017;**37**(8): 1640-1653

[93] Ehsan MA, Zhou Z, Yi Y. Three dimensional integration technology applied to neuromorphic hardware implementation. In: 2015 IEEE International Symposium on Nanoelectronic and Information Systems; IEEE. 2015. pp. 203-206

Chapter 16

Persuasion in Mobile Financial Service: A Case Study with a Bank Savings Mobile Application

Prom Tep Sandrine, Ruer Perrine and Nemery Alexandra

Abstract

Financial institutions are undergoing a technology transformation. The digitization now drives the addition of new services and expectations. In this context, mobile has become a strategic channel to encourage users to adopt specific behaviors and change habits effortlessly. The research question underlying this study focuses on mobile banking applications and how they could support the adoption of savings behaviors. A qualitative study was conducted, in order to evaluate the persuasiveness embedded in a mobile bank saving app. Three experts in human-computer interaction (HCI) assessed the mobile app interfaces through a scientific grid of persuasive criteria to guide their heuristic inspection. Results confirm both a satisfactory level of persuasiveness of the mobile app and the dynamic application of persuasive criteria. The study shows that a mobile app involving certain specific features supports a positive banking customer's experience related to savings. This study contributes to the user experience field, showing that mobile apps can support behavioral change when persuasiveness is embedded in the design process. Using a valid and reliable assessment method to establish the level of persuasiveness of a bank savings mobile app, this study confirms that the persuasion grid is applicable to mobile interfaces.

Keywords: bank savings, mobile app, persuasive technology, user experience, mobile interface, expert heuristic evaluation

1. Introduction

The complexities created by the constant evolution in technology are transforming many sectors, and the banking sector is no exception. Digital progress is changing the way banks and customers interact, with an impact on banking experiences over the last two decades [1]. The banking service has evolved from real face-to-face conversation to virtual discussion (i.e., email, websites, mobile messaging, or chatbot used to accomplish digital transactions).

From another perspective, financial technology (FinTech) is revolutionizing traditional financial practices, such as PayPal, Personal Capital, Kabbage, Apple/Samsung/Google Pay, and Wealthsimple, among others. However, we are witnessing a decline in the financial literacy needed to make informed banking decisions. As developed countries face an increasing level of consumer debt [2], the issue of supporting saving behavior, especially among young consumers, becomes critical.

Banks have a strong desire to retain their customers by offering a wide range of mobile services [1]. This variety of services is being developed with the intent to influence user behavior, for example, by encouraging bank savings or mobile payment solutions. In the area of information technology (IT), we are witnessing the ascendency of persuasion. Persuasion technologies aim to influence a user through a process which changes his or her attitude or behavior [3].

From a managerial point of view, the mobile app serves specific purposes (specific features) such as maintaining a relationship with already acquired customers and building their long-term loyalty through personalization or targeted promotions. A mobile website, on the other hand, is intended as a general tool to serve customers and is therefore more suitable for new customer acquisition. From this perspective, a mobile app lends itself better as a channel to induce behavioral change than as a mobile website.

The main research question is formulated thus: Since payments are increasingly being made through mobiles, what about bank savings? Is a banking mobile app for savings persuasive for customers?

To answer this question, our hypothesis is formulated as follows: to assess a bank mobile app, we can understand the customer intent when we assess the persuasive aspects of a bank saving's mobile app. Our objective is to assess persuasive dimensions of a bank savings mobile application (app) evaluated by experts in the HCI area. The method used is a pilot study in the form of user tests made by three experts in persuasive research. This exploratory work contributes to helping design experts better assess the persuasive aspects of a bank savings mobile app and, in this regard, contribute to achieving designs which could induce the customer to save more money.

The paper's structure is organized as follows: Firstly, we focus on related works dealing with banking technology, bank savings, and persuasion. Secondly, we present the results of applying persuasive heuristic criteria to a dynamic mobile app, identifying the strengths and weaknesses of criteria in assessing a mobile app specialized in bank savings in Canada. Finally, we discuss our results before concluding on further research and the contributions of our study.

2. Literature review

Even though technology is developing so rapidly, human-computer interaction (HCI) does not always follow suit: systems or products do not always consider the end-users in their design. In [4], authors highlight the importance of measuring the usability of a branded app through a "consumer-centric approach" which seeks to include end-users from the beginning and to understand their needs and expectations.

Companies do not take into consideration the importance of HCI when designing their platform and particularly in the context of online savings platforms. Nowadays, banks are investing in digital options such as ATMs, websites, mobile apps, or online chats. Their rivals, FinTech's, on the other hand, are developing digital options which are all online and dematerialize into a smartphone. Some authors wonder how long financial institutions will be able to maintain their lead given the growth of FinTech. These competitors are quicker than banks in offering convenient, reliable, fast, and cost-efficient alternatives to traditional bank payments [5–6].

These bank competitors are influencing their customers and their way of interacting with bank digital solutions. Banks must consider how to encourage their customers by offering bank operations online, what goes through the key factors, strategies, and drawbacks. This can be considered when designing the online solution.

Concerning saving habits in Quebec (Canada), savings have not been a great success over the last decade. In 1990, the average household savings rate was 7.9% compared to 4% in 2015. A Montreal Bank (BMO) survey conducted in 2014 indicated that Quebecers were the least likely to save compared to other Canadian residents [7, 8]. However, all provinces have a low savings rate compared to previous years. In 2017, in Quebec, the most recent statistics present a significant recovery in the household savings rate with an average of 6.3% [9], i.e., Quebecers are still interested in saving money.

From another perspective, technological innovation has been conducive to the creation of persuasive technologies; nevertheless, this does not remove the need to change behavior toward the user's technology [10]. Persuasive technologies aim to change the attitude or behavior of users. The importance of guidelines remains to be shown in order to evaluate persuasion in HCI. The assessment of a persuasive interface is considered to be time-consuming and often less useful. The benefits are the measurement and establishment of problems at the interface level, which help to assess the usability and the persuasiveness of app. To our knowledge, persuasive aspects in banking technology have not yet been validated in the literature.

Criteria involve both static and dynamic aspects to assess the user interface. Static criteria are necessary to promote the acceptance and influence of technology; they are credibility, privacy, personalization, and attractiveness.

	Criteria	**Description**
Static criteria	Credibility	Ability of the interface to inspire confidence and to make the user trust the veracity of its information
	Privacy	Protection of personal data, preservation of personal integrity and security of interaction
	Personalization	Concept of customization aiming to adapt the interface to the user
	Attractiveness	Use of aesthetics (graphic, ...) to capture the attention of the user, to support the interaction and to create a positive emotion
Dynamic criteria	Solicitation	First stage to attract and challenge the user to initiate the relationship
	Initiation	Elements of the media that trigger the persuasive influence
	Commitment	Means that the system continues to involve the user through a process
	Ascendancy	Submission and obedience appear with the completion of the engaging scenario. The ascendency is the deepest form of technological persuasion.

Table 1.
Eight interactive persuasive criteria.

And dynamic criteria encourage users to change their behavior in an organized manner; they are solicitation, initiation, commitment, and ascendancy. The particularity of the dynamic criteria is their temporal aspect. For authors, time is significant because it is a structural element during which the social influence can take place in order to change the user's behavior. In addition, each sub-criterion corresponds to a facet of the multidimensional construct. For instance, the credibility criteria include trustworthiness, expertise, trustfulness, and legitimacy as sub-criteria, following the traditional conceptualization of source credibility in the marketing research literature. The eight main criteria are listed in **Table 1** as defined by the authors in [3].

The particularity of these criteria is they were assessed on mobile responsive website, not for a mobile app. Furthermore, we noticed that interfaces were evaluated with non-dynamic interfaces. We define dynamic as a gradual development, which does not mean a longitudinal study repeated over time. We believe it is relevant to look for the validation of the persuasive criteria for mobile app, particularly with an interactive mobile app.

3. Methodology

Our design research is qualitative and exploratory. We wish to discover whether a mobile app can influence behavior in a savings context and to provide new knowledge in this area. To do so, we aim to examine the persuasive ability by means of a heuristic evaluation in a financial context. We assess the effectiveness of persuasive criteria [3] applied to a Canadian bank savings mobile app interface. With this approach, we aim to determine whether perceived utility (usefulness) and ease of use (usability) are closely related to the level of persuasiveness of the application through practicity. We present our experimental material in Section 3.1, and we describe participants and procedure in Section 3.2.

3.1 Apparatus

Henceforth, we describe the mobile app being assessed in Section 3.1.1 and describe the persuasive criteria used in Section 3.1.2.

3.1.1 The mobile application

In this research, we chose to assess the mobile app Hop 'n S@ve from Desjardins group[1], a financial cooperative in Canada. To access this mobile app, it is necessary to have a bank account available online with the Desjardins group.

This mobile app is qualified as an instant savings tool dedicated to bank savings. The purpose of the tool is to allow anyone to save personal funds instantly, no matter when or where the customer is, for instance, at midnight from home, at 6 pm from a public place, or at 8 am from a car or bus. **Figure 1** shows two interfaces of the Hop 'n S@ve mobile app.

Our interest in this mobile application is its mobile aspect. Even if Hop 'n S@ve is offered for several types of mobile devices (smartphone, digital tablet, smart watch), we focus only on the mobile app for smartphones here.

[1] From Desjardins' website: https://www.desjardins.com/ca/personal/accounts-services/hop-and-save-instant-savings-tool/index.jsp

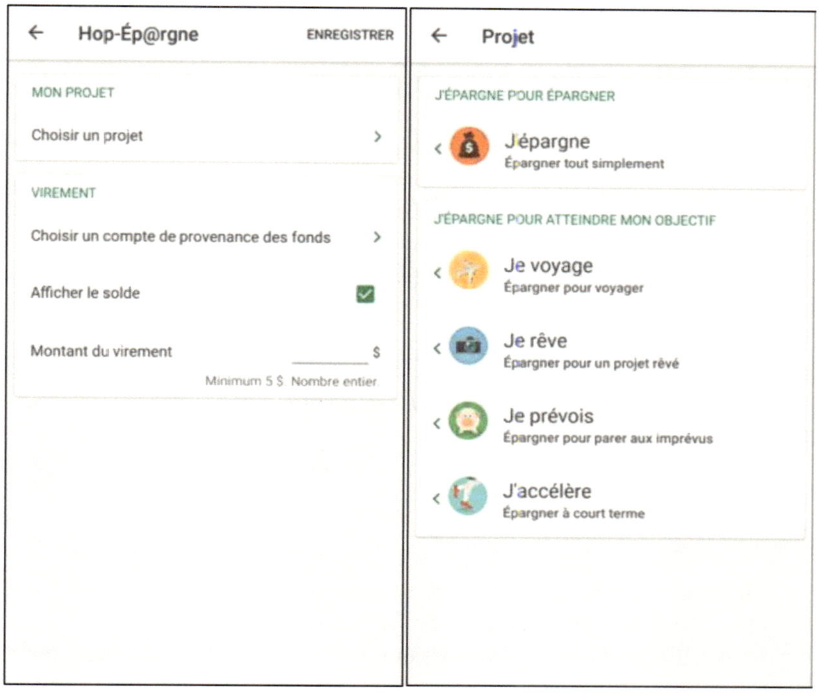

Figure 1.
Interfaces of the Hop 'n S@ve mobile app (screen captures in French as presented during the data collection).

3.1.2 Interactive persuasive criteria

We use the eight interactive persuasive criteria from [3].

These criteria are designed to assess the content of interaction. To do so, they are divided into two dimensions: the static criteria (credibility, privacy, personalization, attractiveness) and the dynamic criteria (solicitation, initiation, commitment, and ascendancy). **Table 1** presents each of these criteria in detail.

This grid was based on a bibliographic analysis (164 documents) and designed for experts in human-computer interaction [11]. The different dimensions of evaluation have been organized according to these two categories in order to facilitate their understanding and reading for the expert's assessors.

3.2 Participants

Given the purpose and context of our research, we defined usability specialists between 18 and 35 years old in the Montreal area. The inclusion criteria required having a bank account in the Desjardins group, living in the Quebec province, and 2 years' experience or more in the HCI area. A total of three specialists in HCI, composed of two women and one man, took part voluntarily for this study. Their experience in HCI was between 5 and 9 years (mean: 7 years).

3.3 Procedure

Instructions given to the experts were to categorize persuasive criteria into the Hop 'n S@ve mobile app. To do so, the protocol followed was to introduce first the persuasive criteria. These ones were proposed in a document with a definition, a description, and illustrated examples (picture, screenshot from app, etc.). Once criteria have been understood, the second phase was to assess the Hop 'n S@ve mobile app. Each participant had to make a fund transfer from the bank account to the savings account to assess persuasive criteria. The experiment with the mobile app is carried out as follows: login to the mobile app, create a saving project, choose a tool amount to save, choose a one-time amount, contribute, and close the session. During this phase, a new document was provided including definition and description of persuasive criteria without example. Each participant was allowing to keep and read the criteria description suggested by the second document.

We collected the following data: the global score of correct identification and the correct identification for each criterion. The correct identification is based on experts' identification during the pretest. Identification was considered as correct if the answer matched with the expected criteria. Quality of the criteria definition was evaluated through the task, that is, if a criterion was defined as correct or false.

4. Result analysis

Participants' aim was to assess and comment persuasive aspects into the mobile app during the use of the persuasive criteria. The global score of correct identification is 59.5%. **Table 2** presents more details on each criterion and the correct identification score.

	Criterion	**Criterion compliance instances identified in the interface**	**Correct**
Static Criteria (69%)	Credibility	-Icon of Desjardins group -Color legitimacy -Desjardins contact information	83%
	Privacy	-Ensured the tool with the FAQ -High level of security with identification and password	67%
	Personalization	-Project creation according to the customer (name, amount) -Different choice (holidays, dream project, home planification, etc.) -First person terms used (i.e. "I", "My")	75%
	Attractiveness	-Visual pictograms used -Simple and clear graphic style corresponding to the banking world - Call to action with the red button "Savings"	51%
Dynamic criteria (50%)	Solicitation	-Pictures, images and vocabulary to reinforce users to interact	67%
	Initiation	-Pre-filled field -Choice of the fixed amount	58%
	Commitment	-Users are congratulated at the end of the process -Progress status information	33%
	Ascendancy	-Notifications -Creation of multiple projects	42%

Table 2.
Correct identification table.

Results show that using the grid makes it possible to determine problems linked to the persuasiveness of interfaces in a bank savings context. Results suggest improving the dynamic criteria, especially the ones called commitment and ascendency.

Results confirm both a satisfactory level of persuasiveness of the mobile app and the dynamic application of persuasive criteria. The study shows that a mobile app involving certain specific features (e.g., features to inform the customer about the various benefits of savings behaviors) supports a positive banking customer's experience regarding savings.

5. Discussion and conclusion

This study was a qualitative evaluation of the persuasiveness of the Hop 'n S@ve mobile app to promote savings behavior among Canadian consumers. The latter is still interested in saving money while not very successful at it in the last decade. Mobile apps have changed consumer behavior even though their purpose is to better manage money. As there are new digital options for banks and new ways of interacting with bank digital solutions, we were interested in understanding the customer's intent as we assess the persuasiveness of a bank saving's mobile app. The growing of persuasion in IT is a good opportunity to examine it in the FinTech context.

To accomplish this, we validate the persuasiveness of the app with experts in the HCI area according to persuasive criteria. Results indicate the Hop 'n S@ve mobile app reaches a high degree of persuasiveness at the interface level. Only commitment criteria may be lacking some persuasiveness in a process leading to the adoption of the targeted specific behavior. It means a mobile app dedicated to bank savings can influence saving behavior.

As far as we know from our literature review, these persuasive criteria were not assessed with a mobile app. The present study confirms the hypothesis that the persuasive criteria grid shall apply to a mobile app in a dynamic context. Hence, this study contributes to the user experience field, showing that mobile apps can support behavioral change when persuasiveness is embedded as part of the design process. Using a valid and reliable assessment method to establish the level of persuasiveness of a bank savings mobile app, this study confirms the persuasion grid applies to mobile interfaces for the banking sector.

Like any research, this study has also encountered some limitations. The first one is the number of expert participants evaluating the app (N = 3). Being an exploratory project, it can only show the tendency trend at best. Our next research will include more participants to confirm this tendency. Also, this evaluation confirms the interest of furthering our research by evaluating with a panel of users of several ages the interest of saving through a mobile app.

The second limit is the fact that the mobile app test was carried out at once. It does not allow for results or answers that could be obtained over a longer period. We could take as examples the commitment or the ascendency criteria. Since they are criteria that are mostly long-term, it was difficult to measure the two dimensions adequately. We have considered them for future research though.

Acknowledgements

We gratefully acknowledge the support of the SSHRC for the funding provided for this research. In addition, the authors wish to acknowledge Mame Matar Adj for the contributions.

Conflict of interest

The authors declare no conflict of interest.

Author details

Prom Tep Sandrine[1*], Ruer Perrine[2] and Nemery Alexandra[3]

1 École des Sciences de la Gestion ESG UQAM Business School, University of Quebec in Montreal, Montreal, QC, Canada

2 Tech3lab, HEC Montreal's UX Research Lab, Montreal, QC, Canada

3 Ubisoft, Montreal, QC, Canada

*Address all correspondence to: promtep.sandrine@uqam.ca

References

[1] Mobile–the ATM in your pocket. Extract from Mobile Finance Services: Banking and Payment market 2007-2011 [Internet]. 2008. Available from: https://wireless-mobilejobs.com/_resx/storage/88ded148-ie388/mobile%20financial%20services%20-%20whitepaper.pdf

[2] Lussardi A, Scheresberg CB. Avery M. Millennial Mobile Payment Users: A Look into their Personal Finances and Financial Behaviors. GFLEC Insights Report. 2018. Available from: http://gflec.org/wp-content/uploads/2018/04/GFLEC-Insight-Report-Millennial-Mobile-Payment-Users-Final.pdf?x87657

[3] Nemery A, Brangier E. Set of guidelines for persuasive interfaces: Organization and validation of the criteria. Journal of Usability Studies. 2014;**9**(3):105-128

[4] Baek TH, Yoo CY. Branded app usability: Conceptualization, measurement, and prediction of consumer loyalty. Journal of Advertising. 2018;**47**(1):70-82

[5] Digital Disruption: How FinTech is Forcing Banking to a Tipping Point. Citi GPS: Global Perspectives and Solutions [Internet]. 2016. Available from: https://ir.citi.com/D%2F5GCKN6uoSvhbvCmUDS05SYsRaDvAykPjb5subGr7f1JMe8w2oX1bqpFm6RdjSRSpGzSaXhyXY%3D

[6] The pulse of FinTech Q4 2016-Global Analysis of Investment in FinTech [Internet]. 2017. Available from: https://assets.kpmg.com/content/dam/kpmg/xx/pdf/2017/02/pulse-of-FinTech-q4-2016.pdf

[7] RadioCanada. Les Québécois bons derniers pour épargner, selon un sondage [Internet]. 2014. Available from: https://ici.radio-canada.ca/nouvelle/650849/epargne-canada-quebecois-derniers-revenu-depenses

[8] RadioCanada. Les Canadiens n'épargnent pas assez, selon une étude de la Banque de Montréal [Internet]. 2015. Available from: https://ici.radio-canada.ca/nouvelle/714753/canadiens-epargne-etude-sondage-banque-montreal

[9] Statistique Canada. Institut de la statistique du Québec. Comptes économiques du Québec. 4e trimestre 2017 [Internet]. 2018. Available from: http://www.stat.gouv.qc.ca/statistiques/economie/comptes-economiques/comptes-revenus-depenses/comptes-2017-04.pdf

[10] Fogg BJ. Creating persuasive technologies: An eight-step design process. In: Proceedings of the 4th International Conference on Persuasive Technology. New York, NY, USA: ACM; 2009. p. 44. DOI: 10.1145/1541948.1542005

[11] Nemery A, Brangier E, Kopp S. Proposition d'une grille de critères d'analyse ergonomiques des formes de persuasion interactive. In: Proceedings of the 22nd Conference on l'Interaction Homme-Machine. New York, NY, USA: ACM; 2010. pp. 153-156